Free-Radical Scavenging and Antioxidant Properties of Melatonin

Free-Radical Scavenging and Antioxidant Properties of Melatonin

Marina Garcia-Macia
Adrián Santos-Ledo

MDPI

Basel • Beijing • Wuhan • Barcelona • Belgrade • Novi Sad • Cluj • Manchester

Marina Garcia-Macia
IBFG
University of Salamanca
Salamanca
Spain

Adrián Santos-Ledo
Human Anatomy and Histology
University of Salamanca
Salamanca
Spain

Editorial Office
MDPI AG
Grosspeteranlage 5
4052 Basel, Switzerland

This is a reprint of articles from the Special Issue published online in the open access journal *Antioxidants* (ISSN 2076-3921) (available at: www.mdpi.com/journal/antioxidants/special_issues/free_radical_scavenging_melatonin).

For citation purposes, cite each article independently as indicated on the article page online and using the guide below:

Lastname, A.A.; Lastname, B.B. Article Title. *Journal Name* **Year**, *Volume Number*, Page Range.

ISBN 978-3-7258-2148-8 (Hbk)
ISBN 978-3-7258-2147-1 (PDF)
https://doi.org/10.3390/books978-3-7258-2147-1

Cover image courtesy of Adrián Santos-Ledo

Contents

About the Editors

Marina Garcia-Macia

Marina Garcia-Macia is a Ramon y Cajal fellow. She is member of the Department of Biochemistry at the University of Salamanca and lead her own group in the Institute of Functional Biology and Genomics. She studies the role of autophagy in several pathological scenarios.

Adrián Santos-Ledo

Adrián Santos-Ledois an Associate Professor in the Department of Human Anatomy and Histology at the University of Slamanca. He studies the role of autophagy during embrionic development and regeneration using zebrafish.

antioxidants

Editorial

Melatonin: A Myriad of Functions to Discover

Adrián Santos-Ledo [1,2,*,†] and Marina García-Macia [3,4,5,*,†]

1 Instituto de Neurociencias de Castilla y León (INCyL), 37007 Salamanca, Spain
2 Department of Human Anatomy and Histology, Universidad de Salamanca, 37007 Salamanca, Spain
3 Institute of Functional Biology and Genomics (IBFG), Universidad de Salamanca-CSIC, 37007 Salamanca, Spain
4 Department of Biochemistry and Molecular Biology, Universidad de Salamanca, 37007 Salamanca, Spain
5 Centre for Biomedical Investigations Network on Frailty and Ageing (CIBERFES), 28029 Madrid, Spain
* Correspondence: santosledo@usal.es (A.S.-L.); marinagarciamacia@usal.es (M.G.-M.)
† These authors contributed equally to this work.

Melatonin is an indoleamine that has captured our attention since 1958 [1]. It was discovered as one of the main regulators of circadian rhythms. At that time, it was simply defined as the sleep–wake regulator hormone, as melatonin is mainly secreted by the pineal gland during the night. Its role in seasonal reproductive functions was also found very soon after. Later, the extra-pineal synthesis of melatonin was discovered (immune system, retina, etc.), together with its production in other organisms, such as bacteria and plants. This fact changed the idea that melatonin was just a hormone that synchronized processes related with the circadian and seasonal rhythms, and brought up the role of melatonin as a very potent antioxidant [2]. The use of oxygen by organisms entails an important cost due to the consequent production of free radicals. To avoid or reduce the detrimental effects of oxidative stress, cells developed specific antioxidant systems, for instance, enzymes such as superoxide dismutase or catalase and, of course, melatonin. This hormone probably appeared in photosynthetic bacteria to prevent oxygen toxicity around 3 billion years ago [3]. Thus, melatonin's first and more primitive function is protecting cells against toxic products, and it was supplemented with a variety of other roles during evolution.

Melatonin's main role as an antioxidant is the focus of many pathological studies, such as in metabolic, degenerative, and cardiovascular disorders, as well as in cancer, where there is impaired redox homeostasis. Melatonin's antioxidant capacity is more effective than that of vitamin E, and it shows intracellular and extracellular activity. It can scavenge ROS directly due to its indole ring, and it can also stimulate the expression of antioxidant enzymes. Furthermore, melatonin can improve mitochondrial homeostasis, which is the main source of free radicals. Despite melatonin's potential as a therapeutic approach for many diseases linked to increased oxidative stress, more studies are required to establish the putative clinical application of this indoleamine. For this reason and to understand the role of melatonin as an antioxidant, we set up this Special Issue.

We have encountered a myriad of articles about the different uses of melatonin as a therapeutic approach in metabolic diseases, cancer, and neurological afflictions. Some authors explored the conserved role of melatonin as an ROS scavenger against sulfur- and nitrogen-mustard-induced toxicity (Contribution 1) and found that melatonin could be used as a medical countermeasure for blister agent poisoning. In line with the antioxidant functions, one of our articles describes, for the first time, the presence of melatonin synthesis in *Archaea* (Contribution 2), which not only informs us about the primitive origin of the molecule, but also about its conservation and robust role as an antioxidant.

Obesity was an interesting target for our authors' studies, particularly the prevention of this medical condition through stimulating thermogenesis (Contribution 3). This implies the use of fat through the mitochondria to produce heat instead of energy (a futile cycle). The use of melatonin as a molecule to stimulate thermogenesis can be a suitable treatment

Citation: Santos-Ledo, A.; García-Macia, M. Melatonin: A Myriad of Functions to Discover. *Antioxidants* **2024**, *13*, 360. https://doi.org/10.3390/antiox13030360

Received: 27 February 2024
Revised: 12 March 2024
Accepted: 15 March 2024
Published: 18 March 2024

for obesity. Our authors showed how melatonin can stimulate these processes to enhance mitochondrial fusion, but in a dose-dependent manner.

Two biomedical fields very interested in melatonin as possible therapy are neuroscience and cancer. Interestingly, the indoleamine seems to have opposing functions, antioxidant and pro-oxidant, depending on the type of studied cell. In this editorial, we provide a thorough review about sepsis-associated encephalopathy, which is a life-threating disfunction caused by infection (Contribution 4). Melatonin was used as a therapeutic substance in many studies, but most of the information came from preclinical studies with animals. Thus, the authors claim the necessity to implement more studies in humans. The manuscripts in our Special Issue differ in the melatonin capabilities in contexts that have common aspects: neurodevelopment and the regeneration of the nervous system. Establishing circadian rhythms is crucial for neurodevelopment. For instance, melatonin enhances important clock proteins also known for their antioxidant capabilities. In particular, individuals with neurodevelopmental disorders such as autism spectrum disorder (ASD), schizophrenia, and bipolar disorder showed lower levels of melatonin secretion and disrupted circadian rhythms (Contribution 5).

Melatonin was used to decipher the relevance of free radicals during the regeneration of optic nerves (Contribution 6). Reactive oxygen species are usually associated with cellular damage, but, currently, they are acquiring more functional roles. When the optic nerve becomes crushed, the regeneration process starts and oligodendrocytes need to succumb and differentiate from new OPCs. Then, they can properly re-myelinate the optic nerve. When melatonin reduces oxidative stress, the regenerative process becomes impaired because damaged oligodendrocytes remain alive.

Melatonin's antioxidant properties have been explored in many studies. However, there are few clues about the pro-oxidant functions of melatonin functioning as an oncostatic. These are recapitulated in the review we present in our Special Issue (Contribution 7). These mechanisms include different pathways that act through melatonin receptors, as well as sirtuins and the anti-Warburg effect. Besides this, the synergy between melatonin and other antitumoral treatments is explored in very aggressive cancers such as triple-negative breast cancer (TNBC) (Contribution 8). New melatonin derivates such as agomelatine, traditionally used as an antidepressant, are being tested as alternative tumoral treatments (Contribution 9). Agomelatine was tested in in vitro colorectal cancer models, and it was able to reduce the proliferation through NF-Kb inhibition. Thus, these new molecules show a double function, antidepressant and antitumoral, which may be due to the modification of the tumor environment or the alteration of the immune response.

This last function is also tackled in our Special Issue with a different approach. The optimization of bird production has deep economic repercussions, as chickens are one of the most consumed meats worldwide. Birds are very sensitive to light and their immune system is very susceptible to different wavelengths, especially bursal B-lymphocytes, which can undergo apoptosis depending on the type of light. This manuscript showed how chickens exposed to a determined light had an increased melatonin level, which reduced B-lymphocyte oxidative stress and prevented apoptosis (Contribution 10).

In summary, with this Special Issue, we try to explore the most recent roles of melatonin as an antioxidant, and we provide a great deal of new information: from melatonin synthesis in a new domain to recent uses of melatonin as a therapy and synergistic treatment for cancer. Thus, we are sure that this indoleamine will be in the spotlight for much more time to come.

Author Contributions: Conceptualization, A.S.-L. and M.G.-M.; Writing-original draft preparation, review and editing, A.S.-L. and M.G.-M. All authors have read and agreed to the published version of the manuscript.

Conflicts of Interest: The authors declare no conflicts of interest.

List of Contributions

1. Ramos, E.; Gil-Martín, E.; De Los Ríos, C.; Egea, J.; López-Muñoz, F.; Pita, R.; Juberías, A.; Torrado, J.; Serrano, D.; Reiter, R.; et al. Melatonin as Modulator for Sulfur and Nitrogen Mustard-Induced Inflammation, Oxidative Stress and DNA Damage: Molecular Therapeutics. *Antioxidants* **2023**, *12*, 397. https://doi.org/10.3390/antiox12020397.
2. Lee, K.; Choi, G.; Back, K. Functional Characterization of Serotonin N-Acetyltransferase in Archaeon Thermoplasma volcanium. *Antioxidants* **2022**, *11*, 596. https://doi.org/10.3390/antiox11030596.
3. Aouichat, S.; Raya, E.; Molina-Carballo, A.; Munoz-Hoyos, A.; Aloweidi, A.; Elmahallawy, E.; Agil, A. Dose-Dependent Effect of Melatonin on BAT Thermogenesis in Zücker Diabetic Fatty Rat: Future Clinical Implications for Obesity. *Antioxidants* **2022**, *11*, 1646. https://doi.org/10.3390/antiox11091646.
4. Sieminski, M.; Szaruta-Raflesz, K.; Szypenbejl, J.; Krzyzaniak, K. Potential Neuroprotective Role of Melatonin in Sepsis-Associated Encephalopathy Due to Its Scavenging and Anti-Oxidative Properties. *Antioxidants* **2023**, *12*, 1786. https://doi.org/10.3390/antiox12091786.
5. Feybesse, C.; Chokron, S.; Tordjman, S. Melatonin in Neurodevelopmental Disorders: A Critical Literature Review. *Antioxidants* **2023**, *12*, 2017. https://doi.org/10.3390/antiox12112017.
6. Pérez-Montes, C.; Jiménez-Cubides, J.; Velasco, A.; Arévalo, R.; Santos-Ledo, A.; García-Macia, M. REDOX Balance in Oligodendrocytes Is Important for Zebrafish Visual System Regeneration. *Antioxidants* **2023**, *12*, 2026. https://doi.org/10.3390/antiox12122026.
7. Florido, J.; Rodriguez-Santana, C.; Martinez-Ruiz, L.; López-Rodríguez, A.; Acuña-Castroviejo, D.; Rusanova, I.; Escames, G. Understanding the Mechanism of Action of Melatonin, Which Induces ROS Production in Cancer Cells. *Antioxidants* **2022**, *11*, 1621. https://doi.org/10.3390/antiox11081621.
8. Estirado, S.; Fernández-Delgado, E.; Viñuelas-Zahínos, E.; Luna-Giles, F.; Rodríguez, A.; Pariente, J.; Espino, J. Pro-Apoptotic and Anti-Migration Properties of a Thiazoline-Containing Platinum(II) Complex in MDA-MB-231 Breast Cancer Cells: The Role of Melatonin as a Synergistic Agent. *Antioxidants* **2022**, *11*, 1971. https://doi.org/10.3390/antiox11101971.
9. Moreno-SanJuan, S.; Puentes-Pardo, J.; Casado, J.; Escudero-Feliu, J.; Khaldy, H.; Arnedo, J.; Carazo, Á.; León, J. Agomelatine, a Melatonin-Derived Drug, as a New Strategy for the Treatment of Colorectal Cancer. *Antioxidants* **2023**, *12*, 926. https://doi.org/10.3390/antiox12040926.
10. Zhang, Y.; Wang, Z.; Dong, Y.; Cao, J.; Chen, Y. Melatonin Nuclear Receptors Mediate Green-and-Blue-Monochromatic-Light-Combinations-Inhibited B Lymphocyte Apoptosis in the Bursa of Chickens via Reducing Oxidative Stress and Nfκb Expression. *Antioxidants* **2022**, *11*, 748. https://doi.org/10.3390/antiox11040748.

References

1. Lerner, A.B. Hormones and Skin Color. *Sci. Am.* **1961**, *205*, 98–108. [CrossRef] [PubMed]
2. Reiter, R.J.; Mayo, J.C.; Tan, D.X.; Sainz, R.M.; Alatorre-Jimenez, M.; Qin, L. Melatonin as an Antioxidant: Under Promises but over Delivers. *J. Pineal Res.* **2016**, *61*, 253–278. [CrossRef] [PubMed]
3. Reiter, R.J.; Rosales-Corral, S.; Tan, D.X.; Jou, M.J.; Galano, A.; Xu, B. Melatonin as a Mitochondria-Targeted Antioxidant: One of Evolution's Best Ideas. *Cell. Mol. Life Sci.* **2017**, *74*, 3863–3881. [CrossRef] [PubMed]

antioxidants

MDPI

Article

REDOX Balance in Oligodendrocytes Is Important for Zebrafish Visual System Regeneration

Cristina Pérez-Montes [1,2,3], Jhoana Paola Jiménez-Cubides [1], Almudena Velasco [1,3,4], Rosario Arévalo [1,3,4], Adrián Santos-Ledo [1,2,3,*,†] and Marina García-Macia [3,5,6,7,*,†]

[1] Instituto de Neurociencias de Castilla y León (INCyL), 37007 Salamanca, Spain; cristina_perez_montes@usal.es (C.P.-M.); jhoanajicu@usal.es (J.P.J.-C.); malmu@usal.es (A.V.); mraa@usal.es (R.A.)
[2] Department of Human Anatomy and Histology, Universidad de Salamanca, 37007 Salamanca, Spain
[3] Instituto de Investigación Biomédica de Salamanca (IBSAL), 37007 Salamanca, Spain
[4] Department of Cell Biology and Pathology, Universidad de Salamanca, 37007 Salamanca, Spain
[5] Institute of Functional Biology and Genomics (IBFG), Universidad de Salamanca/CSIC, 37007 Salamanca, Spain
[6] Department of Biochemistry and Molecular Biology, Universidad de Salamanca, 37007 Salamanca, Spain
[7] Centre for Biomedical Investigations Network on Frailty and Ageing (CIBERFES), 28029 Madrid, Spain
* Correspondence: santosledo@usal.es (A.S.-L.); marinagarciamacia@usal.es (M.G.-M.)
† These authors contributed equally to this work.

Abstract: Zebrafish (*Danio rerio*) present continuous growth and regenerate many parts of their body after an injury. Fish oligodendrocytes, microglia and astrocytes support the formation of new connections producing effective regeneration of the central nervous system after a lesion. To understand the role of oligodendrocytes and the signals that mediate regeneration, we use the well-established optic nerve (ON) crush model. We also used *sox10* fluorescent transgenic lines to label fully differentiated oligodendrocytes. To quench the effect of reactive oxygen species (ROS), we used the endogenous antioxidant melatonin. Using these tools, we measured ROS production by flow cytometry and explored the regeneration of the optic tectum (OT), the response of oligodendrocytes and their mitochondria by confocal microscopy and Western blot. ROS are produced by oligodendrocytes 3 h after injury and JNK activity is triggered. Concomitantly, there is a decrease in the number of fully differentiated oligodendrocytes in the OT and in their mitochondrial population. By 24 h, oligodendrocytes partially recover. Exposure to melatonin blocks the changes observed in these oligodendrocytes at 3 h and increases their number and their mitochondrial populations after 24 h. Melatonin also blocks JNK upregulation and induces aberrant neuronal differentiation in the OT. In conclusion, a proper balance of ROS is necessary during visual system regeneration and exposure to melatonin has a detrimental impact.

Keywords: melatonin; regeneration; ROS; zebrafish; visual system

check for updates

Citation: Pérez-Montes, C.; Jiménez-Cubides, J.P.; Velasco, A.; Arévalo, R.; Santos-Ledo, A.; García-Macia, M. REDOX Balance in Oligodendrocytes Is Important for Zebrafish Visual System Regeneration. *Antioxidants* **2023**, *12*, 2026. https://doi.org/10.3390/antiox12122026

Academic Editors: Ana B Rodríguez Moratinos and Jose Antonio Pariente Llanos

Received: 17 October 2023
Revised: 9 November 2023
Accepted: 20 November 2023
Published: 22 November 2023

1. Introduction

Regenerative medicine aims to regain the lost functionality of an organ due to trauma, injury or disease using organ specific mechanisms. Unfortunately, in most cases, these mechanisms are not properly elucidated [1]. In humans, some tissues, such as blood and epithelia, regenerate efficiently; while others, such as the heart and nervous system, do not. Many groups have devoted themselves to repairing injuries in the nervous system using prosthetics [2], stem cells [3] and gene therapy [4]; their success has been very limited and, so far, there is no efficient regenerative treatment. Unlike mammals, zebrafish (*Danio rerio*) present a robust regenerative capacity, including of the nervous system [5,6]. Furthermore, zebrafish are currently used to model a wide range of human diseases where regenerative therapies could be applied, from heart diseases to Alzheimer's [7,8]. Understanding the regenerative mechanisms in zebrafish would facilitate strategies for regeneration in humans.

The zebrafish visual system resembles that of a human. It is formed by the retina, the optic nerve (ON), formed by the axons of the ganglion cells, and the superior visual center, that in fish is the optic tectum (OT). Gross injuries in the visual system, such as strong light, cryoinjury of the retina and ON crush, cause the degeneration of the ON. Then, the retina loses the connection with the OT, leading to partial and temporal blindness [9]. In zebrafish, many retinal ganglion cells survive the injury and create new axons that are projected contralaterally towards the OT and vision is restored [10]. Glial cells are very important during regeneration and one key difference between mammals and zebrafish is that, in the former, glia do not form a scar but create a regenerative-permissive environment [1,11,12]. Furthermore, in zebrafish during the degeneration and posterior regeneration of oligodendrocytes, the myelinating cells of the central nervous system (CNS) play a crucial role [5]. After injury, many oligodendrocytes die, and new ones arise from oligodendrocyte progenitor cells. These newly formed oligodendrocytes create myelin as the axons regenerate [5]. Mature oligodendrocytes can also survive degeneration and contribute to myelination [13]. However, the role of these latter oligodendrocytes is more controversial, and they have been described to produce aberrant myelination [14]. The role of the oligodendrocytes during regeneration has been well explored in zebrafish larvae [14] and in the adult spinal cord [12] but their exact role in the visual system remains more obscure.

Another open question relates to the cues that are needed for proper regeneration. Certainly, communication between neurons, glia and other cells such as macrophages is key [15]. Several extrinsic and intrinsic signals, such as synaptic sensing, vesicle release and neurotransmitters, have been shown to contribute during regeneration [5,16]. Reactive oxygen species (ROS) have also emerged as relatively new players in this context [17]. ROS are highly reactive chemicals produced from O_2 as normal biproducts of the mitochondrial metabolism and membrane NADPH oxidase. This includes the superoxide ion (O_2^-) that is the precursor of many others such as hydrogen peroxide (H_2O_2). ROS are short-lived but, thanks to their small size, some of them can freely diffuse among cells, playing important roles as signaling molecules [18]. ROS are involved in cell communication, proliferation, cell death and can promote, among other processes, oligodendrogenesis [19]. In zebrafish, ROS are produced after tail injury and promote its regeneration [20] and the presence of antioxidants can be detrimental to an efficient regeneration [21]. CNS is no exception to the production of ROS after injury, but its impact is less well understood [22]. For example, in some contexts, such as demyelinating neurological diseases, excessive ROS are detrimental [23]. Free radicals are a very interesting puzzle since they can exert both beneficial and detrimental effects.

We explore the role of ROS during zebrafish visual system regeneration, using the ON crush model, specifically in fully differentiated oligodendrocytes. To explore the beneficial or detrimental role of oxidative stress, we treated the fish with melatonin, which is a well-known endogenous antioxidant [24,25]. This molecule has been successfully used to reverse cognitive and endocrine deficits produced by alterations of the zebrafish visual system [26].

2. Methods

2.1. Animals

A total of 112 animals were used, half males and half females: 4 animals for each group and treatment in every experiment. All fish used in this work were 12 months old (adults). Fish were kept under a 14 h light–10 h dark cycle and fed twice a day. The water temperature was maintained at 27.8 ± 0.5 °C; pH 7.0 ± 0.2. Animals were fed before surgery and before changing the melatonin solution. We employed two transgenic lines to track mature oligodendrocytes: *Tg(sox10:EGFP)* [27] and *Tg(sox10:tagRFP)* [28]. These lines carry either EGFP or tagRFP under the control of *sox10*. This transcription factor is involved in the differentiation of oligodendrocytes, becoming active when they turn into myelinating cells [29–31]. For the cytometry flux experiments we used the EGFP line because it does not cross with the probes' fluorescence. For the immunohistochemistry experiments we used the tagRFP line because it does not require amplification by immunohistochemistry. All lines were kindly donated by Bruce Appel.

All protocols were performed according to the European Union Directive 86/609/EEC and Recommendation 2007/526/EC, regarding the protection of animals used for experimental and other scientific purposes, enforced in Spanish legislation under the law 6/2013. All protocols were approved by the Bioethics Committee of the University of Salamanca.

2.2. Optic Nerve Crush

ON crush was performed as previously described [32]. Briefly, fish were anesthetized in tricaine methane sulfonate (MS-222) (Sigma-Aldrich, St. Louis, MO, USA) and positioned under a stereomicroscope on an ice-cold surface. A fine forcep was used to partially pull the temporal half of the right eye out of its orbit. Then, the lateral rectus eye muscle was cut with a fine scissor and the eye was pulled out to further expose the ON. With another fine forceps, we crushed the ON, avoiding the ophthalmic artery running along the ON. The fish successfully operated on were return to a fish tank, and allowed to survive for 3, 24 and 72 h after the lesion. The same protocol without crushing was applied to control animals.

2.3. Melatonin Treatment

Exposure to melatonin was performed by immersion, as previously described [26]. Melatonin (98%, Sigma-Aldrich, Saint Louis, MI, USA) was dissolved in ethanol, creating a stock solution of 2.32 mg/mL. Stock solution was wrapped in foil and stored at −20 °C in the dark until use. Melatonin solution at 0.232 mg/L was prepared fresh every time in a 0.5 L tank with standard system water. Control and injured fish with no melatonin were exposed to 0.01% ethanol as the vehicle. One animal was allocated per tank. Thus, the experimental groups were set as control 3, 24 and 72 h (+0.01% ethanol for that period); control melatonin 3, 24 and 72 h (+0.232 mg/L melatonin for that period), regeneration 3, 24 and 72 h (+0.01% ethanol for that period) and regeneration melatonin 3, 24 and 72 h (+0.232 mg/L melatonin for that period).

Zebrafish are diurnal animals linked to circadian rhythms that are modulated by melatonin, so we were careful with the timing of our experiments [25]. Lesion was always performed at 9 am when melatonin production is lowest, followed by exposure to vehicle or melatonin. A fresh solution containing either vehicle or melatonin was prepared every time. In the case of the 3 h group, sacrificed was performed at 12 am. In the case of the 24 h group, the solution was changed once at 9 pm and the fish were sacrificed the following day at 9 am. In the case of the 72 h group, the solution was changed twice every day (9 am and 9 pm) to ensure the stability of the melatonin [26,33], and the fish were sacrificed three days later at 9 am.

2.4. Brain Dissection and Tissue Processing

At 3, 24 and 72 h, fish were endpoint anesthetized in tricaine methane sulfonate before sacrifice, according to Spanish and European laws (2010/63/EU; RD 53/2013; Ley 32/2007; and Orden ECC/566/2015). The head was removed, and the entire brain, including the eyes and ON, was carefully dissected manually with forceps in cold PBS under a microscope.

To prepare the brain for the histochemistry and the immunostaining techniques, the tissue was fixed in paraformaldehyde 4% overnight at 4 °C. The sample was cryoprotected in 10% sucrose in PBS and embedded in a solution containing 1.5% agar and 10% sucrose. Finally, 12 mm horizontal sections, to observe ON and OT simultaneously, were obtained in a cryostat. Sections were stored at −20 °C until use.

For flow cytometry we used a modified protocol [34]. Briefly, both ON, the ipsilateral OT (IOT) and the contralateral OT (COT), were sectioned with surgical scissors and located separately in solution A (EBBS, 2% BSA, 1.3% DNase) in ice. After collecting the tissue from all animals, tissue was transferred to solution B (EBBS medium, 2% BSA, 1.3 DNase, 1% Trypsine). Tissue was incubated for 5 min at 37 °C, then disaggregated with a siliconized pipette and then incubated for 10 more minutes. Samples were centrifuged for 5 min at 3000× *g* at 4 °C, supernatant was removed and 1 mL of solution A was added. Sample was resuspended with a blue tip and centrifuged again. Finally, supernatant was removed and 1 mL of Hank's solution plus 200 µL of KRPG buffer (145 mM NaCl, 5.7 mM, 566 Na_2HPO_4, 4.86 mM KCl, 0.54 mM $CaCl_2$, 1.22 mM $MgSO_4$, 5.5 mM glucose, pH 7.35) were added and created a cell suspension.

2.5. Probes and Flow Cytometry

Mitochondrial ROS were determined with the fluorescent probe MitoSox® (Life Technologies, Carlsbad, CA, USA) [34]. Cell suspensions were incubated with 2 µM of MitoSox® for 30 min at 37 °C in a 5% CO_2 atmosphere in KRPG buffer. MitoSox® fluorescence intensity was assessed by flow cytometry (FACScalibur flow cytometer, BD Biosciences, Franklin Lakes, NJ, USA) and expressed in arbitrary units.

The mitochondrial membrane potential ($\Delta\psi m$) was assessed with MitoProbe DiIC1® (Life Technologies) (50 nM) by flow cytometry (FACScalibur flow cytometer, BD Biosciences) and expressed in arbitrary units [34]. For this purpose, cell suspensions were incubated with a probe 30 min at 37 °C in PBS. Values of $\Delta\psi m$ were obtained after subtraction of the potential value determined in the presence of carbonyl cyanide-4-(trifluoromethoxy) phenylhydrazone (CCCP) (10 µM, 15 min) for each sample.

2.6. Histochemistry

Acetylcholine esterase (AChE) histochemistry was performed as previously described [35]. Briefly, sections were rehydrated in 0.1 M maleate buffer pH 6.0, and then incubated at room temperature for 30 min in a medium containing 65 mM maleate buffer pH 6.0, 0.5 mM sodium citrate, 0.47 mM cupric sulphate, 0.05 mM potassium ferricyanide, 0.37 mM acetylthiocholine iodide and 0.06 mM ethopropazine as an inhibitor of non-specific esterases. AChE activity was enhanced using 0.0125% DAB and 0.0033% H_2O_2 in 0.2 M Tris-HCl buffer, pH 7.6. The reaction was monitored under the microscope and stopped by washing the sections in the same Tris-HCl buffer (around 45 min). Finally, sections were dehydrated in an ethanol series, cleared with xylene and cover slipped with Entellan.

2.7. Immunostaining

Sections were washed several times in PBS and then pre-incubated for 90 min in 5% normal donkey (DK) serum in PBS with 0.2% Triton X-100 (PBSTx) at room temperature (RT). After that, primary antibodies (Table 1) were incubated overnight in 5% normal DK serum with 0.2% PBSTx at 4 °C. Sections were washed in PBS and then incubated for 90 min at RT in darkness with a 1:400 dilution of Alexa 488 and Alexa 647 fluorescent secondary antibodies (Table 2), in a buffer containing 5% normal DK serum in PBS. Next, sections were washed in PBS and then incubated for 10 min in 1:10,000 4′,6-diamidino-2-fenilindol (DAPI) for nucleus staining. Sections were washed thoroughly and mounted with Fluoromount-G® Mounting Medium (Invitrogen, Waltham, MA, USA). The RFP fluorescence in the *sox10:tagRFP* transgenic line was robust enough and we never used an antibody to reinforce it.

Table 1. Primary antibodies.

Antigen	Host	Reference	Dilution	Observations
Choline acetyltransferase (ChAT)	Goat	Sigma-Aldrich; ab144P	1:100	Catalyzes the reversible synthesis of acetylcholine from acetyl CoA and choline at cholinergic synapses
Succinate dehydrogenase complex subunit B (SDHB)	Mouse	Abcam; ab14714	1:200	Complex II of the respiratory chain involved in the oxidation of succinate

Table 2. Secondary antibodies.

Antigen	Host	Reference	Conjugated	Dilution
Anti-Goat	Donkey	Jackson ImmunoResearch (West Grove, PA, USA)	Alexa 488	1:400
Anti-Mouse	Donkey	Jackson ImmunoResearch	Alexa 647	1:400

2.8. Western Blot

OT cells were lysed in a commercial RIPA Lysis buffer (0.1% sodium dodecylsulfate, 1% NP-40, 150 mM NaCl, 1% sodium deoxycholate and 25 mM Tris-HCl, pH 7.6) (G-Biosciences). ON cells were lysed in a noncommercial RIPA buffer (1% sodium dodecylsulfate, 1% NP-40, 150 mM NaCl, 5% sodium deoxycholate and 25 mM Tris-HCl, pH 8). Lysis buffers were supplemented with a protease and phosphatase inhibitor mix (Sigma-Aldrich). Aliquots of cell lysates (10–20 μg of protein from ON or OT, respectively) were subjected to SDS/PAGE 10–12% (vol/vol) acrylamide gel (MiniProtean; Bio-Rad, Hercules, CA, USA) including PageRuler Prestained Protein Ladder (Thermo Fisher Scientific, Waltham, MA, USA). The resolved proteins were transferred electrophoretically to nitrocellulose membranes (0.2 μm, BioRad). Membranes were blocked with 5% (wt/vol) low-fat milk in TTBS (20 mM Tris, 150 mM NaCl, and 0.1% (vol/vol) Tween 20 (pH 7.5) for 1 h. After blocking, membranes were immunoblotted with primary antibodies, anti-p-JNK and JNK (1/1000, Cell Signalling, Danvers, MA, USA) and MBP (Myelin basic protein, 1/500, Abcam, Cambridge, United Kingdom) overnight at 4 °C. After incubation with horseradish peroxidase conjugated anti-rabbit IgG (1/5000, Cell Signalling), membranes were immediately incubated with the Supersignal West Femto (Thermo Fisher Scientific) before exposure using the Fusion FX transilluminator (Vilber GmbH, Eberhardzell, Germany).

2.9. Imaging and Quantification

AChE histochemistry was imaged in a light microscope (Leica Aristoplan; Leica, Wetzlar, Germany) with an Olympus OP- 70 digital camera (Olympus Corporation, Shinjuku City, Tokyo, Japan) coupled to an Olympus Provis AX70 photomicroscope using a 20× objective. All sections were imaged in the same session to avoid variations. To quantify the staining, FIJI 2.9.0 software and the threshold plugin were used. Images were turned into black and white, inversed and turned into 8 bits. Then threshold was established based on control animals to 110/200. Integrated density and area were measured for every section. Three non-consecutive sections were quantified from four different fish and the average is shown in the graphs.

ChAT and SDHB immunostainings were imaged in a Confocal microscope (Leica Stellaris inverted DMi8) using a 40× oil immersion objective for ChAT and 63× oil immersion objective for SDHB. For ChAT, 6 tiles with z-stack (xyz scan) were acquired and automatically assembled by the Leica LASX software. For SDHB, 4 tiles with z-stack were acquired and automatically assembled by the LASX software. The number of z-stacks was determined by observing the limits of each section (14 μm). Images were obtained under constant conditions to minimize image acquisition variation and stored as 1024 × 1024 pixels and 8-bit TIFF files.

Colocalization between SDHB and *sox10:tagRFP* oligodendrocytes was analyzed using FIJI, as previously described [36]. First, background from the *sox10:tagRFP* channel and SDHB channel was subtract with the "subtract background" tool (rolling ball radius 100 pixels). Then, *sox10:tagRFP* oligodendrocytes were selected manually with the "polygon selections" tool and added to ROI Manager. The SDHB threshold was established based on control animals and added to ROI Manager. To analyze the colocalization between the *sox10:tagRFP* and SDHB channels, each *sox10:tagRFP* oligodendrocyte selection was merged with the SDHB threshold using the command "AND" from ROI Manager. Then, merged selections were added to ROI Manager one by one. Finally, area, integrated density and mean gray value were measured for every merged selection added to ROI Manager using the command ''Measure" from the ROI Manager. Eight cells (four from tectal OT and four from the periventricular gray zone, PGZ) from three non-consecutive sections from four different animals in each group were used.

Sox10:tagRFP oligodendrocytes were quantified manually using the Cell Counter plugin from FIJI in the tectal OT and PGZ. Three non-consecutive sections from four different animals in each group were used.

In case of ChAT immunohistochemistry, positive neurons were quantified manually using the Cell Counter plugin from FIJI in the middle portion of the OT. Three non-consecutive sections from four different animals in each group were used. The average of these three sections is shown in the graphs.

The protein abundances of all Western blots per condition were measured by densitometry of the bands quantified with FIJI in the linear phase of the exposure without reaching saturation. At least three biologically independent replicates were always performed, although only one representative Western blot is usually shown in the main figures.

Figures were generated using Adobe Photoshop CS6.

2.10. Statistics

Graphs were generated using GraphPad Prism 5. Data are represented as mean ± standard error of mean (SEM). Student's *t*-test was performed when comparing two groups (for example regeneration vs. control). ANOVA with Tukey´s Multiple Comparison Test was performed to search for differences among all groups (for example, the number of oligodendrocytes at any specific timepoint). The type of statistics and the significance are included in the figure legends.

3. Results

3.1. ROS Are Produced by sox10:EGFP Oligodendrocytes after Optic Nerve Crushing

The role of ROS during regeneration has been shown in several scenarios [20,22]. To understand whether they are also involved during zebrafish visual system regeneration, we used the ON crush model [32]. The right ON was always injured. After letting the fish survive for 3, 24 or 72 h, we extracted the whole brain together with the eyes and dissected three different parts of the visual system: both ON together, the ipsilateral optic tectum (IOT, right one) and the contralateral optic tectum (COT, left one). In zebrafish all fibers from ON cross the midline and reach the opposite side [37]. Thus, COT is the side that lost the projections. We disaggregated the cells and exposed them to Mitosox®, a probe that responds to O_2^- (Figure 1A). Then, we run the cells through flow cytometry, sorting the *sox10:EGFP* cells (fully differentiated oligodendrocytes).

In the three visual areas investigated, sox10:EGFP oligodendrocytes show a greatly increased ROS production at 3 hpl (Figure 1B). The IOT presented the highest peak, with almost six times more ROS production than the control fish, followed by COT that tripled ROS (Figure 1B). These changes were not due to differences in the mitochondrial population state, as indicated by the lack of differences in the membrane potential when measured with MitoProbe DiIC1® (Figure S1A). At 24 hpl, the effect on ROS production passed in the OT but it was maintained in the ON (Figure 1C). *Non-GFP* cells behaved differently (Figure 1D). *Non-GFP* cells at 3 hpl showed a reduction in ROS production but only in the IOT, the side that is not injured and, thus, is responding to the healthy ON (Figure 1D). At 24 hpl, *non-GFP* cells from the three areas showed a reduction in ROS. This analysis shows that there is an important but small window of ROS production in *sox10:EGFP* oligodendrocytes.

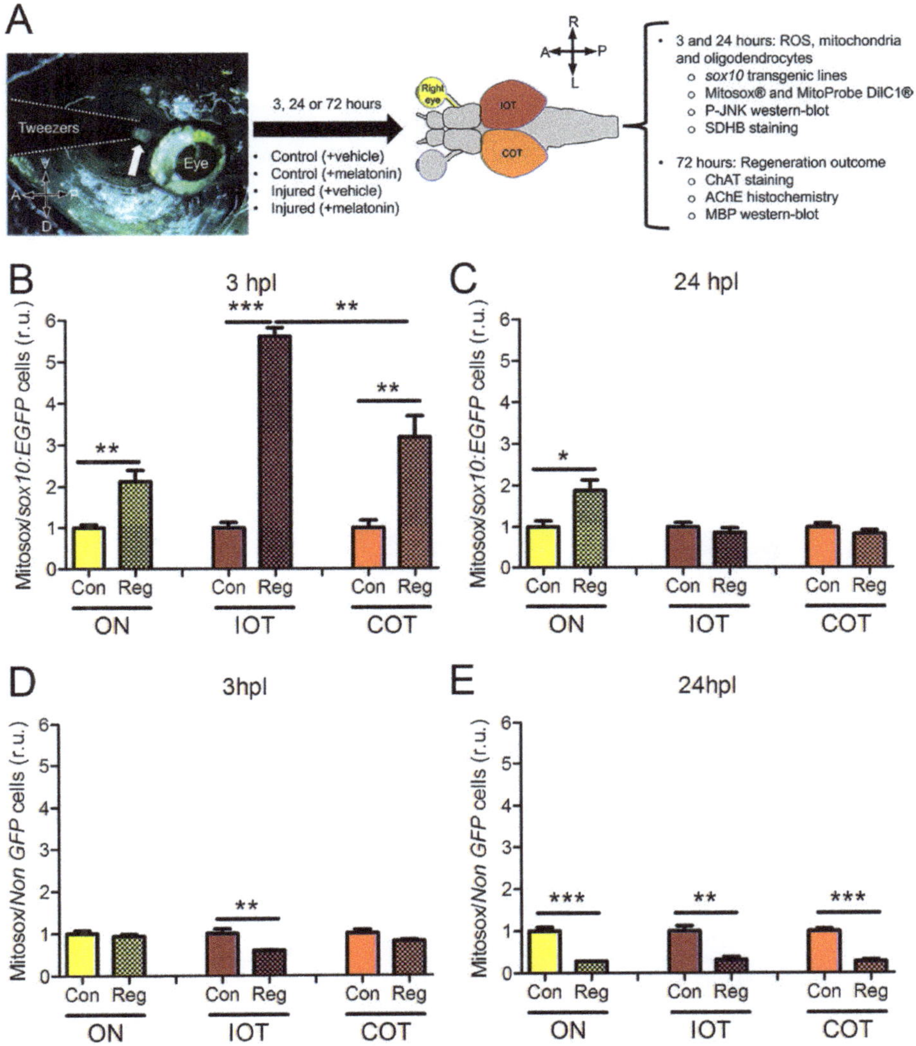

Figure 1. ROS measurement after ON crush. (**A**) Experimental design; the arrow points to the crushed ON from the ventral side. (**B–E**) Mitosox® fluorescence quantified by flow cytometry in *sox10:EGFP* cells at 3 hpl (**B**) and 24 hpl (**C**) or *non-GFP* cells at 3 hpl (**D**) and 24 hpl (**E**). Values are relative to control mean. A: anterior; Con: control; COT: contralateral optic tectum; D: dorsal; hpl: hours post-lesion; IOT: ipsilateral optic tectum; L: left; ON: optic nerve; P: posterior; R: right; Reg: regeneration; r.u.: random units; V: ventral. Student's *t*-test between Reg and Con for each timepoint and region * $p < 0.05$; ** $p < 0.01$; *** $p < 0.001$. n = 4 for each group.

3.2. Melatonin Does Not Always Exert an Antioxidant Effect in sox10:EGFP Oligodendrocytes

To explore whether ROS played a positive or negative role during regeneration of the visual system, we exposed the zebrafish to the natural antioxidant melatonin. Since melatonin was dissolved in ethanol, control animals were also treated with this vehicle. Therefore, for every timepoint we have four experimental groups (see Figure 1A for more details). In the graphs, we show the change in ROS production after exposing the fish to melatonin for all groups. The value shown in the graph (Figure 2) is the result of the

ROS measurement in each animal treated with melatonin minus the average ROS value in animals treated with vehicle in each respective group. Thus, positive values mean higher production of ROS after melatonin treatment (pro-oxidative) and negative ones mean a reduction (antioxidant).

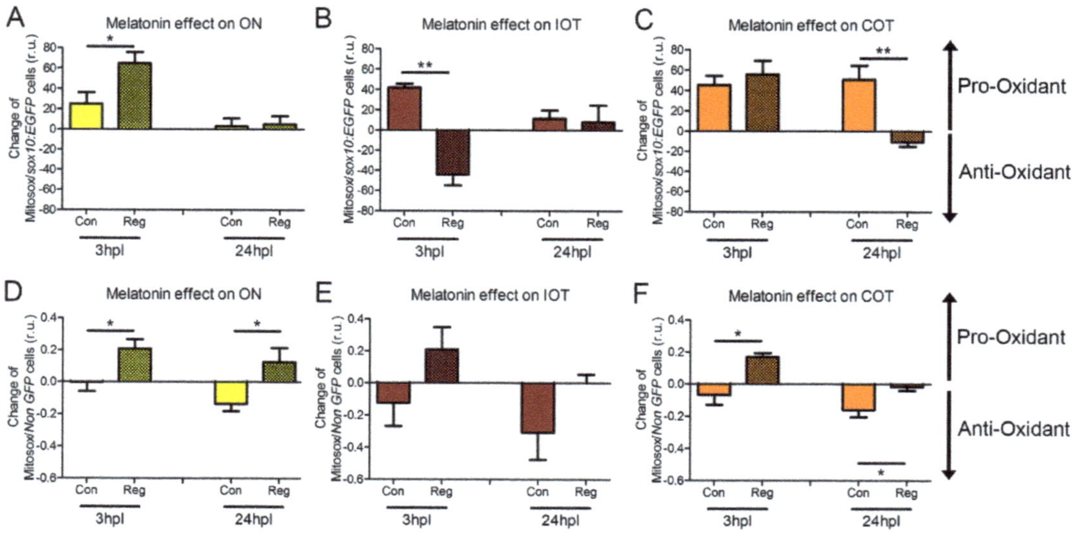

Figure 2. Melatonin impact on ROS production. (**A–F**) Mitosox® fluorescence quantified by flow cytometry in the *sox10:EGFP* cells of ON (**A**), IOT (**B**) and COT (**C**) and in the *non-GFP* cells of ON (**D**), IOT (**E**) and COT (**F**) at 3 and 24 hpl. The value shown in the graph is the result of the ROS measurement in each animal treated with melatonin minus the average ROS value in animals treated with vehicle in each respective group. Con: control; COT: contralateral optic tectum; hpl: hours post-lesion; IOT: ipsilateral optic tectum; Mel: melatonin; ON: optic nerve; Reg: regeneration; r.u.: random units. For (**A–F**), Student's t-test was performed between Reg and Con for each timepoint and region. * $p < 0.05$; ** $p < 0.01$. n = 4 for each group in all experiments.

Surprisingly, exposure to melatonin induced the production of mitochondrial ROS (pro-oxidant) in *sox10:EGFP* oligodendrocytes at 3 hpl in the ON of both control and injured fish, with more production in the latter. (Figure 2A). This effect was not observed at 24 h (Figure 2A). In the IOT, melatonin had an antioxidant effect in the regenerating fish but not in the control fish at 3 h, an effect that was absent at 24 h (Figure 2B). The opposite happened in the COT, the side that is primary responding to the injury: at 3 h, melatonin equally increased ROS production in the COT of the control and regenerating fish but we observed an antioxidant effect in the COT of the regenerating fish at 24 h (Figure 2C).

In all three areas at both timepoints, *non-GFP* cells had a reduced production of ROS after melatonin treatment in control animals (Figure 2D–F). In the regenerating fish, *non-GFP* cells either did not respond (IOT and COT at 24 hpl, Figure 2E,F) or the production was slightly increased (ON at 3 and 24 hpl, and IOT and COT at 3 hpl, Figure 2D–F).

3.3. Oligodendrocytes and Their Mitochondrial Population Are Altered during Regeneration

We observed important and unexpected changes in the ROS state in the OT. To further characterize this aspect and investigate if the production of ROS occurred simultaneously with changes in the mitochondria, first, we quantified the number of full oligodendrocytes present in the horizontal section based on the presence of *sox10:tagRFP* and their morphology. We also stained mitochondria using the SDHB antibody and quantified its fluorescence specifically in these oligodendrocytes of the OT (Figure 3. for 3 hpl and Figure 4 for 24 hpl).

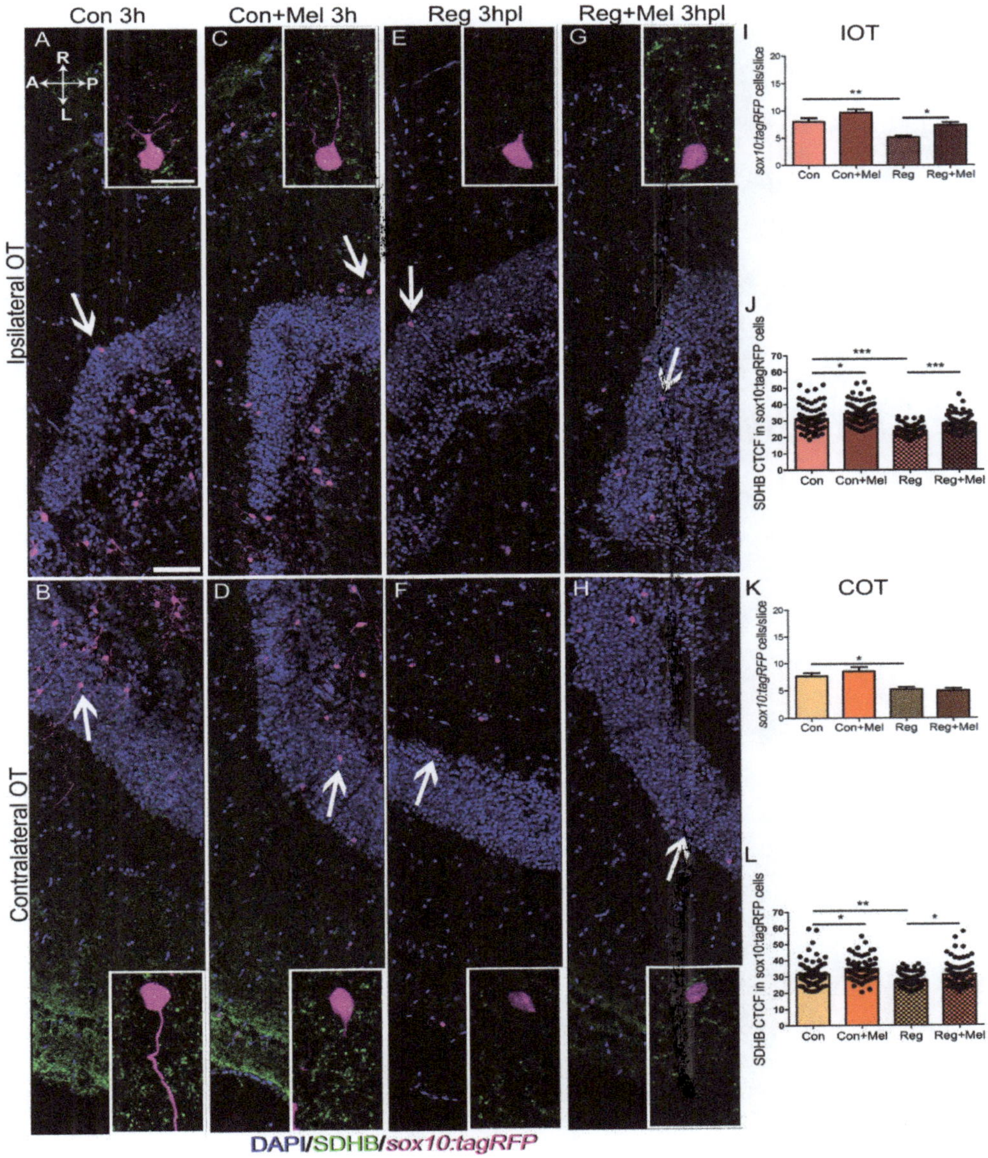

Figure 3. Analysis of oligodendrocyte and the mitochondrial distribution at 3 hpl. Immunostaining for SDHB (green) in a transgenic line carrying *sox10:tagRFP* reporter (magenta). Nuclei are counterstained with DAPI (blue). (**A–H**) Control IOT (**A**), control COT (**B**), control exposed to melatonin IOT (**C**), control exposed to melatonin COT (**D**), regenerating fish IOT (**E**), regenerating fish COT (**F**), regenerating fish exposed to melatonin IOT (**G**), regenerating fish exposed to melatonin COT (**H**). (**I–L**) Quantification of the number of oligodendrocytes per section in the IOT (**I**) and the COT (**K**); CTCF value of SHDB in oligodendrocytes in the IOT (**J**) and the COT (**L**). A: anterior; Con: control; COT: contralateral optic tectum; h: hours; hpl: hours post-lesion; IOT: ipsilateral optic tectum; L: left; Mel: melatonin; P: posterior; R: right; Reg: regeneration; r.u.: random units. ANOVA with Tukey's Multiple Comparison Test * $p < 0.05$; ** $p < 0.01$; *** $p < 0.001$. Scale bar: 100 mm; 10 mm in inset. n = 4 for each group.

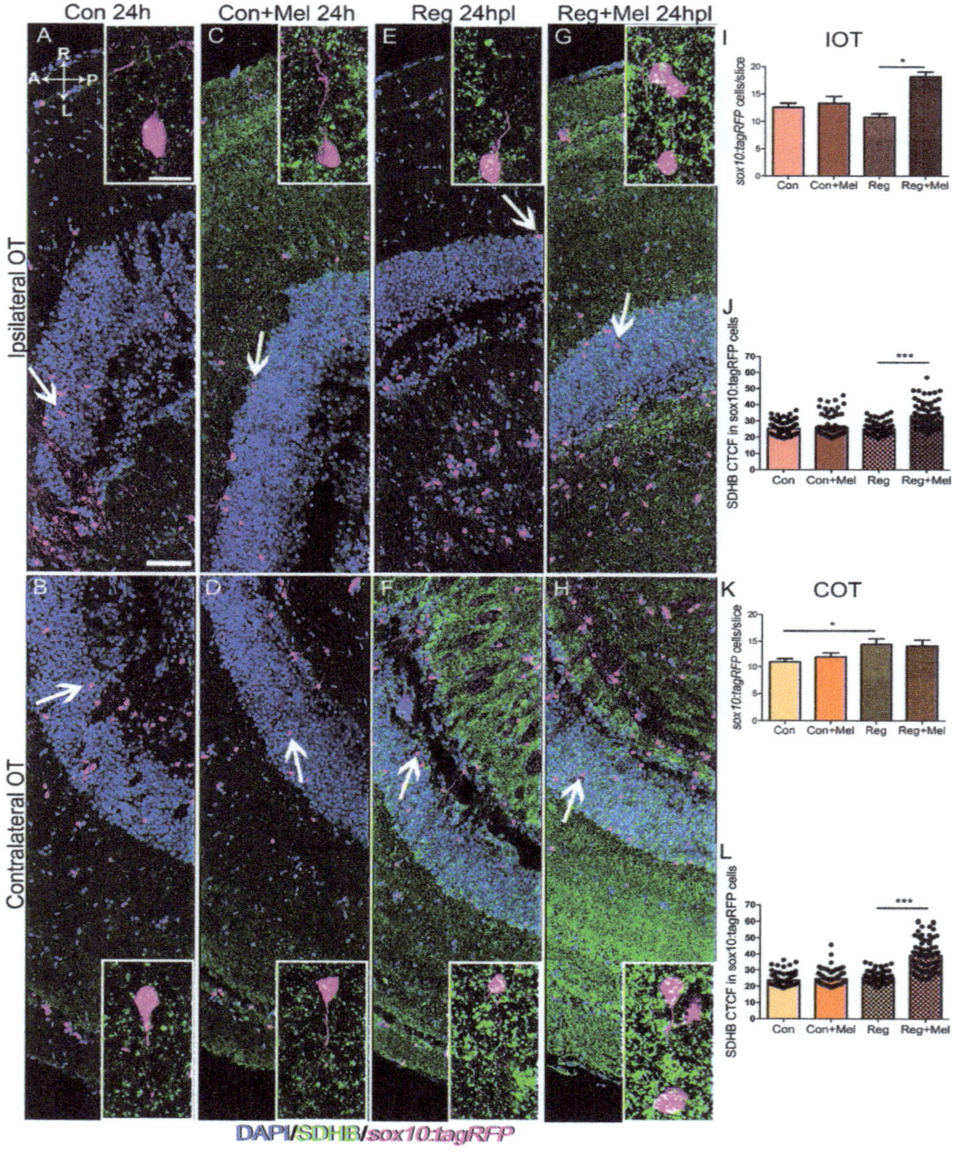

Figure 4. Analysis of oligodendrocyte and the mitochondrial distribution at 24 hpl. Immunostaining for SDHB (green) in a transgenic line carrying *sox10:tagRFP* reporter (magenta). Nuclei are counterstained with DAPI (blue). (**A–H**) Control IOT (**A**), control COT (**B**), control exposed to melatonin IOT (**C**), control exposed to melatonin COT (**D**), regenerating fish IOT (**E**), regenerating fish COT (**F**), regenerating fish exposed to melatonin IOT (**G**), regenerating fish exposed to melatonin COT (**H**). (**I–L**) Quantification of the number of oligodendrocytes per section in the IOT (**I**) and the COT (**K**); CTCF value of SHDB in oligodendrocytes in the IOT (**J**) and the COT (**L**). A: anterior; Con: control; COT: contralateral optic tectum; h: hours; hpl: hours post-lesion; IOT: ipsilateral optic tectum; L: left; Mel: melatonin; P: posterior; R: right; Reg: regeneration; r.u.: random units. ANOVA with Tukey's Multiple Comparison Test * $p < 0.05$; *** $p < 0.001$. Scale bar: 100mm; 10mm in inset. n = 4 for each group.

Oligodendrocytes were observed all over the horizontal section of the IOT (Figure 3A,C,E,G) and the COT (Figure 3B,D,F,H) from the tectal superficial layers to the PGZ, close to the ventricle. Mitochondria (SDHB staining) are shown in green and, as expected, are present in every cell type (Figure 3). Insets allowed us to explore the mitochondrial content of specific oligodendrocytes (insets in Figure 3).

At 3 hpl, we observed a reduction in the number of *sox10:tagRFP* oligodendrocytes in the IOT and the COT (For IOT: Figure 3A vs. Figure 3E, quantified in Figure 3I; for COT: Figure 3B vs. Figure 3F, quantified in Figure 3K). Melatonin blocked this phenotype in the IOT (Figure 3G, quantified in Figure 3I), but not in the COT (Figure 3H, quantified in Figure 3K). Exposure to melatonin produced an increase in SDHB fluoresce in control animals (for IOT: Figure 3A vs. Figure 3C, quantified in Figure 3J; for COT: Figure 3B vs. Figure 3D, quantified in Figure 3L). The oligodendrocytes of both OT hemispheres in regenerating animals showed a decreased fluorescence for mitochondria, suggesting a reduction in the population (for IOT: Figure 3E, quantified in Figure 3J; for COT: Figure 3F quantified in Figure 3L). Interestingly, exposure to melatonin in the injured fish abolished the decrease in SDHB fluorescence in both hemispheres of the OT (for IOT: Figure 3G, quantified in Figure 3J; for COT: Figure 3H quantified in Figure 3L). We did not quantify other cells of the OT due the lack of clear cellular landmarks, but a similar trend could be observed: reduce fluorescence in injured fish, suppression of this phenotype by melatonin and little or no changes in control animals (Figure 3A–H).

By 24 h (Figure 4) the number of oligodendrocytes in the IOT is almost the same in the control and regenerating fish (Figure 4A vs. Figure 4E, quantified in Figure 4I). Melatonin increased the number of oligodendrocytes just in the IOT of lesioned fish (Figure 4C vs. Figure 4G, quantified in Figure 4I). However, regenerating fish presented more *sox10:tagRFP* oligodendrocytes in the COT (Figure 4B vs. Figure 4F, quantified in Figure 4K), with no further increased induced by melatonin (Figure 4F vs. Figure 4H, quantified in Figure 4K). In terms of SDHB fluorescence, an indirect view of the mitochondrial population, we could observe a similar effect. There was no difference within the control group or between the control and regenerating fish (Figure 4A–F, quantified in Figure 4J for IOT and Figure 4L for COT). However, the oligodendrocytes in the injured fish exposed to melatonin presented higher SDHB fluorescence (Figure 4G,H, quantified in Figure 4J for IOT and Figure 4L for COT). These data indicate changes in the oligodendrocytes concomitant to changes in their mitochondrial population.

3.4. Exposure to Melatonin Induces Changes in the OT Undergoing Regeneration

We have described increased ROS production after ON crushing, together with a decrease in the number of *sox10:tagRFP* oligodendrocytes and their mitochondria in the OT at 3 hpl. Both effects were blocked by exposure to melatonin. Thus, we wanted to know if melatonin had a short-term impact on the general regeneration of the visual system. Acetylcholine is one of the most important neurotransmitters in the visual system [38]. The activity of acetylcholine depends on two enzymes, one for its production: choline acetyl transferase (ChAT); and another for its degradation: acetylcholine esterase (AChE). We explored both at 72 hpl through immunohistochemistry and histochemistry, respectively.

ChAT neurons were present in the PGZ projecting towards more superficial areas of the OT. We found around forty of these neurons per section of OT in both hemispheres in control animals (Figure 5A,B, quantified in Figure 5I,J). As expected, exposure to melatonin did not induce any changes in control animals (Figure 5C,D, quantified in Figure 5I,J). At this stage, fish undergoing regeneration did not show any differences either (Figure 5C,D, quantified in Figure 5I,J). However, we found more ChAT neurons in the IOT of fish that were injured and exposed to melatonin (Figure 5G, quantified in Figure 5I) but not in the COT (Figure 5H, quantified in Figure 5J). The IOT is the only hemisphere receiving light input at this timepoint.

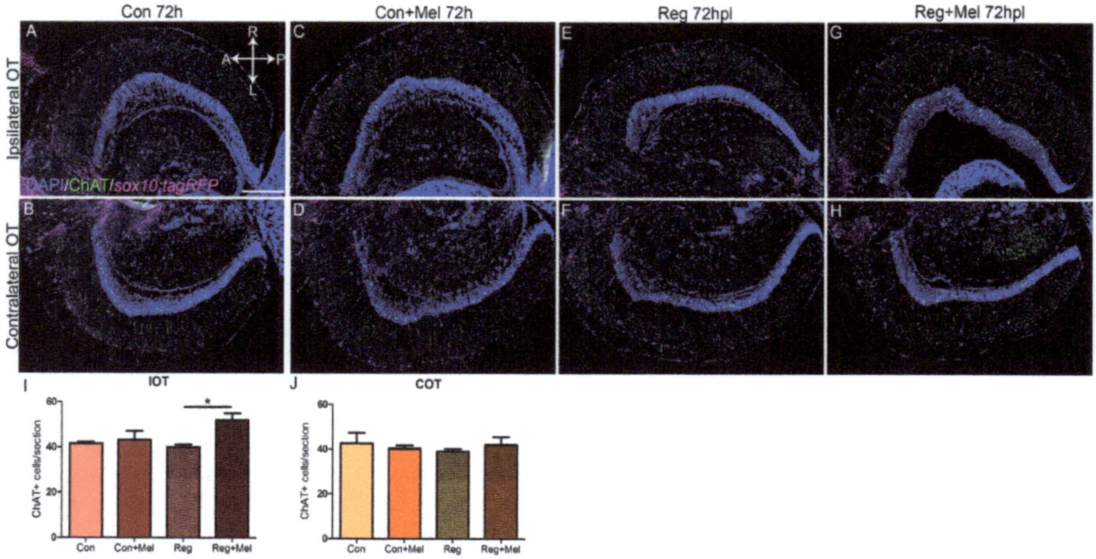

Figure 5. Quantification of ChAT neurons during regeneration at 72 hpl. (**A–H**) Immunohistochemistry against ChAT in the OT at 72 hpl in control (**A**,**B**), control exposed to melatonin (**C**,**D**), injured (**E**,**F**) and injured treated with melatonin (**G**,**H**). (**I**,**J**) Quantification of ChAT neurons per section of IOT (**I**) and COT (**J**). A: anterior; ChAT: choline acetyltransferase; Con: control; COT: contralateral optic tectum; h: hours; hpl: hours post-lesion; IOT: ipsilateral optic tectum; L: left; Mel: melatonin; P: posterior; R: right; Reg: regeneration; r.u.: random units. ANOVA with Tukey's Multiple Comparison Test * $p < 0.05$. Scale bar: 200 mm. n = 4 for each group.

To follow up on the impact of melatonin on the acetylcholine metabolism, we analyzed the activity of the AChE through histochemistry (Figure 6). We did not find significant differences in the staining between control, control exposed to melatonin and injured fish in either side of the OT (Figure 6A–F, quantified in Figure 6I for IOT and in Figure 6J for COT). However, we found a decreased activity of AChE in the COT of lesioned fish treated with melatonin (Figure 6H, quantified in Figure 6J). This is the side that is responding to injury and thus is not receiving light input. We did not find differences in the IOT of regenerating fish treated with melatonin (Figure 6G, quantified in Figure 6I).

3.5. *Melatonin Treatment Hinders Regeneration*

We wondered if the increased ROS production in injured fish (Figure 1B,C) and the exposure to melatonin (Figure 2A–C) had an impact on cell signaling. Thus, we investigated the abundance of JNK, a MAP Kinase involved in stress response and its active form p-JNK [39]. At 3 hpl, regenerating fish had an increased p-JNK/JNK ratio in ON (Figure 7A, quantified in Figure 7D) and the COT (Figure 7C, quantified in Figure 7F), the areas affected by the injury. Melatonin treatment blocked this upregulation in both regions (ON: Figure 7A, quantified in Figure 7D; COT: Figure 7C, quantified in Figure 7F). In the IOT we did not observe a significant increase in p-JNK/JNK (Figure 7B, quantified in Figure 7E); but this ratio was also reduced in this area after melatonin treatment (Figure 7B, quantified in Figure 7E). IOT was the only visual area where melatonin had a clear antioxidant effect at 3 hpl (Figure 2B). Melatonin did not alter the p-JNK/JNK ratio in control animals (Figure 7A–C, quantified in Figure 7D–F). Thus, we observed an activation of the JNK pathway after injury that was quenched by melatonin in all three areas.

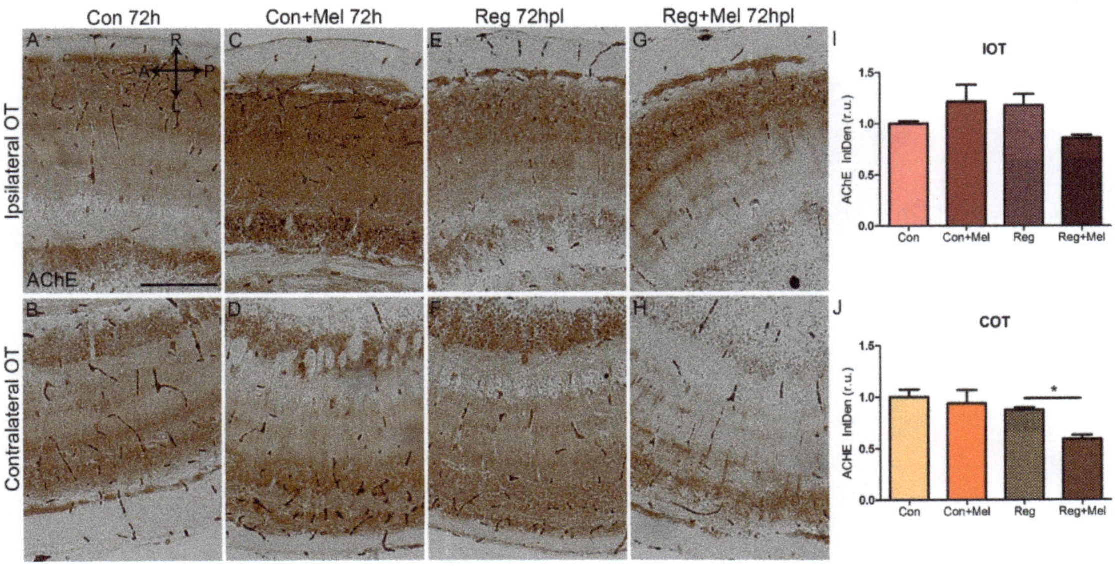

Figure 6. Quantification of AChE during regeneration at 72 hpl. (**A–H**) Histochemistry for AChE in the OT at 72 hpl in control (**A**,**B**), control exposed to melatonin (**C**,**D**), injured (**E**,**F**) and injured treated with melatonin (**G**,**H**). (**I**,**J**) Quantification of AChE staining in the IOT (**I**) and COT (**J**). A: anterior; AChE: acetylcholine esterase; Con: control; COT: contralateral optic tectum; h: hours; hpl: hours post-lesion; IOT: ipsilateral optic tectum; L: left; Mel: melatonin; P: posterior; R: right; Reg: regeneration; r.u.: random units. ANOVA with Tukey's Multiple Comparison Test * $p < 0.05$. Scale bar: 200 mm. Scale bar: 200 mm. n = 4 for each group.

Exposure to melatonin impacts signaling, mitochondrial state and different cell populations. Therefore, we wondered if the general recovery of the injured fish was affected. We then analyzed the presence of MBP in all three visual areas by Western blot at 72 hpl (Figure 7G–L). Melatonin did not produce any impact on the abundance of MBP in control animals (Figure 7G–I, quantified in Figure 7J–L). At this stage, we did not observe any important changes in the abundance of MBP in the regenerating fish (Figure 7G–I, quantified in Figure 7J–L). However, in all areas, injured fish exposed to melatonin presented significantly less MBP (Figure 7G–I, quantified in Figure 7J–L). These results highlight the detrimental impact of melatonin in injured fish.

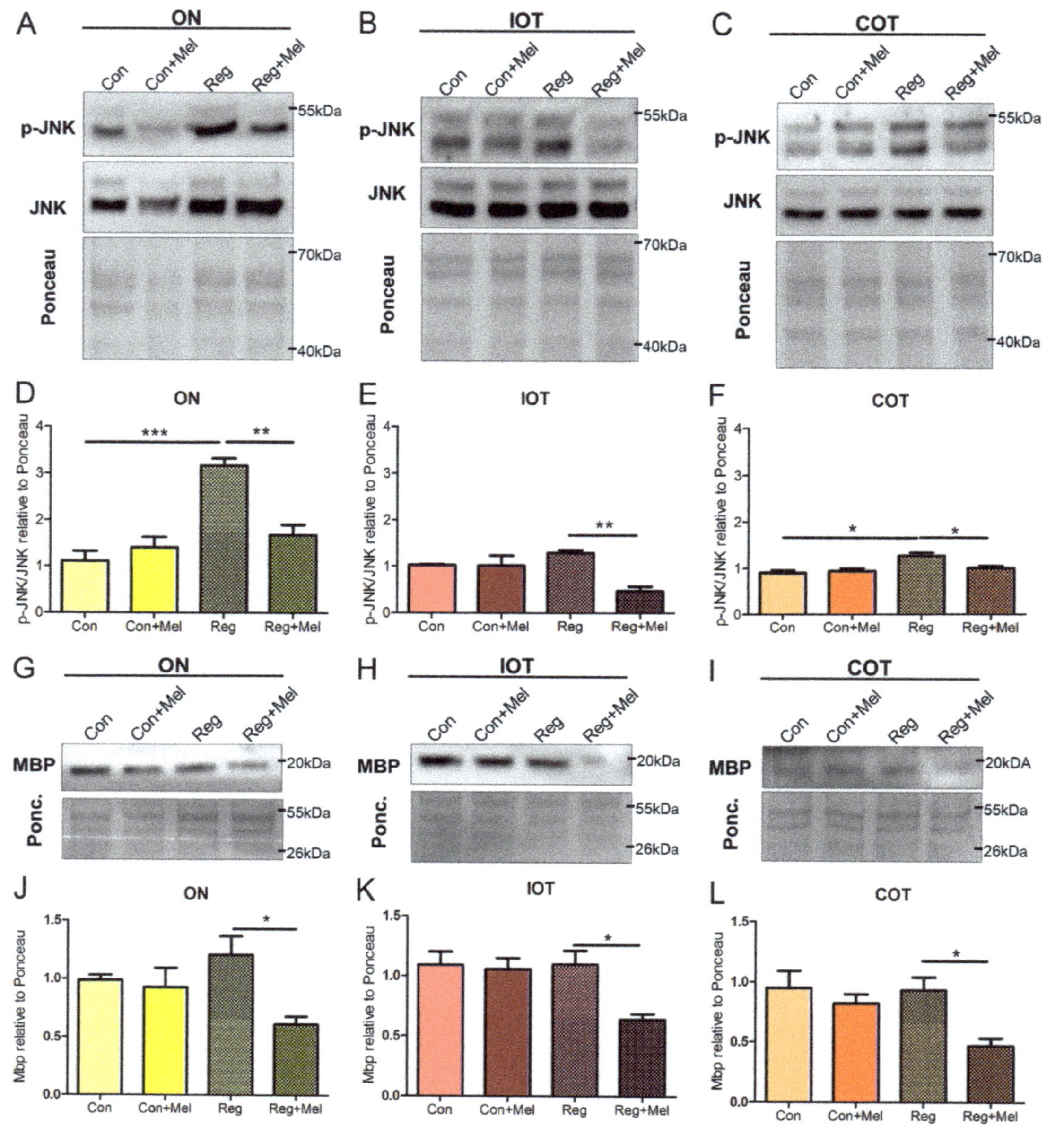

Figure 7. Quantification of MBP at 72 hpl. (**A–F**) Western blot for p-JNK and JNK in tissue from ON (**A**), IOT (**B**), COT (**C**). Quantifications represent value of p-JNK relative to total JNK and relative to total protein (Ponceau), quantified in ON (**D**), IOT (**E**), COT (**F**). (**G–L**) Western blot for MBP in tissue from ON (**G**), IOT (**H**), COT (**I**). Quantifications represent value of MBP relative to total protein (Ponceau) in ON (**J**), IOT (**K**), COT (**L**). Con: control; COT: contralateral optic tectum; hpl: hours post-lesion; IOT: ipsilateral optic tectum; MBP: Myelin Basic Protein; Mel: melatonin; ON: optic nerve; Reg: regeneration; r.u.: random units. ANOVA with Tukey's Multiple Comparison Test was performed to search for differences among all groups * $p < 0.05$; ** $p < 0.01$; *** $p < 0.001$. n = 4 for each group.

4. Discussion

Our results indicate that ROS are produced three hours after injury in different parts of the visual system, with an important contribution from mature oligodendrocytes. Modulating oxidative stress with melatonin induces an aberrant regeneration including diminished production of myelin and disrupted differentiation of the OT. Furthermore, we show that, not only oligodendrocytes in the ON respond to injury, but also those located in the OT. As described during tail regeneration [20], ROS seem to positively contribute to CNS regeneration in zebrafish.

The role of ROS has been explored during the regeneration of open wounds. After the amputation of part of the zebrafish tail, free radicals are produced for at least 24 h and trigger JNK signaling and apoptosis. Blocking either process has detrimental effects and delays regeneration [20,40]. Similar mechanisms have been also reported during wound healing in *Xenopus* [41], *Drosophila* [42] and *C. elegans* [43], although the type of injury was different. Cellular changes induced by ROS, such as recruitment of adherens junctions, are also conserved in different models [42]. Our results show that ROS production by oligodendrocytes peaks in the visual system early, at 3 hpl, including the OT. A previous work also reported the relevance of ROS in the brain using a model of a stab wound in the telencephalon, although they describe the importance of oxidative stress at 24 hpl [22]. In addition to these small timing differences, our results support the role of free radicals as a fast regenerative signal also in the visual system. The source of these ROS has been searched for by several groups. In open wound models, cells close to the wound produce free radicals [42]. Other cells can also produce ROS, for example microglia, another type of supporting glial cell in the nervous system, respond to the oxidative stress produced by a stab at 6 hpl, and then produce ROS themselves [44]. We have shown that fully differentiated oligodendrocytes (*sox10:EGFP*) become a source of ROS after injury at 3 hpl. ROS can control oligodendrocyte differentiation and activity [23,45]. Thus, it is interesting to hypothesize that oligodendrocytes might be the first responders to the injury in the visual system and become a ROS hub that could attract other cells.

To delve into the role of the oligodendrocytes, we tried to cease the production of free radicals with melatonin. However, a clear antioxidant effect was only observed in non-GFP cells in controls, which proved the antioxidant of the melatonin treatment [24]. Fully differentiated oligodendrocytes (*sox10:EGFP*) seem to behave differently to melatonin, as mitochondrial ROS rise. Concomitantly, melatonin produces an increase in mitochondria per oligodendrocyte. This free radical increment in basal conditions may be due to mitochondrial fission [46]. Melatonin had an even more complex effect on the oligodendrocytes under regenerative conditions since it had an antioxidant effect depending on the timepoint, first in the IOT (3 hpl) and later in the COT (24 hpl). This time-dependent melatonin effect has been previously described in other scenarios and reveals a complex interaction with the oligodendrocyte's biology [47]. It is not surprising that oligodendrocytes respond to melatonin since they present both types of receptors [48]. In our scenario, the effect of melatonin seems to depend on the cellular environment, which is very different between both hemispheres, one responding directly to the ON injury (COT) and the other probably responding to excessive workload (IOT) since all eye projections are contralateral in zebrafish [37]. It seems clear, however, that melatonin disrupts the normal process during regeneration, for example delaying the loss of mitochondria in both hemispheres of the OT, a phenotype that has been reported previously [49].

We were interested in evaluating the impact of melatonin on the general regenerative process. Thus, we studied the functionality of the OT (acetylcholine pathway) and the myelination. This neurotransmitter system plays important roles in visual processing and perception in zebrafish [38] and mammals [50]. At the timepoints examined by us, there were no changes in this system during regeneration, but melatonin produced profound effects that were hemisphere specific. ChAT neurons were increased in the IOT, the side that is still receiving light, and AChE activity was decreased in the COT, the side that lost the connection to the eye. When ROS balance is altered, Morin, another antioxidant, has been also shown to decrease AChE activity [51]. More interestingly, melatonin has been described to support survival of ChAT neurons [52,53] and to reduce AChE activity in different rodent models [54,55]. The increased number of ChAT neurons could also be related to the reduction of JNK signaling induced earlier, as shown in vitro [56,57]. So, during regeneration there is a specific cellular environment, likely related to the REDOX misbalance and the presence/absence of light input, where melatonin can push the differentiation of ChAT neurons and block AChE activity.

We have reported the activation of JNK pathway in the ON but also far from the wound, in both OT of injured fish. We have also described the negative impact of melatonin in JNK activation during regeneration. It has been shown that ROS spike leads to JNK and P38 Map Kinase activation [42,58]; and the JNK pathway through c-Jun activation is necessary for normal regeneration [40]. Upregulation of the JNK pathway in our model may be linked to the drop of fully differentiated oligodendrocytes observed in regeneration [59]. Melatonin also hindered the recovery of MBP, a protein involved in axon myelination [29]. As far as we know, there is no report linking melatonin and the myelination process itself, but melatonin promotes oligodendrocytic survival and maturation in different pathological scenarios [48,60]. We have also observed an increased number of oligodendrocytes that survived the injury after exposure to melatonin. Thus, the disruption of JNK by melatonin may lead to the accumulation of aberrant oligodendrocytes, as we observed in IOT, and/or a global abnormal regeneration (OT and ON) [40,54]. Death of mature oligodendrocytes is necessary for proper regeneration since surviving oligodendrocytes mistarget new axons and produce aberrant myelination [14].

Several groups have reported the benefits of melatonin during regeneration, at least in the rodent peripheral system [49,61] or during demyelinating diseases [62,63]. In fact, melatonin may improve the function of Schwann cell (another type of myelinating cell) through a mechanism related to mitochondrial biology and its antioxidant properties [61,64]. However, other authors have reported that ROS are beneficial for the repairment of the peripheral nervous system [65] and during the regeneration of mammalian fat pads [66] and liver [67], and blocking ROS during zebrafish tail regeneration is detrimental [20,21]. All of them tissues with important regenerative capabilities. Then, there could exist a relationship between tissues' capacity to respond to ROS and their regenerative potential. Our results support the beneficial effects of ROS during the regeneration of the zebrafish visual system and the detrimental impact of melatonin. Caution might be advised with the therapeutic potential of this molecule. In some scenarios, melatonin might be beneficial but its impact on neural differentiation is not well understood in pathological conditions. For example, it can exacerbate neurological disability scores in rodents with encephalomyelitis [68]. More experiments are necessary to understand the role of ROS during the regeneration of different tissues and models, and to comprehend the precise mechanisms elicited by melatonin.

5. Conclusions

Oligodendrocytes present important levels of ROS production. ROS are produced during early regeneration in the oligodendrocytes of the ON and the OT. Deregulation of the ROS production leads to aberrant regeneration. Exposure to melatonin is detrimental to the regeneration of the zebrafish visual system.

Supplementary Materials: The following supporting information can be downloaded at: https://www.mdpi.com/article/10.3390/antiox12122026/s1, Figure S1: Analysis of the mitochondrial membrane potential.title (A) Membrane potential measured with MitoProbe DiIC1® by flow cytometry. Con: control; COT: contralateral optic tectum; hpl: hours post-lesion; IOT: ipsilateral optic tectum; ON: optic nerve; Reg: regeneration; r.u.: random units.

Author Contributions: Conceptualization: C.P.-M., R.A., A.S.-L. and M.G.-M. Methodology: C.P.-M., J.P.J.-C., A.S.-L. and M.G.-M. Formal analysis: C.P.-M., A.S.-L. and M.G.-M. Original Draft: C.P.-M.,

R.A., A.S.-L. and M.G.-M. Review and editing: C.P.-M., A.V., R.A., A.S.-L. and M.G.-M. Funding: A.V., A.S.-L. and M.G.-M. All authors have read and agreed to the published version of the manuscript.

Funding: This project was funded by: Lanzadera TCUE21-21_003, C2 program for Universidad de Salamanca; Convocatoria de ayudas de investigación Biomédica 2021, FERP21/003, Fundación Eugenio Rodríguez Pascual; ESCI 2022 Exploratory Research Grants; ESCI22/001, European Society for Clinical Investigation ESCI; Internationalization Project "CL-EI-2021-08-IBFG Unit of Excellence" from Spanish National Research Council (CSIC), funded by the Regional Government of Castile and Leon and co-financed by the European Regional Development Fund (ERDF "Europe drives our growth"). CPM is a USAL fellow Plan Especial Grado Medicina. MGM is a Ramón y Cajal Researcher RYC2021-033684-I, MICIN. We are members of Women in Autophagy (WIA), la Sociedad Española de Bioquímica y Biología Molecular (SEBBM) and la Sociedad Española de Autofagia (SEFAGIA).

Institutional Review Board Statement: All protocols were performed according to the European Union Directive 86/609/EEC and Recommendation 2007/526/EC, regarding the protection of animals used for experimental and other scientific purposes, enforced in Spanish legislation under the law 6/2013. All protocols were approved by the Bioethics Committee of the University of Salamanca.

Informed Consent Statement: Not Applicable.

Data Availability Statement: Data are contained within the article.

Acknowledgments: We would like to thank the Institute for Biomedical Research of Salamanca (IBSAL).

Conflicts of Interest: Authors declare no competing interest.

References

1. Ghaddar, B.; Lübke, L.; Couret, D.; Rastegar, S.; Diotel, N. Cellular Mechanisms Participating in Brain Repair of Adult Zebrafish and Mammals after Injury. *Cells* **2021**, *10*, 391. [CrossRef] [PubMed]
2. Soltan, A.; Barrett, J.M.; Maaskant, P.; Armstrong, N.; Al-Atabany, W.; Chaudet, L.; Neil, M.; Sernagor, E.; Degenaar, P. A head mounted device stimulator for optogenetic retinal prosthesis. *J. Neural Eng.* **2018**, *15*, 065002. [CrossRef] [PubMed]
3. Allsopp, T.E.; Ebneth, A.; Cabrera-Socorro, A. Deploying human pluripotent stem cells to treat central nervous system disorders: Facts, challenges and realising the potential. *Stem Cell Res.* **2019**, *41*, 101581. [CrossRef] [PubMed]
4. Qin, Y.; Li, S.; Li, X.J.; Yang, S. CRISPR-Based Genome-Editing Tools for Huntington's Disease Research and Therapy. *Neurosci. Bull.* **2022**, *38*, 1397–1408. [CrossRef]
5. Masson, M.A.; Nait-Oumesmar, B. Emerging concepts in oligodendrocyte and myelin formation, inputs from the zebrafish model. *Glia* **2023**, *71*, 1147–1163. [CrossRef]
6. Gonzalez, D.; Allende, M.L. Current Advances in Comprehending Dynamics of Regenerating Axons and Axon-Glia Interactions after Peripheral Nerve Injury in Zebrafish. *Int. J. Mol. Sci.* **2021**, *22*, 2484. [CrossRef]
7. Derrick, C.J.; Santos-Ledo, A.; Eley, L.; Henderson, D.; Chaudhry, B. Sequential action of JNK genes establishes the embryonic left-right axis. *Development* **2022**, *149*, dev200136. [CrossRef]
8. Wang, B.J.; Her, G.M.; Hu, M.-K.; Chen, Y.-W.; Tung, Y.-T.; Wu, P.-Y.; Hsu, W.-M.; Lee, H.; Jin, L.-W.; Hwang, S.-P.L.; et al. ErbB2 regulates autophagic flux to modulate the proteostasis of APP-CTFs in Alzheimer's disease. *Proc. Natl. Acad. Sci. USA* **2017**, *114*, E3129–E3138. [CrossRef]
9. Catalani, E.; Cherubini, A.; Del Quondam, S.; Cervia, D. Regenerative Strategies for Retinal Neurons: Novel Insights in Non-Mammalian Model Organisms. *Int. J. Mol. Sci.* **2022**, *23*, 8180. [CrossRef]
10. Stella, S.L.; Geathers, J.S.; Weber, S.R.; Grillo, M.A.; Barber, A.J.; Sundstrom, J.M.; Grillo, S.L. Neurodegeneration, Neuroprotection and Regeneration in the Zebrafish Retina. *Cells* **2021**, *10*, 633. [CrossRef]
11. Jurisch-Yaksi, N.; Yaksi, E.; Kizil, C. Radial glia in the zebrafish brain: Functional, structural, and physiological comparison with the mammalian glia. *Glia* **2020**, *68*, 2451–2470. [CrossRef]
12. Zhou, L.; McAdow, A.R.; Yamada, H.; Burris, B.; Klatt Shaw, D.K.; Oonk, K.; Poss, K.D.; Mokalled, M.H. Progenitor-derived glia are required for spinal cord regeneration in zebrafish. *Development* **2023**, *150*, dev201162. [CrossRef] [PubMed]
13. Duncan, I.D.; Radcliff, A.B.; Heidari, M.; Kidd, G.; August, B.K.; Wierenga, L.A. The adult oligodendrocyte can participate in remyelination. *Proc. Natl. Acad. Sci. USA* **2018**, *115*, E11807–E11816. [CrossRef] [PubMed]
14. Neely, S.A.; Williamson, J.; Klingseisen, A.; Zoupi, L.; Early, J.; Williams, A.C.; Lyons, D.A. New oligodendrocytes exhibit more abundant and accurate myelin regeneration than those that survive demyelination. *Nat. Neurosci.* **2022**, *25*, 415–420. [CrossRef] [PubMed]
15. Zeng, C.W. Macrophage-Neuroglia Interactions in Promoting Neuronal Regeneration in Zebrafish. *Int. J. Mol. Sci.* **2023**, *24*, 6483. [CrossRef]
16. Thornton, M.A.; Hughes, E.G. Neuron-oligodendroglia interactions: Activity-dependent regulation of cellular signaling. *Neurosci. Lett.* **2020**, *727*, 134916. [CrossRef]

17. Olguín-Albuerne, M.; Morán, J. Redox Signaling Mechanisms in Nervous System Development. *Antioxid. Redox Signal.* **2018**, *28*, 1603–1625. [CrossRef]
18. Herb, M.; Gluschko, A.; Schramm, M. Reactive Oxygen Species: Not Omnipresent but Important in Many Locations. *Front. Cell Dev. Biol.* **2021**, *9*, 716406. [CrossRef]
19. Pérez Estrada, C.; Covacu, R.; Sankavaram, S.R.; Svensson, M.; Brundin, L. Oxidative stress increases neurogenesis and oligodendrogenesis in adult neural progenitor cells. *Stem Cells Dev.* **2014**, *23*, 2311–2327. [CrossRef]
20. Gauron, C.; Rampon, C.; Bouzaffour, M.; Ipendey, E.; Teillon, J.; Volovitch, M.; Vriz, S. Sustained production of ROS triggers compensatory proliferation and is required for regeneration to proceed. *Sci. Rep.* **2013**, *3*, 2084. [CrossRef]
21. Wu, Y.; Lim, Y.-W.; Stroud, D.A.; Martel, N.; Hall, T.E.; Lo, H.P.; Ferguson, C.; Ryan, M.T.; McMahon, K.-A.; Parton, R.G. Caveolae sense oxidative stress through membrane lipid peroxidation and cytosolic release of CAVIN1 to regulate NRF2. *Dev. Cell* **2023**, *58*, 376–397.e4. [CrossRef] [PubMed]
22. Anand, S.K.; Sahu, M.R.; Mondal, A.C. Induction of oxidative stress and apoptosis in the injured brain: Potential relevance to brain regeneration in zebrafish. *Mol. Biol. Rep.* **2021**, *48*, 5099–5108. [CrossRef] [PubMed]
23. Spaas, J.; van Veggel, L.; Schepers, M.; Tiane, A.; van Horssen, J.; Wilson, D.M.; Moya, P.R.; Piccart, E.; Hellings, N.; Eijnde, B.O.; et al. Oxidative stress and impaired oligodendrocyte precursor cell differentiation in neurological disorders. *Cell Mol. Life Sci.* **2021**, *78*, 4615–4637. [CrossRef] [PubMed]
24. Santos-Ledo, A.; de Luxán-Delgado, B.; Caballero, B.; Potes, Y.; Rodríguez-González, S.M.; Boga, J.A.; Coto-Montes, A.; García-Macia, M. Melatonin Ameliorates Autophagy Impairment in a Metabolic Syndrome Model. *Antioxidants* **2021**, *10*, 796. [CrossRef]
25. Aranda-Martínez, P.; Fernández-Martínez, J.; Ramírez-Casas, Y.; Guerra-Librero, A.; Rodríguez-Santana, C.; Escames, G.; Acuña-Castroviejo, D. The Zebrafish, an Outstanding Model for Biomedical Research in the Field of Melatonin and Human Diseases. *Int. J. Mol. Sci.* **2022**, *23*, 7438. [CrossRef]
26. Giacomini, A.C.V.V.; Teixeira, K.H.; Marcon, L.; Scolari, N.; Bueno, B.W.; Genario, R.; de Abreu, N.S.; Demin, K.A.; Galstyan, D.S.; Kalueff, A.V.; et al. Melatonin treatment reverses cognitive and endocrine deficits evoked by a 24-h light exposure in adult zebrafish. *Neurosci. Lett.* **2020**, *733*, 135073. [CrossRef]
27. Carney, T.J.; Dutton, K.A.; Greenhill, E.; Delfino-Machín, M.; Dufourcq, P.; Blader, P.; Kelsh, R.N. A direct role for Sox10 in specification of neural crest-derived sensory neurons. *Development* **2006**, *133*, 4619–4630. [CrossRef]
28. Blasky, A.J.; Pan, L.; Moens, C.B.; Appel, B. Pard3 regulates contact between neural crest cells and the timing of Schwann cell differentiation but is not essential for neural crest migration or myelination. *Dev. Dyn.* **2014**, *243*, 1511–1523. [CrossRef]
29. Hughes, A.N.; Appel, B. Microglia phagocytose myelin sheaths to modify developmental myelination. *Nat. Neurosci.* **2020**, *23*, 1055–1066. [CrossRef]
30. Santos-Ledo, A.; Pérez-Montes, C.; DeOliveira-Mello, L.; Arévalo, R.; Velasco, A. Oligodendrocyte origin and development in the zebrafish visual system. *J. Comp. Neurol.* **2023**, *531*, 515–527. [CrossRef]
31. Xiao, Y.; Petrucco, L.; Hoodless, L.J.; Portugues, R.; Czopka, T. Oligodendrocyte precursor cells sculpt the visual system by regulating axonal remodeling. *Nat. Neurosci.* **2022**, *25*, 280–284. [CrossRef] [PubMed]
32. Liu, Q.; Londraville, R.L. Using the adult zebrafish visual system to study cadherin-2 expression during central nervous system regeneration. *Methods Cell Sci.* **2003**, *25*, 71–78. [CrossRef] [PubMed]
33. Cavallo, A.; Hassan, M. Stability of melatonin in aqueous solution. *J. Pineal Res.* **1995**, *18*, 90–92. [CrossRef] [PubMed]
34. Morant-Ferrando, B.; Jimenez-Blasco, D.; Alonso-Batan, P.; Agulla, J.; Lapresa, R.; Garcia-Rodriguez, D.; Yunta-Sanchez, S.; Lopez-Fabuel, I.; Fernandez, E.; Carmeliet, P.; et al. Fatty acid oxidation organizes mitochondrial supercomplexes to sustain astrocytic ROS and cognition. *Nat. Metab.* **2023**, *5*, 1290–1302. [CrossRef] [PubMed]
35. Darvesh, S.; MacDonald, S.E.; Losier, A.M.; Martin, E.; Hopkins, D.A.; Armour, J.A. Cholinesterases in cardiac ganglia and modulation of canine intrinsic cardiac neuronal activity. *J. Auton. Nerv. Syst.* **1998**, *71*, 75–84. [CrossRef]
36. Beccari, S.; Sierra-Torre, V.; Valero, J.; Pereira-Iglesias, M.; García-Zaballa, M.; Soria, F.N.; De Las Heras-Garcia, L.; Carretero-Guillen, A.; Capetillo-Zarate, E.; Domercq, M.; et al. Microglial phagocytosis dysfunction in stroke is driven by energy depletion and induction of autophagy. *Autophagy* **2023**, *19*, 1952–1981. [CrossRef]
37. Corbo, C.P.; Othman, N.A.; Gutkin, M.C.; Alonso, A.e.C.; Fulop, Z.L. Use of different morphological techniques to analyze the cellular composition of the adult zebrafish optic tectum. *Microsc. Res. Tech.* **2012**, *75*, 325–333. [CrossRef]
38. Karaduman, A.; Karoglu-Eravsar, E.T.; Kaya, U.; Aydin, A.; Adams, M.M.; Kafaligonul, H. The optomotor response of aging zebrafish reveals a complex relationship between visual motion characteristics and cholinergic system. *Neurobiol. Aging* **2021**, *98*, 21–32. [CrossRef]
39. Santos-Ledo, A.; Washer, S.; Dhanaseelan, T.; Eley, L.; Alqatani, H.; Chrystal, P.W.; Papoutsi, T.; Henderson, D.J.; Chaudhry, B. Alternative splicing of jnk1a in zebrafish determines first heart field ventricular cardiomyocyte numbers through modulation of hand2 expression. *PLoS Genet.* **2020**, *16*, e1008782. [CrossRef]
40. Ishida, T.; Nakajima, T.; Kudo, A.; Kawakami, A. Phosphorylation of Junb family proteins by the Jun N-terminal kinase supports tissue regeneration in zebrafish. *Dev. Biol.* **2010**, *340*, 468–479. [CrossRef]
41. Love, N.R.; Chen, Y.; Ishibashi, S.; Kritsiligkou, P.; Lea, R.; Koh, Y.; Gallop, J.L.; Dorey, K.; Amaya, E. Amputation-induced reactive oxygen species are required for successful Xenopus tadpole tail regeneration. *Nat. Cell Biol.* **2013**, *15*, 222–228. [CrossRef] [PubMed]

42. Hunter, M.V.; Willoughby, P.M.; Bruce, A.E.E.; Fernandez-Gonzalez, R. Oxidative Stress Orchestrates Cell Polarity to Promote Embryonic Wound Healing. *Dev. Cell* **2018**, *47*, 377–387.e4. [CrossRef] [PubMed]
43. Xu, S.; Chisholm, A.D. *C. elegans* epidermal wounding induces a mitochondrial ROS burst that promotes wound repair. *Dev. Cell* **2014**, *31*, 48–60. [CrossRef]
44. Narra, S.S.; Rondeau, P.; Fernezelian, D.; Gence, L.; Ghaddar, B.; Bourdon, E.; d'Hellencourt, C.L.; Rastegar, S.; Diotel, N. Distribution of microglia/immune cells in the brain of adult zebrafish in homeostatic and regenerative conditions: Focus on oxidative stress during brain repair. *J. Comp. Neurol.* **2023**, *531*, 238–255. [CrossRef]
45. Smith, J.; Ladi, E.; Mayer-Proschel, M.; Noble, M. Redox state is a central modulator of the balance between self-renewal and differentiation in a dividing glial precursor cell. *Proc. Natl. Acad. Sci. USA* **2000**, *97*, 10032–10037. [CrossRef]
46. Chang, J.Y.; Yu, F.; Shi, L.; Ko, M.L.; Ko, G.Y. Melatonin Affects Mitochondrial Fission/Fusion Dynamics in the Diabetic Retina. *J. Diabetes Res.* **2019**, *2019*, 8463125. [CrossRef]
47. Laste, G.; Ripoll Rozisky, J.; Caumo, W.; Lucena da Silva Torres, I. Short- but not long-term melatonin administration reduces central levels of brain-derived neurotrophic factor in rats with inflammatory pain. *Neuroimmunomodulation* **2015**, *22*, 358–364. [CrossRef] [PubMed]
48. Olivier, P.; Fontaine, R.H.; Loron, G.; Van Steenwinckel, J.; Biran, V.; Massonneau, V.; Kaindl, A.; Dalous, J.; Charriaut-Marlangue, C.; Aigrot, M.-S.; et al. Melatonin promotes oligodendroglial maturation of injured white matter in neonatal rats. *PLoS ONE* **2009**, *4*, e7128. [CrossRef]
49. Li, B.; Zhang, Z.; Wang, H.; Zhang, D.; Han, T.; Chen, H.; Chen, J.; Chen, Z.; Xie, Y.; Wang, L.; et al. Melatonin promotes peripheral nerve repair through Parkin-mediated mitophagy. *Free Radic. Biol. Med.* **2022**, *185*, 52–66. [CrossRef]
50. Sajedin, A.; Menhaj, M.B.; Vahabie, A.H.; Panzeri, S.; Esteky, H. Cholinergic Modulation Promotes Attentional Modulation in Primary Visual Cortex—A Modeling Study. *Sci. Rep.* **2019**, *9*, 20186. [CrossRef]
51. Issac, P.K.; Guru, A.; Velayutham, M.; Pachaiappan, R.; Arasu, M.V.; Al-Dhabi, N.A.; Choi, K.C.; Harikrishnan, R.; Arockiaraj, J. Oxidative stress induced antioxidant and neurotoxicity demonstrated in vivo zebrafish embryo or larval model and their normalization due to morin showing therapeutic implications. *Life Sci.* **2021**, *283*, 119864. [CrossRef] [PubMed]
52. Caruso, G.I.; Korde, D.S.; Humpel, C. Melatonin Supports the Survival of Cholinergic Neurons in Organotypic Brain Slices of the Basal Nucleus of Meynert. *Pharmacology* **2023**, *108*, 204–212. [CrossRef]
53. Chen, B.H.; Park, J.H.; Lee, T.-K.; Song, M.; Kim, H.; Lee, J.C.; Kim, Y.-M.; Lee, C.-H.; Hwang, I.K.; Kang, I.J.; et al. Melatonin attenuates scopolamine-induced cognitive impairment via protecting against demyelination through BDNF-TrkB signaling in the mouse dentate gyrus. *Chem. Biol. Interact.* **2018**, *285*, 8–13. [CrossRef] [PubMed]
54. Albazal, A.; Delshad, A.A.; Roghani, M. Melatonin reverses cognitive deficits in streptozotocin-induced type 1 diabetes in the rat through attenuation of oxidative stress and inflammation. *J. Chem. Neuroanat.* **2021**, *112*, 101902. [CrossRef] [PubMed]
55. Liu, P.; Cheng, M.; Guo, J.; Cao, D.; Luo, J.; Wan, Y.; Fang, Y.; Jin, Y.; Xie, S.-S.; Liu, J. Dual functional antioxidant and butyrylcholinesterase inhibitors for the treatment of Alzheimer's disease: Design, synthesis and evaluation of novel melatonin-alkylbenzylamine hybrids. *Bioorg. Med. Chem.* **2023**, *78*, 117146. [CrossRef]
56. Harper, S.J.; Saporito, M.S.; Hewson, L.; Young, L.; Smith, D.; Rigby, M.; Jackson, P.; Curtis, N.; Swain, C.; Hefti, F.; et al. CEP-1347 increases ChAT activity in culture and promotes cholinergic neurone survival following fimbria-fornix lesion. *Neuroreport* **2000**, *11*, 2271–2276. [CrossRef]
57. Liang, X.; Nagai, A.; Terashima, M.; Sheikh, A.M.; Shiota, Y.; Mitaki, S.; Kim, S.U.; Yamaguchi, S. Cystatin C induces apoptosis and tyrosine hydroxylase gene expression through JNK-dependent pathway in neuronal cells. *Neurosci. Lett.* **2011**, *496*, 100–105. [CrossRef]
58. Santabárbara-Ruiz, P.; López-Santillán, M.; Martínez-Rodríguez, I.; Binagui-Casas, A.; Pérez, L.; Milán, M.; Corominas, M.; Serras, F. ROS-Induced JNK and p38 Signaling Is Required for Unpaired Cytokine Activation during Drosophila Regeneration. *PLoS Genet.* **2015**, *11*, e1005595. [CrossRef]
59. Canedo-Antelo, M.; Serrano, M.P.; Manterola, A.; Ruiz, A.; Llavero, F.; Mato, S.; Zugaza, J.L.; Pérez-Cerdá, F.; Matute, C.; Sánchez-Gómez, M.V. Inhibition of Casein Kinase 2 Protects Oligodendrocytes from Excitotoxicity by Attenuating JNK/p53 Signaling Cascade. *Front. Mol. Neurosci.* **2018**, *11*, 333. [CrossRef]
60. Breton, J.M.; Long, K.L.P.; Barraza, M.K.; Perloff, O.S.; Kaufer, D. Hormonal Regulation of Oligodendrogenesis II: Implications for Myelin Repair. *Biomolecules* **2021**, *11*, 290. [CrossRef]
61. Klymenko, A.; Lutz, D. Melatonin signalling in Schwann cells during neuroregeneration. *Front. Cell Dev. Biol.* **2022**, *10*, 999322. [CrossRef]
62. Ghareghani, M.; Scavo, L.; Jand, Y.; Farhadi, N.; Sadeghi, H.; Ghanbari, A.; Mondello, S.; Arnoult, D.; Gharaghani, S.; Zibara, K. Melatonin Therapy Modulates Cerebral Metabolism and Enhances Remyelination by Increasing PDK4 in a Mouse Model of Multiple Sclerosis. *Front. Pharmacol.* **2019**, *10*, 147. [CrossRef] [PubMed]
63. Wen, J.; Ariyannur, P.S.; Ribeiro, R.; Tanaka, M.; Moffett, J.R.; Kirmani, B.F.; Namboodiri, A.M.A.; Zhang, Y. Efficacy of N-Acetylserotonin and Melatonin in the EAE Model of Multiple Sclerosis. *J. Neuroimmune Pharmacol.* **2016**, *11*, 763–773. [CrossRef] [PubMed]
64. Tiong, Y.L.; Ng, K.Y.; Koh, R.Y.; Ponnudurai, G.; Chye, S.M. Melatonin Prevents Oxidative Stress-Induced Mitochondrial Dysfunction and Apoptosis in High Glucose-Treated Schwann Cells via Upregulation of Bcl2, NF-κB, mTOR, Wnt Signalling Pathways. *Antioxidants* **2019**, *8*, 198. [CrossRef] [PubMed]

65. Duregotti, E.; Negro, S.; Scorzeto, M.; Zornetta, I.; Dickinson, B.C.; Chang, C.J.; Montecucco, C.; Rigoni, M. Mitochondrial alarmins released by degenerating motor axon terminals activate perisynaptic Schwann cells. *Proc. Natl. Acad. Sci. USA* **2015**, *112*, E497–E505. [CrossRef] [PubMed]
66. Labit, E.; Rabiller, L.; Rampon, C.; Guissard, C.; André, M.; Barreau, C.; Cousin, B.; Carrière, A.; Eddine, M.A.; Pipy, B.; et al. Opioids prevent regeneration in adult mammals through inhibition of ROS production. *Sci. Rep.* **2018**, *8*, 12170. [CrossRef] [PubMed]
67. Bai, H.; Fang, C.-W.; Shi, Y.; Zhai, S.; Jiang, A.; Li, Y.-N.; Wang, L.; Liu, Q.-L.; Zhou, G.-Y.; Cao, J.-H.; et al. Mitochondria-derived H_2O_2 triggers liver regeneration via FoxO3a signaling pathway after partial hepatectomy in mice. *Cell Death Dis.* **2023**, *14*, 216. [CrossRef]
68. Ghareghani, M.; Dokoohaki, S.; Ghanbari, A.; Farhadi, N.; Zibara, K.; Khodadoust, S.; Parishani, M.; Ghavamizadeh, M.; Sadeghi, H. Melatonin exacerbates acute experimental autoimmune encephalomyelitis by enhancing the serum levels of lactate: A potential biomarker of multiple sclerosis progression. *Clin. Exp. Pharmacol. Physiol.* **2017**, *44*, 52–61. [CrossRef]

antioxidants

Review

Melatonin in Neurodevelopmental Disorders: A Critical Literature Review

Cyrille Feybesse [1,*], Sylvie Chokron [2] and Sylvie Tordjman [1,2,3,*]

1 Pôle Hospitalo-Universitaire de Psychiatrie de l'Enfant et de l'Adolescent (PHUPEA), Centre Hospitalier Guillaume Regnier, 154 rue de Châtillon, 35000 Rennes, France
2 Integrative Neuroscience and Cognition Center (INCC), CNRS UMR 8002, Université Paris Cité, 45 rue des Saints-Pères, 75006 Paris, France; sylvie.chokron@cnrs.fr
3 Faculté de Médecine, Université de Rennes, 2 Avenue du Professeur Léon Bernard, 35000 Rennes, France
* Correspondence: c.feybesse@ch-guillaumeregnier.fr (C.F.); s.tordjman@ch-guillaumeregnier.fr (S.T.); Tel.: +33-2-99-51-06-04 (C.F. & S.T.)

Abstract: The article presents a review of the relationships between melatonin and neurodevelopmental disorders. First, the antioxidant properties of melatonin and its physiological effects are considered to understand better the role of melatonin in typical and atypical neurodevelopment. Then, several neurodevelopmental disorders occurring during infancy, such as autism spectrum disorder or neurogenetic disorders associated with autism (including Smith–Magenis syndrome, Angelman syndrome, Rett's syndrome, Tuberous sclerosis, or Williams–Beuren syndrome) and neurodevelopmental disorders occurring later in adulthood like bipolar disorder and schizophrenia, are discussed with regard to impaired melatonin production and circadian rhythms, in particular, sleep–wake rhythms. This article addresses the issue of overlapping symptoms that are commonly observed within these different mental conditions and debates the role of abnormal melatonin production and altered circadian rhythms in the pathophysiology and behavioral expression of these neurodevelopmental disorders.

Keywords: neurodevelopmental disorders; brain development; melatonin; circadian rhythms; sleep–wake rhythms; sleep disturbance; autism; schizophrenia

check for updates

Citation: Feybesse, C.; Chokron, S.; Tordjman, S. Melatonin in Neurodevelopmental Disorders: A Critical Literature Review. *Antioxidants* **2023**, *12*, 2017. https://doi.org/10.3390/antiox12112017

Academic Editors: Marina Garcia-Macia and Adrián Santos-Ledo

Received: 10 September 2023
Revised: 29 October 2023
Accepted: 31 October 2023
Published: 20 November 2023

1. Introduction

Melatonin (5-Methoxy-N-acetyltryptamine, N-acetyl-5-methoxytryptamine, NSC-113928) is a neurohormone synthesized in the pineal gland, and its secretion is enhanced by darkness and inhibited by light (daylight or artificial light). Indeed, Melatonin is often labeled as being the «darkness hormone» for having its peak secretion during the late evening hours while its production is suppressed by light [1,2]. Melatonin is a pleiotropic neuroendocrine molecule released in the brain at night and plays a crucial role in the synchronization of circadian rhythms, including sleep–wake rhythms, and controlling seasonal rhythms, including reproduction [3]. Melatonin is considered an endogenous synchronizer located in the suprachiasmatic nuclei (SCN) of the hypothalamus, stabilizing bodily rhythms and as a chronobiotic molecule that reinforces variations or adjusts the time period of the central biological clock [4]. It derives from serotonin to form first N-acetylserotonin (NAS) by acetylation through the enzyme arylalkylamine-N-acetyltransferase (AANAT, EC: 2.3.1.87) and then forms melatonin by methylation through the enzyme acetylserotonin-O-methyltransferase (ASMT, EC: 2.1.1.4). Serotonin also contributes to the development of the brain before acting as a neurotransmitter on mature brain; in particular, serotonin has a role on dendritic development and branching in the hippocampus and cortex [5]. Melatonin synthesis is regulated periodically in the SCN. This clock determines the circadian rhythm of melatonin secretion. Melatonin is a powerful natural antioxidant providing beneficial and protective effects against oxidative stress [6]; melatonin is found in mitochondria at concentrations higher than the ones in the blood, suggesting that melatonin could be viewed as a mitochondrial antioxidant [7]. A specific section on the antioxidant properties of melatonin is developed in this article. In addition, melatonin receptors can be found in the regions associated with the master circadian clock [8]. Melatonin has an important role in the circadian cycle, which is the temporal organization of physiological, cellular, neural, biochemical, and behavioral processes. It helps the body to anticipate the different phases of the day in a proactive way [9]. Melatonin affects various temporal processes via mainly high G protein-coupled melatonin receptors 1 and 2 (MT1, MT1) widely distributed across brain and peripherical tissues (for a review on melatonin receptors, see Tordjman et al. [10]). Many melatonin receptors are found in neuroendocrine and acoustic-vocal integration areas [11]. Animal models with diurnal birds or breeding green treefrogs changed vocal behaviors through melatonin interactions with receptors influencing both inhibitory and excitatory signaling; melatonin mediates the regulation of neural excitability in vocal-motor circuits. Scholars are now aware of the role melatonin can have on social communication and its development [12]. In the same line, as discussed below, melatonin is important for child development even before birth.

2. Melatonin and Typical Neurodevelopment

Since the beginning of the normal pregnancy process, melatonin secretion is involved with the oocyte quality and the parturition course. The night-time concentrations start increasing after 24 weeks of gestation and reach significantly higher levels by 32 weeks. Melatonin receptors are widespread in the embryo and fetus since early stages. There is solid evidence that this neurohormone has a role in fetal neuroprotection, as normal sleep patterns are involved in human neurodevelopment even at this stage. It is noteworthy that melatonin is the regulating factor of the fetal sleep patterns development set in late pregnancy. Voiculescu et al. [13] found strong evidence that melatonin has a positive effect on the outcome of compromised pregnancies. Melatonin concentrations progressively increase in maternal blood during pregnancy, reaching their maximum at term. Researchers have also found melatonin in amniotic fluid [14]. In addition, chronic disruption leads to reproductive dysfunction and appears to be an important factor in the development of offspring diseases in adulthood (this relates to the concept of fetal programming). Melatonin decreases in conditions associated with serious outcomes for the fetus and appears to be involved in preeclampsia and intrauterine growth restriction [15]. Animal models of fetal growth restriction in newborn lambs showed that maternal administration of melatonin reduced fetal hypoxia, improved neurodevelopment, and decreased brain injury and oxidative stress [16]. Taken together, these studies suggest that the effects of melatonin on the development of human fetuses appear to be not limited to circadian rhythmicity.

After birth, melatonin is secreted into the general circulation and in the cerebrospinal fluid, allowing this neurohormone to circulate throughout the body and the brain [17]. In the brain, extra-pineal melatonin behaves similarly to neurotrophic molecules [18,19] and is capable of modulating cell survival, proliferation, and differentiation by signaling pathways that can be triggered in response to stimulation of membrane and intracellular receptors. Thus, melatonin plays a crucial role in brain neuroplasticity and neurodevelopment via neurotrophic factors, promoting its growth and survival [20,21].

In addition, natural maternal melatonin is also a powerful free radical scavenger and an antioxidant protecting the baby and fetus within the maternal-placental-fetal unit [22]. Indeed, maternal melatonin deprivation during gestation or lactation has been shown to delay the infant's physical maturation and neurobehavioral development. Melatonin found in the gastrointestinal tract of newborns is of maternal origin knowing that melatonin easily penetrates the placenta during the preterm period; melatonin is secreted into the mother's milk after birth and may be involved in the production of meconium [2,23–25]. Melatonin also follows a circadian rhythm in human breast milk, and studies report undetectable levels of melatonin during the day and high levels at night [22,26]. Several studies reporting that melatonin rhythms set around 3 months of age in typical development allow us to understand better why infants begin at this period of life to have more regular sleep–wake cycles combined with regular night-time sleep lasting 6 to 8 h [23,27]. Similarly, the infant's circadian cortisol rhythm is only set up at around 3 months of age [28]. The development of circadian rhythms occurs fully between 49-and 52-weeks post-conception and corresponds to the developmental phase when increased periods of deep sleep during the night are consolidated as infants have fewer nocturnal awakenings [23]. Without maternal melatonin, infants establish circadian rhythms mainly by neurological maturation [29]. Infants born prematurely or facing circumstances related to normal intrauterine development show a significant delay in pineal rhythmicity. In the life span, the highest melatonin levels are found in children younger than 4 years [30]. Infants benefit from increased cell proliferation differentiation and survival rates of novel neurons in the hippocampus when melatonin is administrated after birth [31]. It also plays a role in terms of having effects on excitation/inhibition balance by changes in neurotransmitter levels [32]. The balance between excitation and inhibition in synaptic inputs of neural circuits must be tight to avoid the pathogenesis of neurodevelopmental disorders [33]. Melatonin levels decline progressively with age, although circadian rhythms tend to be highly consistent day to day at any age [30]. Biological aging is a natural process leading to the disruption of circadian rhythms; aging is associated with the dampening of circadian gene expression, as aging is associated with an increase in oxidative stress [34].

The antioxidant properties of melatonin and its physiological effects have first to be considered in the next section to understand better the role of melatonin in typical and atypical neurodevelopment, as discussed in the following sections.

3. Antioxidant Properties and Physiological Effects

The antioxidant action of melatonin involved in the cardiovascular, immune, gastrointestinal, oncostatic, and brain-protective effects of melatonin is presented in this section. The protective effects of melatonin are summarized in Figure 1 and are developed below.

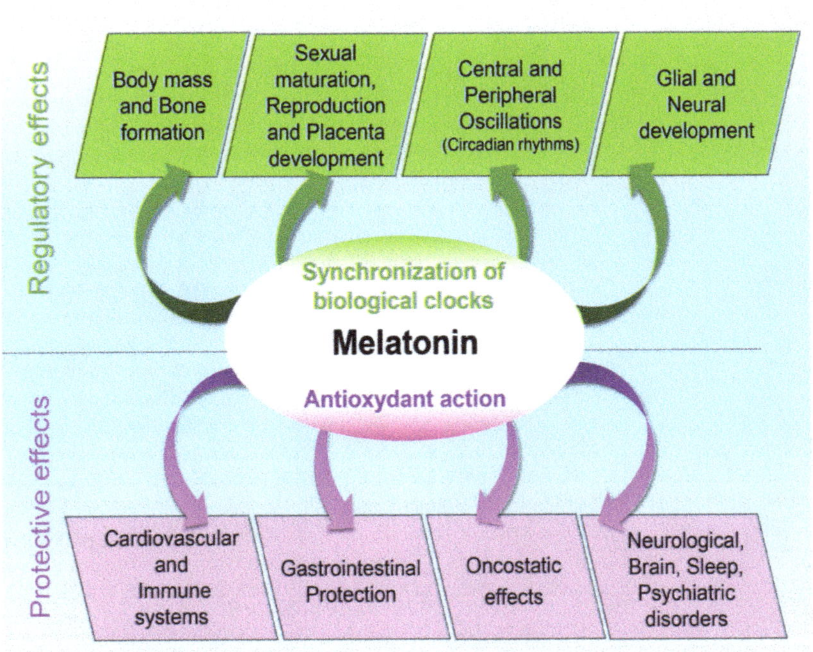

Figure 1. Antioxidant action and protective effects of melatonin.

Melatonin regulates blood pressure and autonomic cardiovascular and immune function, in addition to other physiological processes such as free radical detoxification and antioxidant effects via MT3 receptors, protecting the brain from oxidative stress [35–43].

Also, the antioxidant action of melatonin protects the gastrointestinal tract from ulcerations by (1) reducing hydrochloric acid secretion and oxidative effects of bile acids on the intestinal epithelium; (2) increasing microcirculation and bicarbonate secretion from duodenal mucosa via MT2 receptors (this alkaline secretion is an important mechanism for duodenal protection against gastric acid); and (3) fostering epithelial regeneration [4,44].

Melatonin has direct immuno-enhancement effects in both humans and animals, which is relevant to its function in immunological regulation [45,46]. The production of cytokines, and more precisely, certain interleukins (IL-2, IL-6, and IL-12), is selectively stimulated by melatonin [47]. Melatonin also improves T helper immune responses [46,48]. Additionally, the antioxidant properties of melatonin contribute to its immunostimulatory effects [47] and act indirectly by lowering the production of nitric oxide, which helps to reduce the inflammatory response [49].

Furthermore, given that oxidative stress is implicated in the origin, promotion, and development of carcinogenesis, the oncostatic protective effects of melatonin have been documented and linked to its anti-oxidative action [50,51].

Regarding brain protection, there is growing experimental evidence showing therapeutic benefits of melatonin for prematurity as well as for neurodegenerative conditions like Alzheimer's disease, Parkinson's disease, Huntington's disease, and amyotrophic lateral sclerosis (for a review, see Polimeni et al. [52]). To determine the precise therapeutic concentrations required according to the specific disease, age of individuals, and brain lesion, as well as to examine the short- and long-term effects of melatonin on physiological, functional, and cognitive outcomes, further studies and clinical trials are requested in preterm neonates as well as aging adults. Finally, in addition to its therapeutic benefits for sleep problems, melatonin is of major interest regarding its antioxidant action, increasing brain protection against oxidative stress and inflammation, in general for atypical development, and in particular for neurodevelopmental disorders.

4. Melatonin and Atypical Neurodevelopment

The establishment and maintenance of several circadian rhythms, such as the one involved in the secretion of melatonin and sleep–wake rhythms, depend upon the interaction of light perceived by the retina and the suprachiasmatic nucleus [23]. Some stressful and traumatic situations experienced by pregnant women decrease maternal melatonin production, and this can have an impact on the internal rhythms and post-natal development of the fetus [24]. Brain anomalies such as the reduction of the hippocampus volume can occur during early childhood or before birth due to the impact of stress in this brain region [25]. In the same way, children in their primary infancy who are coping with stressful situations or traumas like insecure attachment, separation from the mother, or abuse can consequently develop brain anomalies [25]. These traumas have long-term effects on cognitive functioning [22]. Among children diagnosed with developmental disabilities, many of them (the frequencies range from 25% to 85%) also show sleep problems [26].

Tauman et al. [53] show relationships between low melatonin production in the first weeks of life and impaired psychomotor development by measuring nocturnal urinary excretion of 6-sulfatoxymelatonin. Melatonin production is related to brain functioning and has effects on neurological development, given its impact on the onset of circadian rhythms, as shown by a study on REM sleep (i.e., the rapid eye movement during sleep) [54]. The relationships between low 6-sulfatoxymelatonin excretion and developmental delay appear very early, including reports at 16 weeks of age [33].

5. Melatonin and Neurodevelopmental Disorders

Neurodevelopmental disorders encompass intellectual disability, autism spectrum disorder (ASD), or neurogenetic disorders associated with ASD, as well as schizophrenia or bipolar disorder occurring later in life, as these two last conditions are more and more considered as neurodevelopmental disorders [55]. These different disorders can be seen lying on a neurodevelopmental continuum and having considerable comorbidity, as observed in patients showing an overall deficit of melatonin production [56]. A high prevalence of altered circadian rhythms, including sleep–wake rhythms, was observed in individuals with these neurodevelopmental disorders [57], strengthening the interest to focus on melatonin metabolism in these disorders given the role of melatonin in sleep–wake rhythms, synchronization of circadian rhythms, and neural development.

5.1. Relationships between Melatonin and Neurodevelopmental Disorders in Infancy

Melatonin production and circadian rhythms have been consistently associated with mental disorders that occur in primary infancy. Early neurodevelopmental disorders, such as autism spectrum disorder (ASD), have been associated with a dysregulation of circadian cycles, especially the circadian cycle of melatonin production. It is noteworthy that alterations in circadian sleep–wake rhythms are frequently observed in these neurodevelopmental disorders with abnormalities in melatonin secretion.

ASD is a behavioral syndrome with altered sensory motor development and sleep–wake rhythms [58–60]. Key behavioral features of ASD are characterized by impairments in social communication and restricted interests with repetitive patterns of behaviors [61–63]. This condition is often associated with common comorbidities such as intellectual disability, epilepsy, and severe sleep disorders [64]. Family home movies of infants who were subsequently diagnosed with ASD showed motor and emotional asynchrony between infants before 12 months of age and their parents [65,66]. These early signs are not specific to autism but offer indicators of atypical development, which become more evident in the second year of life [65]. In later stages of development, several signs were also reported, such as abnormal eye contact and other social communication impairments in learning through imitation (people's faces, gestures, or vocal signals), social reciprocity, joint attention, and orienting to name or body language [67,68].

Melatonin is a common pharmacologic treatment used to deal with sleep disturbance due to circadian phase delay. The melatonin treatment provides a significant decrease in sleep latency and night awakenings and an increase in sleep quality and sleep efficiency (for a review, see Tordjman et al. [58,68]). However, Moon et al. [69] indicated that evidence of the therapeutic benefits of melatonin on psychiatric disorders is robust only in autism, attention deficit hyperactivity disorder (ADHD), and neurocognitive disorders. Sleep disturbances, such as falling asleep or having night awakenings, are relatively common among children with ASD, and their prevalence is higher when compared with children with typical development. This contributes to a variety of disturbances in their daily lives, such as behavioral problems, self-injurious behaviors, and other-injurious behaviors, and emotional problems like depression or anxiety. Finally, sleep deprivation among children with ASD also has negative ecological consequences since it affects parents' or caregivers' overall mental health. These difficulties can be related to odd bedtime routines and bedtime resistance [70].

Several studies reported that individuals with ASD showed lower melatonin levels in urine, plasma, and pineal gland than control groups [64]. Furthermore, several studies provided evidence of relationships between melatonin deficit and social communication impairments that are prevalent in neurodevelopmental disorders. In ASD, a lack of melatonin production was associated with language impairments [71,72]. The Tordjman et al. studies showed that abnormally low nocturnal melatonin excretion is significantly associated with severe autistic social communication impairments, especially verbal communication and social imitative play impairments in children and adolescents with ASD [73,74]. Moreover, this deficit in melatonin may be involved in ASD development through desynchronized, disrupted, and abnormal circadian rhythms but also through several physiological pathways, including a lack of antioxidant protective effects (as seen in Section 3 on antioxidant properties and physiological effects, melatonin protects the brain from oxidative stress and its antioxidant action decreases the production of nitric oxide which helps in turn to decrease the inflammatory response), and impairments in neurotransmission, synaptic plasticity and metabolic pathways [75]. It is noteworthy that nitro-oxidative stress, immune-inflammatory, neurotransmission, synaptic plasticity, and metabolic pathways are also under the control of the circadian clock [75]. In addition, researchers found metabolic disorders and neurochemical imbalances in the melatonin/serotonin system of children with ASD or Down syndrome [76]. It was suggested that the autistic, well-replicated hyperserotonemia [73] could cause a loss of serotonin terminals [77], possibly involved in autistic behaviors, given that serotonin production was associated with poor social interaction, emotional detachment, and aggression towards others [78]. Concerning Down syndrome, associations were observed with a serotonin deficit in the postmortem brain, cerebrospinal fluid, and blood [5].

Furthermore, altered melatonin circadian rhythms and impaired melatonin secretion are also reported in several neurogenetic disorders associated with autism, such as Smith–Magenis syndrome, Angelman syndrome, Rett's syndrome, Tuberous sclerosis, or Williams–Beuren syndrome. A summary concerning the melatonin abnormalities found in these neurogenetic disorders associated with autism is presented in Table 1 and includes particular sleep problems, melatonin impairments, and the response to melatonin therapy observed in these developmental neurogenetic disorders.

Table 1. Melatonin abnormality and therapy in neurogenetic disorders associated with autism.

Neurogenetic Disorder	Neurogenetic Disorder Frequency, Estimated Rate (%) of Autism in the Disorder, and Estimated Rate (%) of the Disorder in Autism	Age of Diagnosis	Phenotype (Including Autistic Behaviors and Intellectual Functioning)	Sleep Problems	Melatonin Abnormality	Response to Melatonin Therapy
Smith-Magenis syndrome (SMS) Chromosome 17p11.2 microdeletion encompassing retinoic acid-induced 1 (RAI1) or a mutation in the *RAI1* gene [79–82]	- Frequency: 1 in 15,000–25,000 individuals [82,83] - Estimated rate of autism in SMS: 50–100 [84] - Estimated rate of SMS in autism: NA [84]	Many of the features of SMS are subtle in infancy and early childhood and become more recognizable with advancing age. Despite increased clinical awareness of SMS as well as improved cytogenetic technologies, many children are not definitively diagnosed until early childhood or even school age [85]	Facial dysmorphism, peripheral neuropathy, hypotonia, early feeding problems. Tantrums, self-injurious and stereotyped behaviors, sameness, developmental delay in vocalizations but possible social contact. Normal intellectual functioning to moderate intellectual disability [84,86,87]	Disrupted sleep patterns with shortened sleep cycles are characteristic of SMS and begin typically during the months after birth. Reports of excessive daytime sleepiness, increased sleep latency, frequent nocturnal and early morning awakenings due to an inverted circadian rhythm of melatonin [88]	Inverted circadian rhythm of melatonin secretion [88]	Melatonin therapy is used to regulate sleep problems. Combined with exogenous PRM (prolonged release melatonin), blockade of endogenous melatonin production during the day by the adrenergic antagonist acebutolol can improve impaired sleep and behaviors, and increase melatonin concentrations [89] Patients aged 3–18 years were given PRM (4 to 6 mg/day) as a single evening dose over a treatment duration of 6–72 months. Within 3 months, parents report improvement in sleep duration, sleep latency, number of midnight awakenings and sleep quality. No serious adverse events [88]
Angelman syndrome (AS) Maternal 15q11-q13 deletion, paternal uniparental disomy, mutations of *UBE3A* that encodes ubiquitin protein ligase (UBE3A) [90–94]	- Frequency: 1 in 12,000–20,000 individuals [95] - Estimated rate of autism in AS: 48–80 [84] - Estimated rate of AS in autism: 1 [84]	Developmental delays, between about 6 and 12 months of age, are usually the first signs, and seizures begin often between the age of 2 and 3 years old [96]	Facial dysmorphism, microcephaly, seizures (>1 year), ataxia and walking disturbance, Attention Deficit with Hyperactivity Disorder (ADHD), paroxysmal laughter, tantrums No language, stereotypies, sameness. Severe intellectual disability [84,96–100]	Severe sleep disturbances are common in Angelman syndrome, and are included in the diagnostic criteria [96]	The melatonin secretion profile of patients with Angelman syndrome is impaired, leading to a variety of sleep problems, most prominently in the areas of sleep-wake patterns and sleep duration [100]	Melatonin therapy significantly advanced sleep onset by 28 min, decreased sleep latency by 32 min, increased total sleep time by 56 min, and reduced the number of nights with awakenings from 3.1 to 1.6 nights per week [101]

Table 1. *Cont.*

Neurogenetic Disorder	Neurogenetic Disorder Frequency, Estimated Rate (%) of Autism in the Disorder, and Estimated Rate (%) of the Disorder in Autism	Age of Diagnosis	Phenotype (Including Autistic Behaviors and Intellectual Functioning)	Sleep Problems	Melatonin Abnormality	Response to Melatonin Therapy
Tuberous sclerosis complex (TSC), synonym: Bourneville disease (*TSC1*, 9q34) (*TSC2*, 16p13.3) Pathogenic variants in *TSC1* and *TSC2* genes: 31% and 69%, respectively [102]	- Frequency: 1 in 6000 individuals [103] - Estimated rate of autism in TSC: 25–60 [84] - Estimated rate of TSD in autism: 1–4 [84]	The average age at diagnosis of TSC is 7.5 years with 81% of patients diagnosed before the age of 10. Diagnosis may be difficult because symptoms are not present in all patients, and none are pathognomonic [104]	Autosomal dominant neurocutaneous disorder with ectodermal anomalies clinically diagnosed, renal lesions, seizures, learning disorder. Severe autistic syndrome. Variable intellectual disability [84,103,105–107]	Sleep problems are considered one of the most common behavioral manifestations in children with TSC [108]	Significant differences between TSC melatonin secretion profiles and control ones. Melatonin rhythm but not its amplitude was related to the total number of seizures [109]	Treatment improvement in sleep latency, total sleep time, and sleep fragmentation reported with melatonin at 5 mg dose [110]
Rett's syndrome (RS) Mutation in the *MECP2* gene coding for the methyl CpG binding protein 2 and located at Xq28 [111,112]	- Frequency: 1 in 10,000–15,000 live female births [113] - Estimated rate of autism in RS: 61–100 [84,114,115] - Estimated rate of RS in autism: <1 in female [84]	Because of the apparent normal developmental course in early childhood, diagnosis may be delayed [116]	Developmental course: - Stagnation stage in girls (6–18 months); - Regression stage (12–36 months) with head growth deceleration, appearance of progressive motor symptoms (gait and truncal apraxia, ataxia, decreasing mobility) and respiratory symptoms (hyperventilation, breath holding, apnea); - Pseudo-stationary stage (2–10 years); - Late motor deterioration (>10 years). Autistic behaviors: stereotyped hand movements, absence of language, loss of social engagement. Severe intellectual Disability [84]	Sleep problems are common in Rett's syndrome but there is some variation with age and mutation type [116]	Impaired secretion of melatonin [117]	Exogenous melatonin improved the sleep-wake cycle and sleep onset. The effect was maintained over 2 years without any adverse effects [117,118]

Table 1. *Cont.*

Neurogenetic Disorder	Neurogenetic Disorder Frequency, Estimated Rate (%) of Autism in the Disorder, and Estimated Rate (%) of the Disorder in Autism	Age of Diagnosis	Phenotype (Including Autistic Behaviors and Intellectual Functioning)	Sleep Problems	Melatonin Abnormality	Response to Melatonin Therapy
Williams Beuren syndrome (WBS) 7q11.23 deletion including 26 to 28 genes (typically *CLIP2*, *ELN*, *GTF2I*, *GTF2IRD1*, and *LIMK1*) [119,120]	- Frequency: 1 in 7500–10,000 [119] - Estimated rate of autism in WBS: <10 [84] - Estimated rate of WBS in autism: <1 [84]	The mean age at initial concerns is 0.98 year (Standard Deviation: 1.24), and the mean age at diagnosis may be delayed to 3.66 years (Standard Deviation: 4.13) [121]	Facial dysmorphism, short stature, heart and endocrine malformations, hypercalcemia, feeding problems, hyperacusis, visual spatial deficit, risk for attention deficit. Autistic syndrome but overfriendliness with social disinhibition and overtalkativeness. Mild to moderate intellectual disability [122–127]	Sleep disorders are common in individuals with Williams syndrome [128]	The WBS group had shallower drops in cortisol and less pronounced increase in melatonin at bedtime compared to the control group [129,130]	Melatonin was the most frequently reported medication taken for sleep problems in WBS, with 91% of parents reporting benefits for their child with WBS, and very few, if any, side effects [131,132]

5.2. Association of Melatonin with Mental Disorders Emerging in Early Adulthood

Neuroscientists are trying to understand schizophrenia through a new approach involving neurodevelopmental maturation. Schizophrenia onset usually occurs in adolescence or early adulthood but might be related to vulnerability in infancy. Studies in individuals with a first episode of schizophrenia reported a decrease in grey matter volume for most of the examined brain regions and cerebellar area [133]. This reduced volume of grey matter can get even smaller and extend to other surrounding regions in chronic cases [134]. Individuals with schizophrenia suffer from changes in brain microstructure, physiology, and connectivity of widely acting neurotransmitter systems, resulting in affective, cognitive, and psychotic symptoms. Schizophrenia leads to major impairments regarding the more complex cognitive performances, the so-called higher-order cognitive functions like, for example, verbal episodic memory or executive functioning [135]. Individuals with schizophrenia show errors in integrative information processing that are hypothesized to result in mis-connectivity or dysconnectivity, leading to a dysfunction of multiple brain circuits [134]. Melatonin is viewed as an important biological marker of the circadian cycle and as a psychiatric therapeutic agent [136]. Individuals with schizophrenia show lower levels of nocturnal melatonin secretion compared to a healthy control group, poor sleep efficiency, and disrupted circadian rhythms [137,138]. The Galván-Arrieta et al. study on olfactory neuronal precursors in schizophrenia and typical development suggested that a deficit in melatonin may lead to impaired neurodevelopment in schizophrenia [139]. Brain autopsies revealed abnormal elevated HIOMT activity in schizophrenia due to abnormally low activity of an enzyme prior to HIOMT involved in the biosynthesis of melatonin [140]. The decreased endogenous secretion of melatonin can persist even if there are improvements in sleep and positive effects with psychotic agents [141]. Finally, the Beckmann et al. study [142] showed no abnormal melatonin concentrations in the cerebrospinal fluid of individuals with schizophrenia compared to healthy controls. However, the authors conclude that other possibilities, such as changes in biological rhythms related to variations in melatonin activity and its influence on other neuroendocrine functions, may have a role in the pathophysiology of schizophrenia.

Similarly, bipolar disorder can exhibit prodromal manifestation prior to illness onset, underlying eventual similarities of neurodevelopmental abnormalities possibly involved in the pathogenesis of bipolar disorder [143]. Bipolar disorder is mainly associated with a shift in mood, energy, and activity. Patients show sleep alteration, circadian cycle disturbances, emotional deregulation, cognitive impairment, and increased risk for comorbidities [144]. The psychopathology of bipolar disorder (BD) is associated with altered sleep/wake rhythms, thermoregulation, cortisol secretion, and melatonin secretion. BD patients show abnormal rhythmic activity that is more functionally impacted during inter-episode periods (the passage from a mania state to a depressive one) [145]. Circadian cycles are dysregulated regarding the sleep–wake rhythms, especially the number of hours of sleep [146]. Manic symptoms were associated with less robust circadian rhythms leading to a decreased need for sleep, but also with other symptoms such as thought disorder, increased rate and amount of speech, and increased motor activity and energy [147]. It is noteworthy that a manic phase with a decreased need for sleep is considered a critical marker of the appearance of a depressive phase [1]. During the depression phase, hypersomnia is prevalent in 100% of patients and is also followed by a delayed sleep onset with night-time awakenings [148,149]. It has been highlighted that bipolar disorder treatments could benefit from a better understanding of circadian cycles in bipolar disorder [1]. Therapeutic administration of agomelatine (a selective agonist of MT1/MT2 receptors and a selective antagonist of 5-HT2C/5-HT2B receptors) is of special interest given that agomelatine is involved in the resynchronization of interrupted circadian rhythms with therapeutic benefits on sleep patterns, resynchronizing circadian rhythms in individuals with depression (a few studies has also been published on ASD, ADHD, and anxiety) [150].

6. Transnosographic Approach on the Role of Melatonin in Neurodevelopmental Disorders

Making an association between neurodevelopmental disorders and the pineal gland became more significant since the discovery of melatonin in the 1950s [148]. First, impairments in melatonin secretion have been associated with a significant decrease in sleep efficiency, notably in elderly individuals with continuity at different ages [6,10]. Second, melatonin seems to have a protective role in neurodevelopmental disorders with effects on early synaptic plasticity and neurotransmitter levels [33]. As indicated previously, circadian cycles have an impact on our bodies, shaping the timing and rhythms of various physiological and behavioral processes [1]. Children having an absence or alteration of circadian rhythms may have difficulties adapting to changes in their internal or external environment [151].

There is compelling evidence indicating that impairments in the endogenous circadian system, and especially in the sleep–wake rhythms (e.g., a delayed excretion of melatonin with a later onset of circadian rhythms), might precede the appearance of clinical symptoms for a variety of psychiatric disorders such as major depressive disorder, anxiety disorders or schizophrenia [152]. It suggests the existence of a biological dysfunction of rhythmicity and synchrony of rhythms in neurodevelopmental disorders. Reduced melatonin activity would then create a timing dysfunction of biological clocks with physiological and psychological disturbances leading to autistic social communication impairments. This timing dysfunction of biological clocks in early infancy could then lead to clinical psychopathological neurodevelopmental disorders later and even to psychotic or borderline psychotic states when the individual is emerging into adulthood. A disrupted sleep–wake cycle was associated as a major component for mood, anxiety, and psychotic disorders in adolescence [153]. Even if neurodevelopmental disorders display considerable genetic as well as environmental heterogeneity, severe mental conditions such as schizophrenia or bipolar conditions have increasing support from neuroscientific disciplines as originating from their origins from disturbed development of the nervous system based on the neurodevelopmental hypothesis [56,154,155]. New nosology conceptions should consider the genetic overlap between schizophrenia and psychopathologies associated with neurodevelopmental disorders that manifest in childhood [156]. Those conditions show some similarities in their phenotypes. All of them present significant cognitive impairment, tend to be more common among males, and are associated with developmental delay as well as neurological and motor abnormalities [155]. Symptoms and behaviors observed in autism are also found in adults with schizophrenia [157]. For example, autism and schizophrenia share intellectual disability, social communication impairment, or social withdrawal. One common factor that might be associated with these symptoms is a past or current deficiency of melatonin production.

There is a higher risk for ASD in children when the parents have mood disorders. The prevalence of ASD is higher in the offspring of parents with BD, then in the offspring of mothers with affective disorders, and finally in the offspring of parents with depressive disorders [158]. One possible explanation for these associations might be a possible shared genetic etiology between BD, depressive disorders, and ASD [159]. Reported family co-aggregation of BD and schizophrenia, as the common co-occurrence of ASD and BD, indicates a potential neurodevelopmental pathway [143]. The appearance of these neurodevelopmental disorders in families with ASD, BD, or schizophrenia was higher than in control families [160]. Further evidence to link the role of neurodevelopment in psychiatric disorders occurring later in life is the finding that neonatal lesions produce schizophrenia-like behaviors that emerge in post-adolescence [161].

In the Singh et al. meta-analysis, genetic data regarding rare coding variants in whole-exome sequences of 4133 schizophrenia cases and 9274 controls, de novo mutations in 1077 trios and copy number variants from 6882 cases and 11,255 controls, provided evidence that individuals with schizophrenia share rare damaging variants. Cases of schizophrenia with intellectual disability suggest that intellectual disability might be a dimension shared by neurodevelopmental disorders. These results support an overlap of genetic risk between schizophrenia and other neurodevelopmental disorders [162,163]. Several studies have concluded that schizophrenia is associated with cognitive impairments that probably result from a long-term neurodevelopmental evolution, even if it might be etiologically variable for each individual [135]. Risk factors associated with this disorder exert primarily their effects during developmental periods, leading to a detrimental maturation of the brain. The dysconnectivity in brain circuits found in schizophrenia is seen more and more as the result of abnormal brain development [134].

7. Conclusions

Psychiatric nosography suggests possible links between the pathogenesis of neurodevelopmental disorders and altered circadian rhythms. Altered circadian rhythms may participate in the development of these neurodevelopmental disorders but can also elicit and worsen psychopathological symptoms associated with this range of conditions. Melatonin production would consequently play a considerable role in sleep disorders and cognitive and social communication impairments. Neurodevelopmental disorders can greatly benefit from melatonin therapeutic administration and developmental behavior interventions that focus on rhythm synchronization [10,151]. Melatonin has been used with therapeutic efficacy for its resynchronizing effects to treat free-running rhythm disorder and delayed phase syndrome, among other circadian rhythm disorders [136]. Circadian disruption is a common prodrome for schizophrenia and bipolar disorder, as they are also for other mood disorders, but the nature of the relationships between melatonin, circadian rhythms, and psychopathology is still poorly understood today [164]. Further research is warranted to ascertain better the mechanisms underlying the effects of abnormal melatonin production and altered circadian rhythms on the pathogenesis and behavioral expression of neurodevelopmental disorders such as autism spectrum disorder or neurogenetic disorders associated with autism, schizophrenia, and bipolar disorder. Finally, given the neuroprotective and neurotrophic role of melatonin, it is a major issue to understand the relationships between the pathophysiology of melatonin metabolism and the development and severity of certain mental disorders discussed in the present paper.

Author Contributions: Conceptualization, C.F. and S.T.; writing—original draft preparation, C.F., S.C. and S.T.; supervision, S.T. All authors have read and agreed to the published version of the manuscript.

Funding: This research received no external funding.

Acknowledgments: We thank Tali Nir for their assistance in the elaboration of Table 1.

Conflicts of Interest: The authors declare no conflict of interest.

References

1. Abreu, T.; Braganca, M. The bipolarity of light and dark: A review on bipolar disorder and circadian cycles. *J. Affect. Disord.* **2015**, *185*, 219–229. [CrossRef] [PubMed]
2. Abdelgadir, I.S.; Gordon, M.A.; Akobeng, A.K. Melatonin for the management of sleep problems in children with neurodevelopmental disorders: A systematic review and meta-analysis. *Arch. Dis. Child.* **2018**, *103*, 1155–1162. [CrossRef] [PubMed]
3. Jang, S.W.; Liu, X.; Pradoldej, S.; Ye, K. N-acetylserotonin activates TrkB receptor in a circadian rhythm. *Proc. Natl. Acad. Sci. USA* **2010**, *107*, 3876–3881. [CrossRef]
4. Pandi-Perumal, S.R.; Srinivasan, V.; Maestroni, G.J.M.; Cardinali, D.P.; Poeggeler, B.; Hardeland, R. Melatonin: Nature's most versatile biological signal? *FEBS J.* **2006**, *273*, 2813–2838. [CrossRef]
5. Whitaker-Azmitia, P.M. Serotonin and brain development: Role in human developmental diseases. *Brain Res. Bull.* **2001**, *56*, 479–485. [CrossRef]
6. Ferlazzo, N.; Andolina, G.; Cannata, A.; Costanzo, M.G.; Rizzo, V.; Currò, M.; Ientile, R.; Caccamo, D. Is Melatonin the Cornucopia of the 21st Century? *Antioxidants* **2020**, *9*, 1088. [CrossRef] [PubMed]
7. Reiter, R.J.; Mayo, J.C.; Tan, D.X.; Sainz, R.M.; Alatorre-Jimenez, M.; Qin, L. Melatonin as an antioxidant: Under promises but over delivers. *J. Pineal Res.* **2016**, *61*, 253–278. [CrossRef]
8. Kandalepas, P.C.; Mitchell, J.W.; Gillette, M.U. Melatonin signal transduction pathways require E-Box-mediated transcription of Per1 and Per2 to reset the SCN clock at dusk. *PLoS ONE* **2016**, *11*, e0157824. [CrossRef]

9. Dibner, C.; Schibler, U.; Albrecht, U. The mammalian circadian timing system: Organization and coordination of central and peripheral clocks. *Annu. Rev. Physiol.* **2010**, *72*, 517–549. [CrossRef]
10. Tordjman, S.; Chokron, S.; Delorme, R.; Charrier, A.; Bellissant, E.; Jaafari, N.; Fougerou, C. Melatonin: Pharmacology, functions, and therapeutic benefits. *Curr. Neuropharmacol.* **2017**, *15*, 434–443. [CrossRef]
11. Feng, N.Y.; Bass, A.H. 'Singing' fish rely on circadian rhythm and melatonin for the timing of nocturnal courtship vocalization. *Curr. Biol.* **2016**, *26*, 2681–2689. [CrossRef] [PubMed]
12. Wilczynski, W.; Lutterschmidt, D.I. Biological rhythms: Melatonin shapes the space–time continuum of social communication. *Curr. Biol.* **2016**, *26*, R892–R895. [CrossRef] [PubMed]
13. Voiculescu, S.; Zygouropoulos, N.; Zahiu, C.D.; Zagrean, A.M. Role of melatonin in embryo fetal development. *J. Med. Life* **2014**, *7*, 488–492. [PubMed]
14. Kivela, A.; Kauppila, A.; Lappaluoto, J.; Vakkuri, O. Serum and amniotic fluid melatonin during human labor. *J. Clin. Endocrinol. Metab.* **1989**, *69*, 1065–1068. [CrossRef]
15. Nakamura, Y.; Tamura, H.; Kashida, S.; Takayama, H.; Yamagata, Y.; Karube, A.; Sugino, N.; Kato, H. Changes of serum melatonin level and its relationship to feto-placental unit during pregnancy. *J. Pineal Res.* **2001**, *30*, 29–33. [CrossRef]
16. Supramanian, V.G.; Jenjin, G.; Loose, J. Chronic fetal hypoxia activin: A concentration in the late-pregnant sheep. *BJOG Int. J. Obstet. Gyneacol.* **2006**, *113*, 102–109. [CrossRef]
17. Tricoire, H.; Moller, M.; Chemineau, P.; Malpaux, B. Origin of cerebrospinal fluid melatonin and possible function in the integration of photoperiod. *Reprod. Suppl.* **2003**, *61*, 311–321. [CrossRef]
18. Hardeland, R.; Pandi-Perumal, S.R.; Cardinali, D.P. Melatonin. *Int. J. Biochem. Cell Biol.* **2006**, *38*, 313–316. [CrossRef]
19. Reiter, R.J. Melatonin: The Chemical Expression of Darkness. *Mol. Cell. Endocrinol.* **1991**, *79*, C153–C158. [CrossRef]
20. Prakash, Y.; Thompson, M.A.; Meuchel, L.; Pabelick, C.M.; Mantilla, C.B.; Zaidi, S.; Martin, R.J. Neurotrophins in Lung Health and Disease. *Expert Rev. Respir. Med.* **2010**, *4*, 395–411. [CrossRef]
21. Thoenen, H. Neurotrophins and Neuronal Plasticity. *Science* **1995**, *270*, 593–598. [CrossRef] [PubMed]
22. Motta-Teixeira, L.C.; Machado-Nils, A.V.; Battagello, D.S.; Diniz, G.B.; Andrade-Silva, J.; Silva, S.; Matos, R.A.; do Amaral, F.G.; Xavier, G.F.; Bittencourt, J.C.; et al. The absence of maternal pineal melatonin rhythm during pregnancy and lactation impairs offspring physical growth, neurodevelopment, and behavior. *Horm. Behav.* **2018**, *105*, 146–156. [CrossRef] [PubMed]
23. Kennaway, D.J.; Goble, F.C.; Stamp, G.E. Factors influencing the development of melatonin rhythmicity in humans. *J. Clin. Endocrinol. Metab.* **1996**, *81*, 1525–1532. [CrossRef] [PubMed]
24. Torres-Farfan, C.; Mendez, N.; Abarzua-Catalan, L.; Vilches, N.; Valenzuela, G.J.; Seron-Ferre, M. A circadian clock entrained by melatonin is ticking in the rat fetal adrenal. *Endocrinology* **2011**, *152*, 1891–1900. [CrossRef] [PubMed]
25. Kwiatkowski, F.; Abrial, C.; Gachon, F.; Chevrier, R.; Curé, H.; Chollet, P. Stress, cancer et rythme circadien de la mélatonine. *Pathol. Biol.* **2004**, *53*, 269–272. [CrossRef]
26. Hollway, J.A.; Aman, M.G. Pharmacological treatment of sleep disturbance in developmental disabilities. A review of the literature. *Res. Dev. Disabil.* **2011**, *32*, 939–962. [CrossRef]
27. Joseph, D.; Chong, N.W.; Shanks, M.E.; Rosato, E.; Taub, N.A.; Petersen, S.A. Getting rhythm: How do babies do it? *Arch. Dis. Child Fetal Neonatal* **2014**, *100*, F50–F54. [CrossRef]
28. Price, D.A.; Close, G.C.; Fielding, B.A. Age of appearance of circadian rhythm in salivary cortisol values in infancy. *Arch. Dis. Child.* **1983**, *58*, 454–456. [CrossRef]
29. Paprocka, J.; Kijonka, M.; Rzepka, B.; Sokol, M. Melatonin in hypoxic-ischemic brain injury in term and preterm babies. *Int. J. Endocrinol.* **2019**, *2019*, 9626715. [CrossRef]
30. Touitou, Y. Human aging and melatonin. Clinical relevance. *Exp. Gerontol.* **2001**, *36*, 1083–1100. [CrossRef]
31. Ramirez-Rodriguez, G.; Ortiz-Lopez, L.; Dominguez-Alonso, A.; Benitez-King, G.A.; Kempermann, G. Chronic treatment with melatonin stimulates dendrite maturation and complexity in adult hippocampal neurogenesis of mice. *J. Pineal Res.* **2011**, *50*, 29–37. [CrossRef] [PubMed]
32. Wan, Q.; Man, H.Y.; Liu, F.; Braunton, J.; Niznik, H.B.; Pang, S.F.; Brown, G.M.; Wang, Y.T. Differential modulation of GABAA receptor function by Mel1a and Mel1b receptors. *Nat. Neurosci.* **1999**, *2*, 401–403. [CrossRef] [PubMed]
33. Yenen, A.S.; Çak, H.T. Melatonin and circadian rhythm in autism spectrum disorders. *Turk Psikiyatr. Derg.* **2020**, *31*, 201–211. [CrossRef]
34. Bocheva, G.; Slominski, R.M.; Janjetovic, Z.; Kim, T.K.; Böhm, M.; Steinbrink, K.; Reiter, R.J.; Kleszczynski, K.; Slominski, A.T. Protective role of melatonin and its metabolites in skin aging. *Int. J. Mol. Sci.* **2022**, *23*, 1238. [CrossRef] [PubMed]
35. Reiter, R.J.; Tan, D.X.; Maldonado, M.D. Melatonin as an antioxidant: Physiology versus pharmacology. *J. Pineal Res.* **2005**, *39*, 215–216. [CrossRef]
36. Jockers, R.; Maurice, P.; Boutin, J.A.; Delagrange, P. Melatonin receptors, heterodimerization, signal transduction and binding sites: What's new? *Br. J. Pharmacol.* **2008**, *154*, 1182–1195. [CrossRef] [PubMed]
37. Arendt, J. Melatonin and human rhythms. *Chronobiol. Int.* **2006**, *29*, 21–37. [CrossRef]
38. Arangino, S.; Cagnacci, A.; Angiolucci, M.; Vacca, A.M.B.; Longu, G.; Volpe, A.; Melis, G.B. Effects of melatonin an vascular reactivity, catecholamine levels, and blood pressure in healthy men. *Am. J. Cardiol.* **1999**, *83*, 1417. [CrossRef]
39. Cagnacci, A.; Arangino, S.; Angiolucci, M.; Maschio, E.; Melis, G.B. Influences of melatonin adminstration on the circulation of women. *Am. J. Physiol.* **1998**, *274*, R335–R338.

40. Doolen, S.; Krause, D.N.; Dubocovich, M.L.; Duckles, S.P. Melatonin mediates two distinct responses in vascular smooth muscle. *Eur. J. Pharmacol.* **1998**, *345*, 67–76. [CrossRef]
41. Iuvone, P.M.; Tosini, G.; Pozdeyev, N.; Haque, R.; Klein, D.C.; Chaurasia, S.S. Circadian clocks, clock networks, arylalkylamine N-acetyltransferase, and melatonin in the retina. *Prog. Retin. Eye Res.* **2005**, *24*, 433–456. [CrossRef] [PubMed]
42. Nishiyama, K.; Yasue, H.; Moriyama, Y.; Tsunoda, R.; Ogawa, H.; Yoshimura, M.; Kugiyama, K. Acute effects of melatonin administration on cardiovascular autonomic regulation in healthy men. *Am. Heart J.* **2001**, *141*, E9. [CrossRef] [PubMed]
43. Scheer, F.A.; Van Montfrans, G.A.; Van Someren, E.J.; Mairuhu, G.; Buijs, R.M. Daily night-time melatonin reduces blood pressure in male patients with essential hypertension. *Hypertension* **2004**, *43*, 192–197. [CrossRef] [PubMed]
44. Bubenik, G.A. Gastrointestinal melatonin: Localization, function, and clinical relevance. *Dig. Dis. Sci.* **2002**, *47*, 2336–2348. [CrossRef] [PubMed]
45. Pandi-Perumal, S.R.; Esquifino, A.I.; Cardinali, D.P. The role of melatonin in immunoenhancement: Potential application in cancer. *Int. J. Exp. Pathol.* **2006**, *87*, 81–87. [CrossRef]
46. Carrillo-Vico, A.; Guerrero, J.M.; Lardone, P.J.; Reiter, R.J. A review of the multiple actions of melatonin on the immune system. *Endocrine* **2005**, *27*, 189–200. [CrossRef]
47. Srinivasan, V.; Pandi-Perumal, S.R.; Maestroni, G.J.M.; Esquifino, A.I.; Hardeland, R.; Cardinali, D.P. Role of melatonin in neurodegenerative diseases. *Neurotox. Res.* **2005**, *7*, 293–318. [CrossRef]
48. Regodon, S.; Martin-Palomino, P.; Fernadez-Montesinos, R.; Herrera, J.L.; Carrascosa-Salmoral, M.P.; Piriz, S.; Vadillo, S.; Guerrero, J.M.; Pozo, D. The use of melatonin as a vaccine agent. *Vaccine* **2005**, *23*, 5321–5327. [CrossRef]
49. Hardeland, R.; Pandi-Perumal, S.R. Melatonin, a potent agent in antioxidative defense: Actions as a natural food constituent, gastrointestinal factor, drug and prodrug. *Nutr. Metab.* **2005**, *2*, 22. [CrossRef]
50. Karbownik, M.; Lewinski, A.; Reiter, R.J. Anticarcinogenic actions of melatonin which involve antioxidative processes: Comparison with other antioxidants. *Int. J. Biochem. Cell. Biol.* **2001**, *33*, 735–753. [CrossRef]
51. Klaunig, J.E.; Xu, Y.; Isenberg, J.S.; Bachowski, S.; Kolaja, K.L.; Jiang, J.; Stevenson, D.E.; Walborg, E.F., Jr. The role of oxidative stress in chemical carcinogenesis. *Environ. Health Perspect.* **1998**, *106*, 289–295. [CrossRef] [PubMed]
52. Polimeni, G.; Esposito, E.; Bevelacqua, V.; Guarneri, C.; Cuzzocrea, S. Role of melatonin supplementation in neurodegenerative disorders. *Front. Biosci.* **2014**, *19*, 429–446. [CrossRef] [PubMed]
53. Tauman, R.; Zisapel, N.; Laudon, M. Melatonin production in infants. *Pediatr. Neurol.* **2002**, *26*, 379–382. [CrossRef]
54. Boeve, B.F.; Silber, M.H.; Ferman, T.J. Melatonin for treatment of REM sleep behavior disorder in neurologic disorders: Results in 14 patients. *Sleep Med.* **2003**, *4*, 281–284. [CrossRef] [PubMed]
55. Fitzgerald, M. *The Future of Psychiatry and Neurodevelopmental Disorders: A Paradigm Shift*; Intech Open: London, UK, 2019; ISBN 978-1-78923-826-6.
56. Morris-Rosendahl, D.; Crocq, M. Neurodevelopmental disorders—The history and future of a diagnostic concept. *Dialogues Clin. Neurosci.* **2020**, *22*, 65–72. [CrossRef] [PubMed]
57. Charrier, A.; Olliac, B.; Roubertoux, P.; Tordjman, S. Clock genes and altered sleep-wake rhythms: Their role in the development of psychiatric disorders. *Int. J. Mol. Sci.* **2017**, *18*, 938. [CrossRef] [PubMed]
58. Tordjman, S.; Cohen, D.; Anderson, G.M.; Botbol, M.; Coulon, N.; Roubertoux, P. Reframing autism as a behavioral syndrome and not a specific mental disorder: Implications of genetic and phenotypic heterogeneity. *Neurosci. Biobehav. Rev.* **2018**, *89*, 132–150. [CrossRef]
59. Lalanne, S.; Fougerou-Leurent, C.; Anderson, G.M.; Schroder, C.; Nir, T.; Chokron, S.; Delorme, R.; Claustrat, B.; Bellissant, E.; Kermarrec, S.; et al. Melatonin: From pharmacokinetics to clinical use in autism spectrum disorder. *Int. J. Mol. Sci.* **2021**, *22*, 1490. [CrossRef]
60. Kloeckner, A.; Justard, C.; Bullinger, A.; Nicoulaud, L.; Tordjman, S.; Cohen, D. Intérêt d l'abord sensorimoteur dans les pathologies autistiques sévères: Introduction aux travaux d'André Bullinger. *Neuropsychol. Enf. Adolesc.* **2009**, *57*, 154–159. [CrossRef]
61. Tordjman, S.; Celume, M.P.; Denis, L.; Motillon, T.; Keromnes, G. Reframing schizophrenia and autism as bodily self-consciousness disorders leading to a deficit of theory of mind and empathy with social communication impairments. *Neurosci. Biobehav. Rev.* **2019**, *103*, 401–413. [CrossRef]
62. Grandgeorge, M.; Hausberger, M.; Tordjman, S.; Deleau, M.; Lazartigues, A.; Lemonnier, E. Environmental factors influence language development in children with autism spectrum disorders. *PLoS ONE* **2009**, *4*, e4683. [CrossRef] [PubMed]
63. Keromnes, G.; Chokron, S.; Celume, M.P.; Berthoz, A.; Botbol, M.; Canitano, L.; Du Boisgueheneuc, F.; Jaafari, N.; Lavenne-Collot, N.; Martin, B.; et al. Exploring self-consciousness from self- and other-image recognition in the mirror: Concepts and evaluation. *Front. Psychol.* **2019**, *10*, 719. [CrossRef] [PubMed]
64. Benabou, M.; Rolland, T.; Leblond, C.S.; Millot, G.A.; Huguet, G.; Delorme, R.; Leboyer, M.; Pagan, C.; Callebert, J.; Maronde, E.; et al. Heritability of the melatonin synthesis variability in autism spectrum disorders. *Sci. Rep.* **2017**, *7*, 17746. [CrossRef] [PubMed]
65. Maestro, S.; Casella, C.; Milone, A.; Muratori, F.; Palacio-Espasa, F. Study of the onset of autism through home movies. *Psychopathology* **1999**, *32*, 292–300. [CrossRef] [PubMed]
66. Saint-Georges, C.; Cassel, R.S.; Cohen, D.; Chetouani, M.; Laznik, M.C.; Maestro, S.; Muratori, F. What studies of family home movies can teach us about autistic infants: A literature review. *Res. Autism Spectr. Disord.* **2010**, *4*, 355–366. [CrossRef]

67. Stern, D. *Le Monde Interpersonnel du Nourrisson*; PUF: Paris, France, 1989.
68. Tordjman, S.; Najjar, I.; Bellissant, E.; Anderson, G.M.; Barburoth, M.; Cohen, D.; Jaafari, N.; Schischmanoff, O.; Fagard, R.; Lagdas, I.; et al. Advances in the research of melatonin in autism spectrum disorders: Literature review and new perspectives. *Int. J. Mol. Sci.* **2013**, *14*, 20508–20542. [CrossRef]
69. Moon, E.; Kim, K.; Partonen, T.; Linnaranta, O. Role of melatonin in the management of sleep and circadian disorders in the context of psychiatric illness. *Curr. Psychiatry Rep.* **2022**, *24*, 623–634. [CrossRef]
70. Fang, Y.; Zhao, C.; Xiang, H.; Zhao, X.; Zhong, R. Melatonin inhibits formation of mitochondrial permeability transition pores and improves oxidative phosphorylation of frozen-thawed ram sperm. *Front. Endocrinol.* **2020**, *10*, 896. [CrossRef]
71. Nir, I.; Meir, D.; Zilber, N.; Knobler, H.; Hadjez, J.; Lerner, Y. Brief report: Circadian melatonin, thyroid-stimulating hormone, prolactin, and cortisol levels in serum of young adults with autism. *J. Autism Dev. Disord.* **1995**, *25*, 641–654. [CrossRef]
72. Hu, V.W.; Sarachana, T.; Kim, K.S.; Nguyen, A.; Kulkarni, S.; Steinberg, M.E.; Luu, T.; Lai, Y.; Lee, N.H. Gene expression profiling differentiates autism case–controls and phenotypic variants of autism spectrum disorders: Evidence for circadian rhythm dysfunction in severe autism. *Autism Res.* **2009**, *2*, 78–97. [CrossRef]
73. Tordjman, S.; Anderson, G.M.; Bellissant, E.; Botbol, M.; Charbuy, H.; Camus, F.; Graignic, R.; Kermarrec, S.; Fougerou, C.; Cohen, D.; et al. Day and nighttime excretion of 6-sulphatoxymelatonin in adolescents and young adults with autistic disorder. *Psychoneuroendocrinology* **2012**, *37*, 1990–1997. [CrossRef] [PubMed]
74. Tordjman, S.; Anderson, G.M.; Pichard, N.; Charbuy, H.; Touitou, Y. Nocturnal excretion of 6-sulphatoxymelatonin in children and adolescents with autistic disorder. *Biol. Psychiatry* **2005**, *57*, 134–138. [CrossRef] [PubMed]
75. Abdul, F.; Sreenivas, N.; Kommu, J.V.S.; Banerjee, M.; Berk, M.; Maes, M.; Leboyer, M.; Debnath, M. Disruption of circadian rhythm and risk of autism spectrum disorder: Role of immune-inflammatory, oxidative stress, metabolic and neurotransmitter pathways. *Rev. Neurosci.* **2021**, *33*, 93–109. [CrossRef]
76. Bonnot, O.; Herrera, P.M.; Tordjman, S.; Walterfang, M. Secondary psychosis induced by metabolic disorders. *Front. Neurosci.* **2015**, *9*, 177. [CrossRef]
77. Rose Meyer, R. A review of the serotonin transporter and prenatal cortisol in the development of autism spectrum disorders. *Mol. Autism* **2013**, *4*, 37. [CrossRef] [PubMed]
78. Eslinger, P.J.; Anders, S.; Ballarini, T.; Boutros, S.; Krach, S.; Mayer, A.V.; Zahn, R. The neuroscience of social feelings: Mechanisms of adaptive social functioning. *Neurosci. Biobehav. Rev.* **2021**, *128*, 592–620. [CrossRef] [PubMed]
79. Smith, A.C.; McGavran, L.; Robinson, J.; Waldstein, G.; Macfarlane, J.; Zonona, J.; Reiss, J.; Lahr, M.; Allen, L.; Magenis, E. Interstitial deletion of (17) (P11.2p11.2) in nine patients. *Am. J. Hum. Genet.* **1986**, *24*, 393–414. [CrossRef]
80. Vostanis, P.; Harrington, R.; Prendergast, M.; Farndon, P. Case reports of autism with interstitial deletion of chromosome 17 (p11.2 p11.2) and monosomy of chromosome 5 (5pter > 5p15.3). *Psychiatr. Genet.* **1994**, *4*, 109–111. [CrossRef]
81. Rinaldi, B.; Villa, R.; Sironi, A.; Garavelli, L.; Finelli, P.; Bedeschi, M.F. Smith-Magenis Syndrome: Clinical Review, Biological Background and Related Disorders. *Genes* **2022**, *13*, 335. [CrossRef]
82. Elsea, S.H.; Girirajan, S. Smith-Magenis syndrome. *Eur. J. Hum. Genet.* **2008**, *16*, 412–421. [CrossRef]
83. GHR. Smith Magenis Syndrome. 2018. Available online: https://rarediseases.info.nih.gov/diseases/8197/smith-magenis-syndrome (accessed on 2 November 2023).
84. Robert, C.; Pasquier, L.; Cohen, D.; Fradin, M.; Canitano, R.; Damaj, L.; Odent, S.; Tordjman, S. Role of genetics in the etiology of Autistic Spectrum Disorder: Towards a hierarchical diagnostic strategy. *Int. J. Mol. Sci.* **2017**, *18*, 618. [CrossRef] [PubMed]
85. Gropman, A.L.; Duncan, W.C.; Smith, A.C.M. Neurologic and developmental features of the Smith-Magenis Syndrome (del 17p11.2). *Pediatr. Neurol.* **2006**, *34*, 337–350. [CrossRef] [PubMed]
86. Laje, G.; Morse, R.; Richter, W.; Ball, J.; Pao, M.; Smith, A.C.M. Autism spectrum features in smith-magenis syndrome. *Am. J. Med. Genet.* **2010**, *154*, 456–462. [CrossRef] [PubMed]
87. Smith, A.C.M.; Dykens, E.; Greenberg, F. Behavioral Phenotype of Smith-Magenis Syndrome (del 17 p11.2). *J. Med. Genet. Neuropsychiatry* **1998**, *81*, 179–185. [CrossRef]
88. De Leersnyder, H.; de Blois, M.C.; Claustrat, B.; Romana, S.; Albrecht, U.; von Kleist-Retzow, J.-C.; Delobel, B.; Viot, G.; Lyonnet, S.; Vekemans, M.; et al. Inversion of the circadian rhythm of melatonin in the Smith-Magenis syndrome. *J. Pediatr.* **2001**, *139*, 111–116. [CrossRef]
89. De Leersnyder, H. Inverted rhythm of melatonin secretion in Smith-Magenis syndrome: From symptoms to treatment. *Trends Endocrinol. Metab.* **2006**, *17*, 291–298. [CrossRef]
90. Chamberlain, S.J.; Chen, P.F.; Ng, K.Y.; Bourgois-Rocha, F.; Lemtiri-Chlieh, F.; Levine, E.S.; Lalande, M. Induced pluripotent stem cell models of the genomic imprinting disorders Angelman and Prader-Willi syndromes. *Proc. Natl. Acad. Sci. USA* **2010**, *107*, 17668–17673. [CrossRef]
91. Mabb, A.M.; Judson, M.C.; Zylka, M.J.; Philpot, B.D. Angelman syndrome: Insights into genomic imprinting and neurodevelopmental phenotypes. *Trends Neurosci.* **2011**, *34*, 293–303. [CrossRef]
92. Erickson, C.A.; Wink, L.K.; Baindu, B.; Ray, B.; Schaefer, T.L.; Pedapati, E.V.; Lahiri, D.K. Analysis of peripheral amyloid precursor protein in Angelman syndrome. *Am. J. Med. Genet.* **2016**, *170*, 2334–2337. [CrossRef]
93. Li, W.; Yao, A.; Zhi, H.; Kaur, K.; Zhu, Y.; Jia, M.; Zhao, H.; Wang, Q.; Jin, S.; Zhao, G.; et al. Angelman syndrome protein Ube3a regulates synaptic growth and endocytosis by inhibiting BMP signaling in Drosophila. *PLoS Genet.* **2016**, *12*, e1006062. [CrossRef]

94. Williams, C.A.; Driscoll, D.J.; Dagli, A.I. Clinical and genetic aspects of Angelman syndrome. *Genet. Med.* **2010**, *12*, 385–395. [CrossRef] [PubMed]
95. NORD. Angelman Syndrome. 2018. Available online: https://rarediseases.org/rare-diseases/angelman-syndrome/ (accessed on 8 November 2023).
96. Williams, C.A.; Beaudet, A.L.; Clayton-Smith, J.; Knoll, J.H.; Kyllerman, M.; Laan, L.A.; Magenis, E.; Moncla, A.; Schinzel, A.A.; Summers, J.A.; et al. Conference Report. Angelman Syndrome 2005: Updated consensus for diagnostic criteria. *Am. J. Med. Genet.* **2006**, *140A*, 413–418. [CrossRef] [PubMed]
97. Clayton-Smith, J. Clinical research on Angelman syndrome in the United Kingdom: Observations on 82 affected individuals. *Am. J. Med. Genet.* **1993**, *46*, 12–15. [CrossRef]
98. Summers, J.A.; Allison, D.B.; Lynch, P.S.; Sandler, L. Behaviour problems in Angelman syndrome. *J. Intellect. Disabil. Res.* **1995**, *39*, 97–106. [CrossRef] [PubMed]
99. Walz, N.C. Parent report of stereotyped behaviors, social interaction, and developmental disturbances in individuals with Angelman syndrome. *J. Autism Dev. Disord.* **2007**, *37*, 940–947. [CrossRef] [PubMed]
100. Paprocka, J.; Kijonka, M.; Wojcieszek, P.; Pęcka, M.; Emich-Widera, E.; Sokół, M. Melatonin and Angelman syndrome: Implications and mathematical model of diurnal secretion. *Int. J. Endocrinol.* **2017**, *2017*, 5853167. [CrossRef]
101. Braam, W.; Didden, R.; Smits, M.G.; Curfs, L.M.G. Melatonin for chronic insomnia in Angelman syndrome: A randomized placebo-controlled trial. *J. Child Neurol.* **2008**, *23*, 649–654. [CrossRef]
102. Curatolo, P.; Bombardieri, R.; Jozwiak, S. Tuberous sclerosis. *Lancet* **2008**, *372*, 657–668. [CrossRef]
103. GHR. Tuberous Sclerosis Complex. 2018. Available online: https://rarediseases.info.nih.gov/diseases/7830/tuberous-sclerosis (accessed on 17 November 2023).
104. Staley, B.A.; Vail, E.A.; Thiele, E.A. Tuberous sclerosis complex: Diagnostic challenges, presenting symptoms, and commonly missed signs. *Pediatrics* **2011**, *127*, e117–e125. [CrossRef]
105. Bolton, P.F.; Park, R.J.; Higgins, J.N.; Griffiths, P.D.; Pickles, A. Neuro-epileptic determinants of autism spectrum disorders in tuberous sclerosis complex. *Brain* **2002**, *125*, 1247–1255. [CrossRef]
106. Muto, Y.; Sasaki, H.; Sumitomo, M.; Inagaki, H.; Kato, M.; Kato, T.; Miyai, S.; Kurahashi, H.; Shiroki, R. Genotype-phenotype correlation of renal lesions in the tuberous sclerosis complex. *Hum. Genome Var.* **2022**, *9*, 5. [CrossRef] [PubMed]
107. Jeste, S.S.; Varcin, K.J.; Hellemann, G.S.; Gulsrud, A.C.; Bhatt, R.; Kasari, C.; Wu, J.Y.; Sahin, M.; Nelson, C.A. Symptom profiles of autism spectrum disorder in tuberous sclerosis complex. *Neurology* **2016**, *87*, 766–772. [CrossRef] [PubMed]
108. Bruni, O.; Cortesi, F.; Giannotti, F.; Curatolo, P. Sleep disorders in tuberous sclerosis: A polysomnographic study. *Brain Dev.* **1995**, *17*, 52–56. [CrossRef] [PubMed]
109. Hancock, E.; O'Callaghan, F.; English, J.; Osborne, J.P. Melatonin excretion in normal children and in tuberous sclerosis complex with sleep disorder responsive to melatonin. *J. Child Neurol.* **2005**, *20*, 21–25. [CrossRef] [PubMed]
110. Hancock, E.; O'Callaghan, F.; Osborne, J.P. Effect of melatonin dosage on sleep disorder in tuberous sclerosis complex. *J. Child Neurol.* **2005**, *20*, 78–80. [CrossRef]
111. Amir, R.E.; Van Den Veyver, I.B.; Wan, M.; Tran, C.Q.; Francke, U.; Zoghbi, H.Y. Rett syndrome is cause by mutations in X-linked MECP2, encoding methil-CpG-binding protein 2. *Nat. Genet.* **1999**, *23*, 185–188. [CrossRef]
112. Gadalla, K.; Bailey, M.E.; Cobb, S.R. MeCP2 and Rett syndrome: Reversibility and potential avenues for therapy. *Biochem. J.* **2011**, *439*, 1–14. [CrossRef]
113. NORD. Rett Syndrome. 2015. Available online: https://rarediseases.org/rare-diseases/rett-syndrome/ (accessed on 8 November 2023).
114. Richards, C.; Jones, C.; Groves, L.; Moss, J.; Oliver, C. Prevalence of autism spectrum disorder phenomenology in genetic disorders: A systematic review and meta-analysis. *Lancet Psychiatry* **2015**, *2*, 909–916. [CrossRef]
115. Ramocki, M.B.; Peters, S.U.; Tavyev, Y.J.; Zhang, F.; Carvalho, C.M.; Schaaf, C.P.; Richman, R.; Fang, P.; Glaze, D.G.; Lupski, J.R.; et al. Autism and other neuropsychiatric symptoms are prevalent in individuals with MeCP2 duplication syndrome. *Ann. Neurol.* **2009**, *66*, 771–782. [CrossRef]
116. Young, D.; Nagarajan, L.; De Klerk, N.; Jacoby, P.; Ellaway, C.; Leonard, H. Sleep problems in Rett syndrome. *Brain Dev.* **2007**, *29*, 609–616. [CrossRef]
117. Miyamoto, A.; Oki, J.; Takahashi, S.; Okuno, A. Serum melatonin kinetics and long-term melatonin treatment for sleep disorders in Rett syndrome. *Brain Dev.* **1999**, *21*, 59–62. [CrossRef]
118. McArthur, A.; Budden, S. Sleep dysfunction in Rett syndrome: A trial of exogenous melatonin treatment. *Dev. Med. Child Neurol.* **1998**, *40*, 186–192. [CrossRef] [PubMed]
119. GHR. Williams Beuren Syndrome. 2018. Available online: https://rarediseases.info.nih.gov/espanol/13102/sindrome-de-williams (accessed on 12 November 2023).
120. Edelmann, L.; Prosnitz, A.; Pardo, S.; Bhatt, J.; Cohen, N.; Lauriat, T.; Ouchanov, L.; González, P.J.; Manghi, E.R.; Bondy, P.; et al. An atypical deletion of the Williams-Beuren syndrome interval implicates genes associated with defective visuospatial processing and autism. *J. Med. Genet.* **2007**, *44*, 136–143. [CrossRef] [PubMed]
121. Huang, L.; Sadler, L.; O'Riordan, M.A.; Robin, N.H. Delay in diagnosis of Williams syndrome. *Clin. Pediatr.* **2002**, *41*, 257–261. [CrossRef]

122. Morris, C.; Janssens, A.; Shilling, V.; Allard, A.; Fellowes, A.; Tomlinson, R.; Williams, J.; Coon, J.T.; Rogers, M.; Beresford, B.; et al. Meaningful health outcomes for paediatric neurodisability: Stakeholder prioritisation and appropriateness of patient reported outcome measures. *Health Qual. Life Outcomes* **2015**, *13*, 87. [CrossRef]
123. Gagliardi, C.; Bonaglia, M.C.; Selicorni, A.; Borgatti, R.; Giorda, R. Unusual cognitive and behavioural profile in a Williams syndrome patient with atypical 7q11.23 deletion. *J. Med. Genet.* **2003**, *40*, 526–530. [CrossRef] [PubMed]
124. Lincoln, A.J.; Searcy, Y.M.; Jones, W.; Lord, C. Social interaction behaviors discriminate young children with autism and Williams syndrome. *J. Am. Acad. Child Adolesc. Psychiatry* **2007**, *46*, 323–331. [CrossRef] [PubMed]
125. Tordjman, S.; Anderson, G.M.; Botbol, M.; Toutain, A.; Sarda, P.; Carlier, M.; Saugier-Veber, P.; Baumann, C.; Cohen, D.; Lagneaux, C.; et al. Autistic disorder in patients with Williams-Beuren syndrome: A reconsideration of the Williams-Beuren syndrome phenotype. *PLoS ONE* **2012**, *7*, e30778. [CrossRef] [PubMed]
126. Tordjman, S.; Anderson, G.M.; Cohen, D.; Kermarrec, S.; Carlier, M.; Touitou, Y.; Saugier-Veber, P.; Lagneaux, C.; Chevreuil, C.; Verloes, A. Presence of autism, hyperserotonemia, and severe expressive language impairment in Williams-Beuren syndrome. *Mol. Autism* **2013**, *4*, 29. [CrossRef]
127. Martens, M.A.; Wilson, S.J.; Reutens, D.C. Research Review: Williams syndrome: A critical review of the cognitive, behavioral, and neuroanatomical phenotype. *J. Child Psychol. Psychiatry* **2008**, *49*, 576–608. [CrossRef]
128. Santoro, S.D.; Giacheti, C.M.; Rossi, N.F.; Guissoni Campos, L.M.; Pinato, L. Correlations between behavior, memory, sleep-wake and melatonin in Williams-Beuren syndrome. *Physiol. Behav.* **2016**, *159*, 14–19. [CrossRef] [PubMed]
129. Sniecinska-Cooper, A.M.; Iles, R.K.; Butler, S.A.; Jones, H.; Bayford, R.; Dimitriou, D. Abnormal secretion of melatonin and cortisol in relation to sleep disturbances in children with Williams syndrome. *Sleep Med.* **2015**, *16*, 94–100. [CrossRef] [PubMed]
130. Sniecinska-Cooper, A.M.; Iles, R.K.; Butler, S.A.; Jones, H.; Bayford, R.; Dimitriou, D. Response to the letter "Sleep characteristics of children with Williams syndrome in relation to saliva melatonin and cortisol". *Sleep Med.* **2015**, *16*, 1177. [CrossRef] [PubMed]
131. Martens, M.A.; Seyfer, D.L.; Andridge, R.R.; Coury, D.L. Use and effectiveness of sleep medications by parent report in individuals with Williams syndrome. *J. Dev. Behav. Pediatr.* **2017**, *38*, 765–771. [CrossRef]
132. Reilly, C.; Senior, J.; Murtagh, L. ASD, ADHD, mental health conditions and psychopharmacology in neurogenetic syndromes: Parent survey. *J. Intellect. Disabil. Res.* **2015**, *59*, 307–318. [CrossRef]
133. Farrow, T.F.D.; Whitford, T.J.; Williams, L.M.; Gomes, L.; Harris, A.W.F. Diagnosis-related regional gray matter loss over two years in first episode of schizophrenia and bipolar disorder. *Biol. Psychiatry* **2005**, *58*, 713–723. [CrossRef]
134. Nath, M.; Wong, T.P.; Srivastava, L.K. Neurodevelopmental insights into circuit dysconnectivity in schizophrenia. *Biol. Psychiatry* **2021**, *104*, 110047. [CrossRef]
135. Schaefer, J.; Giangrande, E.; Weinberger, D.R.; Dickinson, D. The global cognitive impairment in schizophrenia: Consistent over decades and around the world. *Schizo. Res.* **2013**, *150*, 42–50. [CrossRef]
136. Morera-Fumero, A.L.; Abreu-Gonzalez, P. Role of melatonin in schizophrenia. *Int. J. Mol. Sci.* **2013**, *14*, 9037–9050. [CrossRef]
137. Monteleone, P.; Natale, M.; La Rocca, A.; Maj, M. Decreased nocturnal secretion of melatonin in drug-free schizophrenics: No change after subchronic treatment with antipsychotics. *Neuropsychobiology* **1997**, *36*, 159–163. [CrossRef]
138. Rao, M.L.; Gross, G.; Strebel, B.; Halaris, A.; Huber, G.; Bräunig, P.; Marler, M. Circadian rhythm of tryptophan, serotonin, melatonin, and pituitary hormones in schizophrenia. *Biol. Psychiatry* **1994**, *35*, 151–163. [CrossRef] [PubMed]
139. Galván-Arrieta, T.; Trueta, C.; Cercós, M.G.; Valdés-Tovar, M.; Alarcón, S.; Oikawa, J.; Zamudio-Meza, H.; Benítez-King, G. The role of melatonin in the neurodevelopmental etiology of schizophrenia: A study in human olfactory neuronal precursors. *J. Pineal Res.* **2017**, *63*, e12421. [CrossRef] [PubMed]
140. Smith, J.A.; Mee, T.J.; Padwick, D.J.; Spokes, E.G. Human post-mortem pineal enzyme activity. *Clin. Endocrinol.* **1981**, *14*, 75–81. [CrossRef] [PubMed]
141. Petursson, H.; Lader, M.H. Withdrawal from long-term benzodiazepine treatment. *Br. Med. J.* **1981**, *283*, 643–645. [CrossRef] [PubMed]
142. Beckmann, H.; Wetterberg, L.; Gattaz, W.F. Melatonin immunoreactivity in cerebrospinal fluid of schizophrenic patients and healthy controls. *Psychiatry Res.* **1984**, *11*, 107–110. [CrossRef]
143. Kloiber, S.; Rosenblat, J.D.; Husain, M.I.; Ortiz, A.; Berk, M.; Quevedo, J.; Vieta, E.; Maes, M.; Birmaher, B.; Soares, J.C.; et al. Neurodevelopmental pathways in bipolar disorder. *Neurosci. Biobehav. Rev.* **2020**, *112*, 213–226. [CrossRef]
144. Leboyer, M.; Kupfer, D.J. Bipolar disorder: New perspectives in health care and prevention. *J. Clin. Psychiatry* **2010**, *71*, 1689–1695. [CrossRef]
145. Salvatore, P.; Ghidini, S.; Zita, G.; De Panfilis, C.; Lambertino, S.; Maggini, C.; Baldessarini, R.J. Circadian activity rhythm abnormalities in ill and recovered bipolar I disorder patients. *Bipolar Disord.* **2008**, *10*, 256–265. [CrossRef]
146. Lewy, A.J. Circadian misalignment in mood disturbances. *Curr. Psychiatry Rep.* **2009**, *11*, 459–465. [CrossRef]
147. Gonzalez, R.; Tamminga, C.A.; Tohen, M.; Suppes, T. The relationship between affective state and the rhythmicity of activity in bipolar disorder. *J. Clin. Psychiatry* **2014**, *75*, 317–322. [CrossRef]
148. López-Munoz, F.; Molina, J.D.; Rubio, G.; Alamo, C. An historical view of the pineal gland and mental disorders. *J. Clin. Neurosci.* **2011**, *18*, 1028–1037. [CrossRef] [PubMed]
149. Pandi-Perumal, S.R.; Moscovitch, A.; Srinivasan, V.; Spence, D.W.; Cardinali, D.P.; Brown, G.M. Bidirectional communication between sleep and circadian rhythms and its implications for depression: Lessons from agomelatine. *Prog. Neurobiol.* **2009**, *88*, 264–271. [CrossRef] [PubMed]

150. Savino, R.; Polito, A.N.; Marsala, G.; Ventriglio, A.; Di Salvatore, M.; De Stefano, M.I.; Valenzano, A.; Marinaccio, L.; Bellomo, A.; Cibelli, G.; et al. Agomelatine: A potential multitarget compound for neurodevelopmental disorders. *Brain Sci.* **2023**, *13*, 734. [CrossRef] [PubMed]
151. Tordjman, S.; Davlantis, K.S.; Georgieff, N.; Geoffray, M.M.; Speranza, M.; Anderson, G.M.; Xavier, J.; Botbol, M.; Oriol, C.; Bellissant, E. Autism as a disorder of biological and behavioral rhythms: Toward new therapeutic perspectives. *Front. Pediatr.* **2015**, *3*, 1. [CrossRef] [PubMed]
152. Lamont, E.W.; Legault-Coutu, D.; Cermakian, N.; Boivin, D.B. The role of circadian clock genes in mental disorders. *Dialogues Clin. Neurosci.* **2007**, *9*, 333–342. [CrossRef] [PubMed]
153. Logan, R.W.; McClung, C.A. Rhythms of life: Circadian disruption and brain disorders across the lifespan. *Nat. Rev. Neurosci.* **2019**, *20*, 49–65. [CrossRef]
154. Murray, R.M.; Lewis, S.W. Is schizophrenia a neurodevelopmental disorder? *Br. Med. J.* **1987**, *295*, 681–682. [CrossRef]
155. Owen, M.J.; O'Donovan, M.C.; Thapar, A.; Craddock, N. Neurodevelopmental hypothesis of schizophrenia. *Br. J. Psychiatry* **2011**, *198*, 173–175. [CrossRef]
156. Rutter, M.; Kim-Cohen, J.; Maughan, B. Continuities and discontinuities in psychopathology between childhood and adult life. *J. Child Psychol. Psychiatry* **2006**, *47*, 276–295. [CrossRef]
157. Stahlberg, O.; Soderstrom, H.; Rastam, M.; Gillberg, C. Bipolar disorder, schizophrenia, and other psychotic disorders in adults with childhood onset AD/HD and/or autism spectrum disorders. *J. Neural. Transm.* **2004**, *111*, 891–902. [CrossRef]
158. Ayano, G.; Maravilla, J.C.; Alati, R. Risk of autistic spectrum disorder in offspring with parental mood disorders: A systematic review and meta-analysis. *J. Affect. Disord.* **2019**, *248*, 185–197. [CrossRef] [PubMed]
159. Lee, S.H.; Ripke, S.; Neale, B.M.; Faraone, S.V.; Purcell, S.M.; Perlis, R.H.; Mowry, B.J.; Thapar, A.; Goddard, M.E.; Witte, J.S. Genetic relationship between five psychiatric disorders estimated from genome-wide SNPs. *Nat. Genet.* **2013**, *45*, 984–994. [CrossRef]
160. Cannon, M.; Caspi, A.; Moffitt, T.E.; Harrington, H.; Taylor, A.; Murray, R.M.; Poulton, R. Evidence for early-childhood, pan-developmental impairment specific to schizophreniform disorder: Results from a longitudinal birth cohort. *Arch. Gen. Psychiatry* **2002**, *59*, 449–456. [CrossRef] [PubMed]
161. Lipska, B.K.; Jaskiw, G.E.; Weinberger, D.R. Postpubertal emergence of hyperresponsiveness to stress and to amphetamine after neonatal excitotoxic hippocampal damage: A potential animal model of schizophrenia. *Neuropsychopharmacology* **1993**, *9*, 67–75. [CrossRef]
162. Singh, T.; Walters, J.T.; Johnstone, M.; Curtis, D.; Suvisaari, J.; Torniainen, M.; Rees, E.; Lyzgbe, C.; Blackwood, D.; Mclntosh, A.; et al. The contribution of rare variants to risk of schizophrenia in individuals with and without intellectual disability. *Nat. Genet.* **2017**, *49*, 1167–1173. [CrossRef]
163. Le Roy, I.; Roubertoux, P.L.; Jamot, L.; Maarouf, F.; Tordjman, S.; Mortaud, S.; Blanchard, C.; Martin, B.; Guillot, P.V.; Duquenne, V. Neuronal and behavioral differences between Mus musculus domesticus (C57BL/6JBy) and Mus musculus castaneus (CAST/Ei). *Behav. Brain Res.* **1998**, *95*, 135–142. [CrossRef]
164. Walker, W.H.; Walton, J.C.; DeVries, A.C.; Nelson, R.J. Circadian rhythm disruption and mental health. *Transl. Psychiatry* **2020**, *10*, 28. [CrossRef]

antioxidants

MDPI

Review

Potential Neuroprotective Role of Melatonin in Sepsis-Associated Encephalopathy Due to Its Scavenging and Anti-Oxidative Properties

Mariusz Sieminski *, Karolina Szaruta-Raflesz, Jacek Szypenbejl * and Klaudia Krzyzaniak

Department of Emergency Medicine, Medical University of Gdansk, 80-214 Gdansk, Poland; karolina.szaruta@gmail.com (K.S.-R.); klaudiak@gumed.edu.pl (K.K.)

* Correspondence: mariusz.sieminski@gumed.edu.pl (M.S.); jacek.szypenbejl@gumed.edu.pl (J.S.)

Abstract: Sepsis is defined as life-threatening organ dysfunction caused by a dysregulated host response to infection. The brain is one of the organs involved in sepsis, and sepsis-induced brain injury manifests as sepsis-associated encephalopathy (SAE). SAE may be present in up to 70% of septic patients. SAE has a very wide spectrum of clinical symptoms, ranging from mild behavioral changes through cognitive disorders to disorders of consciousness and coma. The presence of SAE increases mortality in the population of septic patients and may lead to chronic cognitive dysfunction in sepsis survivors. Therefore, therapeutic interventions with neuroprotective effects in sepsis are needed. Melatonin, a neurohormone responsible for the control of circadian rhythms, exerts many beneficial physiological effects. Its anti-inflammatory and antioxidant properties are well described. It is considered a potential therapeutic factor in sepsis, with positive results from studies on animal models and with encouraging results from the first human clinical trials. With its antioxidant and anti-inflammatory potential, it may also exert a neuroprotective effect in sepsis-associated encephalopathy. The review presents data on melatonin as a potential drug in SAE in the wider context of the pathophysiology of SAE and the specific actions of the pineal neurohormone.

Keywords: melatonin; sepsis-associated encephalopathy; sepsis; antioxidant; neuroinflammation

Citation: Sieminski, M.; Szaruta-Raflesz, K.; Szypenbejl, J.; Krzyzaniak, K. Potential Neuroprotective Role of Melatonin in Sepsis-Associated Encephalopathy Due to Its Scavenging and Anti-Oxidative Properties. *Antioxidants* **2023**, *12*, 1786. https://doi.org/10.3390/antiox12091786

Academic Editors: Marina Garcia-Macia and Adrián Santos-Ledo

Received: 24 July 2023
Revised: 8 September 2023
Accepted: 20 September 2023
Published: 21 September 2023

1. Introduction—Sepsis and Sepsis-Associated Encephalopathy

1.1. Sepsis

Sepsis is a life-threatening complex of biochemical and clinical abnormalities caused by infection. The actual definition of sepsis was devised in 2016 and describes this syndrome as "life-threatening organ dysfunction caused by dysregulated host response to infection" [1]. The diagnostic workup should start with the clinical suspicion of infection (e.g., fever, symptoms of upper respiratory airway infection, dysuria) followed by screening with the Quick Sequential Organ Failure Assessment (qSOFA) scale. In cases of positive screening, a wider assessment of multi-organ failure should be performed with the Sequential Organ Failure Assessment (SOFA) scale. A score of $\geq$2 points in the SOFA is equivalent to a diagnosis of sepsis [1,2].

Sepsis is a severe problem for healthcare systems, with a global incidence of 189/100,000 person-years in the population of hospitalized adults and with mortality of 26.7% [3]. Rhee et al. found that sepsis was diagnosed in 6% of patients admitted to hospitals in the USA, with mortality of 15% and with 6% of patients requiring palliative care after discharge [4]. The survivors of sepsis frequently struggle with troublesome chronic consequences—in 74% of them, a new co-morbidity is diagnosed within the first year after hospital discharge, and one third of them become dependent on nursing care after surviving sepsis [5]. It is of special note that in 18.5% of sepsis survivors, a diagnosis of cognitive dysfunction was made. These data show the importance of any intervention that potentially might reduce mortality and post-septic disability and have a neuroprotective effect.

1.2. Sepsis-Associated Encephalopathy

Sepsis is defined as the failure of organs resulting from a pathological inflammatory reaction to infection. The central nervous system (CNS) is one of the most vulnerable organs due to its circulatory and metabolic needs. The systemic inflammatory process causes dysfunction of the blood–brain barrier, allowing the influx of proinflammatory mediators to the CNS, resulting in the spreading of inflammation across the cerebral structure. Sepsis-related hypotonia decreases cerebral perfusion, which leads to a deficiency of metabolic substrates for neural tissue. Therefore, symptoms of involvement of the CNS develop rapidly during sepsis, being the first clinical symptoms of sepsis in some cases. Sepsis-associated encephalopathy (SAE) is a term describing brain failure caused by sepsis in the absence of direct brain injury or infection of the central nervous system. It must be remembered that there is no formal clinical definition of SAE or formal diagnostic criteria. Biochemical, neurophysiological or radiological markers of SAE are lacking as well. Therefore, the diagnosis of SAE is made on clinical grounds exclusively. The diagnosis is based upon the clinical observation of cognitive deterioration temporarily related to symptoms of sepsis, with the exclusion of direct brain injury and neuroinfection.

SAE is a diffuse cerebral dysfunction [6]. The clinical spectrum of SAE ranges from discrete cognitive deficits, e.g., reduced attention or diminished readiness for social interactions (so called sickness behavior) [7] through delirium to coma [8]. Most commonly, acute SAE is observed, but its subacute (lasting for weeks) or chronic (lasting for months) forms may be observed. The exact pathophysiological mechanism of SAE remains unknown, with many potential pathways leading to this phenomenon. The most important pathways leading to the development of SAE are dysfunction of the BBB, neuroinflammation, oxidative stress and cerebral ischemia. The development of SAE is a consequence of the dysfunction of the brainstem, frontal cortex and amygdala, as well as of the failure of neuroendocrine centers: the hypothalamus and pituitary gland [9]. Dysfunction of the CNS is reflected by pathological patterns on an electroencephalogram (EEG), which is characterized by the presence of triphasic waves, theta waves or burst suppression [10].

1.3. Clinical Picture of SAE

Sepsis-associated encephalopathy may be categorized as a cognitive and communicative disorder, at least in its mild forms. Patients present problems with concentration, reduced attention and problems with learning and remembering new facts [11]. These cognitive deficits are accompanied by irritability, depressed mood, anxiety and withdrawal from social activities—a spectrum of symptoms defined as sickness behavior [7]. The clinical progression of SAE leads to the appearance of confusion and further features of delirium with disorders of consciousness. These symptoms may develop acutely or subacutely, with the patient being disoriented, agitated and aggressive, with hallucinations and a tendency to wander and speak loudly [6,12], as occurs in hyperactive delirium. Patients with SAE presenting as hypoactive delirium are less "visible": they are quiet, apathetic and pathologically somnolent and show problems with the perception of environmental stimuli [12]. It is of note that this "silent" phenotype of delirium is more common in SAE, which makes detection of the disorder more challenging [13]. Although SAE is defined by a lack of focal brain lesions, the appearance of focal neurological deficits is described in approximately one fifth of patients and seizures may be present in 10% of subjects [14]. The progression of sepsis and development of septic shock with multi-organ failure increase the severity of SAE, finally leading to a coma.

Disturbed functioning of the central nervous system is reflected with a variety of changes in the EEG. Typical changes for septic encephalopathy are the domination of theta or delta rhythms, the presence of triphasic waves and burst suppression [15,16]. Epileptic discharges and pseudo-epileptic discharges are recorded in nearly 10–50% of patients with sepsis, with most seizures being non-convulsive [10,13,17].

The clinical diagnosis of SAE is paralleled with various findings in neuroimaging studies. Volumetric analyses revealed that SAE leads to brain atrophy, with a visible reduction in the volume of the white matter, cortex hippocampus and amygdala [18]. Disturbances of the cerebral blood flow have also been descried in septic patients [19]. Disseminated ischemic lesions are most often found in septic patients [14]. A noticeable percentage of septic patients develop radiological features of posterior reversible encephalopathy syndrome (PRES) [20].

1.4. Epidemiology of SAE

As was mentioned earlier, there are no strict, widely agreed and validated diagnostic criteria for SAE. One of the consequences of this fact is the lack of precise data on the incidence and prevalence of SAE in the population of septic patients. The results of epidemiological studies significantly differ according to the definition of SAE used. The most commonly cited figure of the 70% prevalence of SAE among septic patients comes from a study by Young et al. published in 1990. This prospective study included 69 septic patients and indeed the presence of encephalopathy was found in 70% of them [21]. It must be noted that this study was performed before contemporary definitions of sepsis were devised and therefore data from recent studies should be closer to the actual state. Chen et al., in their retrospective analysis performed in one center (defining SAE as the presence of cognitive and neuropsychiatric disorders in septic patients with no history or chronic neurological disorders or features of metabolic encephalopathy), found a prevalence of 43.6% of SAE. Patients with encephalopathy were older, with no significant differences in gender distribution, had higher scores on the APACHE II and SOFA scales and were at higher risk of in-hospital death [22]. These results are in concordance with those reported by Feng et al. These authors found an incidence of SAE of 42.3% septic patients, and subjects with SAE had higher scores of APACHE II and SOFA and higher mortality rates [23]. Sonneville et al. performed a retrospective multi-center study and found a slightly higher prevalence—53% of septic patients had SAE [24]. Therefore, it may be assumed that about half of all septic patients suffer from encephalopathy and that this condition is related to a more serious course of sepsis and higher mortality.

A frequently neglected fact is that acute sepsis-related brain injury may transform into a chronic form, becoming a cause of neurological and cognitive decline. A prospective study performed by Iwashyna et al. showed that in 10% of sepsis survivors, moderate and severe cognitive deficits were diagnosed de novo [25]. Another study showed that in a population of elderly subjects who survived 3 years after the diagnosis of sepsis, there were approximately 15% of patients with moderate and severe cognitive deficits [26]. In other studies, the prevalence of cognitive disorders reached over 20% [27]. Recently, it was also observed that in a middle-aged population, sepsis increases the rate of cognitive decline [28].

The high prevalence of sepsis-associated encephalopathy, its negative impact on prognosis and its chronic cognitive consequences (especially in the context of an increasing number of sepsis survivors) make disentangling the mechanisms of SAE and finding therapies with the potential to reduce the consequences of SAE urgently needed.

1.5. Melatonin

Melatonin is a neurohormone widely spread in nature, found both in plant and animal species. The most important source of melatonin in humans is the pineal gland, where this neurohormone is produced with the circadian rhythm, but multiple local sources of melatonin also exist, like the retina [29], immune cells [30], skin [31] or the epithelium of the gastrointestinal tract [32]. The most noticeable physiological role of melatonin in humans is the control of the circadian rhythm [33,34], but it exerts numerous other effects. Melatonin is a potent antioxidant [35] and anti-inflammatory factor [36]. It also has an impact on energy metabolism [37] and on the cardiovascular system [38].

The above-mentioned properties of melatonin make it an important factor influencing the inflammatory response. This is why so much attention has been given recently to the role of melatonin in sepsis. The aim of our review is to analyze the available literature on the potential neuroprotective effect of melatonin in sepsis-associated encephalopathy (SAE).

2. Mechanisms of Sepsis-Associated Encephalopathy

2.1. Pathomechanisms of SAE

The pathomechanism of SAE is not completely understood. It is certain that its etiology is multifactorial, with blood–brain barrier disruption, neuroinflammation, ischemia and oxidative stress playing crucial roles.

Sepsis-associated encephalopathy is diagnosed only in situations where the central nervous system is not infected. The systemic inflammatory response continues outside the brain, being hidden behind the blood–brain barrier (BBB). The BBB is a highly integrated "wall" built of endothelial cells, pericytes, astrocytes and microglial cells. Thanks to the BBB, most molecules may pass into the brain only through controlled transcellular transport—this also includes inflammatory mediators. The BBB also limits the migration of peripheral cells of the immune system to the CNS [39]. Therefore, an increase in the permeability of the BBB may be considered as the first step towards the involvement of the brain in sepsis and the development of SAE. Indeed, it has been found in a post-mortem study that proteins (occludin, claudin-5, ZO-1) forming so-called tight junctions between endothelial cells within the BBB were practically absent in the brains of septic patients. This means that during sepsis, the permeability of the BBB is significantly increased [40]. This allows the influx of proinflammatory cytokines into the central nervous system, which is followed by leukocyte infiltration—the systemic inflammatory response invades the brain. The next step is the activation of localized microglia within CNS immunocompetent cells [41]. It has been shown in humans that the activation of microglial cells is related to the development of delirium [42]. Microglial cells activated by proinflammatory cytokines entering the CNS through the pathologically permeable BBB present a proinflammatory phenotype [43] and start to produce their own proinflammatory cytokines—mainly tumor necrosis factor alpha (TNF alfa), interleukin-1 beta (Il-1b) and interleukin-6. The presence of these cytokines, along with the release of reactive oxygen species, nitric oxide and glutamate, further stimulates the neuroinflammatory process, closing a vicious circle of neuroinflammation, leading to the progressing damage of neural cells and to cell death through a mechanism called pyroptosis [44,45]. This injury of neural tissue leads to clinical symptoms of encephalopathy and delirium.

Parallel to the neuroinflammatory process, the septic brain struggles with acute disturbances within the supply chain of metabolic substrates. Sepsis and septic shock are characterized by perturbances in systemic circulation, hypotonia being the most prominent of them [46]. Simultaneously, it was shown that the autoregulation of cerebral arteries is less efficient in septic patients [47–49]. An active inflammatory process is also a pro-thrombotic condition. Regardless of the dysregulated blood flow, the cerebral arteries are blocked with clots in septic patients. Thus, another factor increasing the severity of sepsis-related brain injury is disseminated focal ischemia of the brain, leading to cellular death or significant metabolic perturbances, as the oxygenation of the brain tissue is significantly reduced in sepsis [50]. Therefore, the features of brain ischemia were present in all septic patients in one neuropathological study [51].

Neuroinflammation, BBB disruption and ischemia lead to the malfunction of neural cells, which in turn is the cause of pathological neurotransmission. The most widely studied is the cholinergic pathway—its anti-inflammatory action is suppressed in sepsis [52]. Sepsis-related deficiency in dopaminergic transmission is also suggested by some publications [53,54]. Proinflammatory cytokines (Il-1B) may increase the biological effect of GABA-ergic transmission, leading to pathological somnolence or cognitive disorders [55].

Summing up, disruption of the BBB allows the influx of proinflammatory cytokines into the CNS, which leads to microglial cells' activation and the initiation of a neuroinflammatory vicious cycle (activated cells produce more cytokines, which activate other cells, etc.). The neuroinflammation causes damage and death in neural cells, which is one of the biological substrates in the symptomatology of SAE. Further dysfunction of the brain is caused by sepsis-related hemodynamical instability and hypercoagulability, leading to ischemia. Neuroinflammation and ischemia disturb the normal balance of neurotransmitters, which is another cause of disorders of cognition and consciousness typical of SAE. The most important processes leading to SAE are presented in Figure 1.

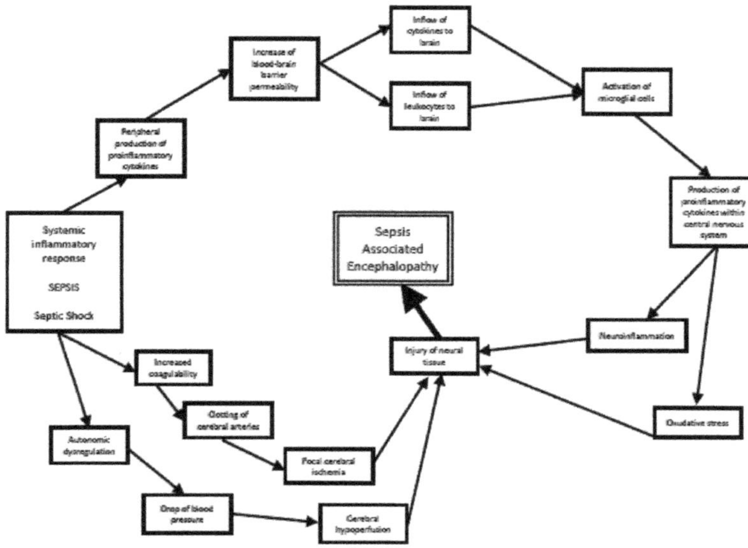

Figure 1. Processes leading to development of SAE.

2.2. The Role of Oxidative Stress and Free Radicals in Sepsis-Associated Encephalopathy

All the above-mentioned processes constitute a serious metabolic challenge to the cells of the central nervous system. The mitochondria of neurons must work in circumstances of increased energetic needs resulting from the high activity of the inflammatory process and of a decreased supply of oxygen and glucose, resulting from ischemia of the CNS. This leads to oxidative stress and mitochondrial dysfunction [56]. In vivo models of lipopolysaccharide-evoked sepsis-associated encephalopathy show that neuroinflammation leads to mitochondrial damage in endothelial cells through the pathological activation of dynamic-related protein 1 (Drp1), which was correlated with increasing mitochondrial oxidative stress and the loss of mitochondrial membrane potential. These phenomena were correlated with the reduced expression of proteins responsible for the formation of tight junctions between endothelial cells (ZO-1 and occludin) and thus leading to the increased permeability of the BBB. Mitochondrial damage was also detected in neuronal cells, with a decrease in oxidative phosphorylation, increased glycolysis and a reduction in the mitochondrial membrane potential and production of ATP. The final consequence of this process is the death of neurons [57]. Other animal studies have also shown the increased production of reactive oxygen species (ROS), reduced production of ATP and intense apoptosis [58]. Activated microglial cells also generate reactive oxygen species (ROS) and reactive nitrogen species (RNS), which finally leads to neuronal injury and death [59]. Oxidative stress, mitochondrial dysfunction and the presence of ROS and RNS trigger proapoptotic cellular caspase-dependent pathways, leading to the death of cells, which is the histological substrate for the clinical symptoms of SAE and its chronic consequences [60]. This is why any therapeutical intervention with the potential to reduce oxidative stress and scavenge reactive species may be neuroprotective and limit the chronic cognitive deficits in survivors of sepsis. This was already shown in animal models with anti-oxidative therapies, e.g., with molecular hydrogen [61] or tetramethylpyrazine [58].

3. Clinical Role of Melatonin

Melatonin is frequently mentioned as a potential drug in septic patients. The antioxidant, anti-inflammatory and neuroprotective properties of this hormone, which are described in detail below, make it a very interesting potential therapeutic factor in patients with sepsis and SAE.

3.1. Melatonin

Melatonin is a neurohormone synthesized mainly in the pineal gland. It is a tryptophan derivative, N-acetyl-5-methoxytryptamine. Apart from the pineal gland, it is also produced in the retina, skin, gastrointestinal tract and in cells of the immunological system [62]. It is produced and released in the circadian rhythm, with the lowest concentrations during the day (<2 pg/mL) and highest during the night (approximately 100 pg/mL) [63]. It has been also discovered that in certain conditions, the local tissue concentration of melatonin may be significantly higher than in plasma [64]. It acts through two G-protein-coupled receptors, MT1 and MT2, most abundantly met in the pars tubercles of the pineal gland, retina and hypothalamus [65]. Nevertheless, it must be noted that most of the anti-inflammatory and antioxidant actions of melatonin are independent of MT1 and MT2 and are exerted through the activation of NRF2-related pathways, as was shown by Janjetovic et al. [66].

3.2. Main Clinical Roles of Melatonin

The main physiological role of melatonin is the regulation of the circadian rhythm. In the case of humans, this means promoting sleep during the dark part of the day. The production and secretion of melatonin is inhibited by light through the stimulation of the retina and activation of the suprachiasmatic nucleus [67]. Its production is increased in darkness, when melatonin may fully exert its physiological role, which is switching the organism to "sleep mode" [68]. The localization of melatonin receptors suggests that the neurohormone acts through the modulation of the function of neural and neuroendocrine centers but also through direct impacts on the systems and organs of the body [62,67]. The chronobiological role of melatonin is expressed through driving the rhythm of the functioning of peripheral tissues and organs, adjusting them to the day–night cycle [69]. This role is crucial in terms of sepsis and the development of SAE. Sleep/wake rhythm disorders are frequently reported in septic patients and their presence negatively impacts prognosis [70]. On the other hand, a strong and probably causal relation between disrupted circadian rhythms and the development of delirium/encephalopathy in critically ill patients has been observed [71]. A low concentration of melatonin was discovered in septic patients with disordered circadian rhythms, which suggests a potential therapeutical role of this neurohormone at least as a circadian rhythm keeper [72]. It must be noted that the usage of melatonin (as a hypnotic drug) in intensive care units has increased in some countries [73]. As "switching to sleep mode" requires the modulation of the activity of the autonomic nervous system, it was proven that melatonin has potential to reduce the sympathetic drive [74]. This potential to modulate the activity of the autonomic nervous system was elegantly shown in patients who underwent pinealectomy—a therapy with exogenic melatonin restored the sympathetic–parasympathetic balance in these subjects [75]. This ability to harmonize both parts of the autonomic nervous system may be especially important in septic patients, in which the disruption of sympathovagal modulation may even precede the development of a systemic inflammatory response [76]. It can be speculated that the use of melatonin may be parallel to the stimulation of the vagal nerve, as successfully implemented in animal models of sepsis [77]. The impact of melatonin on the autonomic nervous system manifests also through the influence of melatonin on the values of blood pressure. Melatonin reduces vasoconstriction and blood pressure, and it is proposed as an adjunct drug in the treatment of hypertension [78]. Melatonin has also a significant impact on metabolism, being an important protective factor against obesity and the development of diabetes mellitus [37,79,80]. Melatonin also plays a role in modulating the inflammatory response, having noticeable anti-inflammatory potential, as it can prevent cell death caused by pyroptosis [81,82]. The above-listed functions of melatonin (autonomic and blood pressure control, metabolism control, modulation of inflammatory response) make it a very important neurohormone and potentially a therapeutic agent in septic patients.

3.3. Scavenging and Antioxidant Potential of Melatonin

As mentioned above, sepsis and SAE are related to the increased production and activity of free radicals. Neuroinflammation, being the most important putative factor of SAE, leads to the very intense production of oxidants within the central nervous system, with normal or decreased abilities for their consumption. This situation fulfills the definition criteria for oxidative stress [83]. Oxidative stress and high concentrations and activity of free radicals lead to the damage of vital molecules such as DNA, phospholipids and proteins, which in turn exert pro-apoptotic action and cause damage to biological membranes, with the fragmentation and death of cells. Cells and tissues can be protected from oxidative stress by molecules called antioxidants (molecules promoting processes that diminish the production of free radicals and intensify the consumption of oxidants) and scavengers (molecules capable of trapping free radicals directly) [62].

Melatonin has significant antioxidant potential. First, it can scavenge some of the free radicals directly. It has been shown that melatonin can react with hydroxyl (OH) radicals [84,85]. Melatonin may also serve as a substrate of reaction, leading to the elimination of peroxyl radicals [86,87] as well as nitric oxide [88].

Apart from the direct scavenging of free radicals, melatonin is capable of suppressing their production through the chelation of metal ions (e.g., Cu, Fe) necessary for the synthesis of oxidants. It was proven that melatonin chelates metal ions—melatonin was able to create complexes with copper, iron, cadmium or aluminum, blocking the participation of these ions in oxidant-producing reactions [89]. Therefore, melatonin is able to protect tissues from oxidative, metal-catalyzed damage [90,91].

Moreover, melatonin may participate in the repair of oxidative molecular damage. Colares et al. have recently proven that treatment with melatonin reduces the damage of DNA in cirrhotic rats [92]. Therapy with melatonin also intensifies the repair of DNA in neoplasmatic cells [83]. Melatonin was shown to reduce oxidative DNA damage in neural tissue caused by ischemia or trauma [93,94].

Lastly, melatonin activates antioxidant enzymes and influences signaling pathways involved in the generation of free radicals. Gou et al. described the protective effect of melatonin on neural tissue in hypoxic–ischemic brain damage exerted through the modulation of pathways leading to the activation of glutathione peroxidase—an antioxidant enzyme [95]. The melatonin-driven activation of antioxidant enzymes is a well-established fact [96].

It is also noteworthy that the metabolites of melatonin are potent antioxidants that increase the therapeutical potential of the pineal neurohormone. These metabolites may counteract environmental stresses, including oxidative stress, as was described in a review by Slominski et al. [97]. The main metabolites of melatonin, like N^1-acetyl-N^2-formyl-5-methoxykynuramine (AFMK), N-acethyl-5-methoxykynuramine (AMK), 6-hydroxymelatonin and 4-hydroxymelatonin, act as direct scavengers, metal-chelating agents or molecules capable of repairing oxidative damage, as was presented in a review by Galano and Reiter [98].

As this short summary shows, both endo- and exogenic melatonin may serve as a potent antioxidant protectant, reducing the oxidative tissue damage evoked by sepsis and neuroinflammation.

4. Role of Melatonin in Sepsis and in Sepsis-Associated Encephalopathy

4.1. Therapeutic Potential of Melatonin in Sepsis

Sepsis is a dysregulated response to infection, resulting from the overproduction of proinflammatory cytokines, extreme oxidative stress and the progressing failure of the organs and systems of the organism. As was shown in numerous studies, melatonin has the potential to exert a multidirectional therapeutical impact, protecting the tissues against the consequences of sepsis.

As was discussed above, melatonin is a powerful antioxidant. Zhen et al., in a recent study, found that melatonin could increase the activity of the antioxidant enzyme superoxide dismutase in the myocardium of septic rats, which led to less oxidative tissue damage [99]. Rahim et al. published similar results based upon an animal model of sepsis, showing that melatonin increases antioxidative defense through NRF2 activation [100]. Melatonin is able to activate NRF2, which in turn increases the expression of antioxidative enzymes (like SOD), which may prevent multiorgan failure in sepsis. Kang et al. proved that melatonin may prevent the development of lung injury through this pathway [101]. The melatonin-induced activation of the NRF2 signaling pathway was also shown to exert neuroprotective action in a model of the lipopolysaccharide-induced injury of neural tissue (analogous to models of septic encephalopathy) [102]. Melatonin is also capable of upregulating the activity of sirtuins (especially SIRT1 and SIRT 3) and thus leads to increased antioxidant activity of superoxide dismutase [103]. In this specific experiment, melatonin was shown to protect the small intestine from sepsis-evoked oxidative injury.

One typical pathological phenomenon in sepsis is mitochondrial dysfunction leading to energetic failure [104]. It was shown that melatonin may reverse sepsis-related pathological changes within mitochondria. The infusion of melatonin was shown to improve mitochondrial respiration and to reduce oxidative stress. This effect was paralleled with the reduction of organ failure markers [105]. Escames et al. found that melatonin exerted a bidirectional therapeutical effect upon mitochondria in septic mice: the pineal neurohormone inhibited mitochondrial nitric oxide synthase, reducing oxidative stress and simultaneously restoring the production of ATP [106].

Melatonin influences also the interplay between pro- and anti-inflammatory cytokines. Carillo-Vico et al. found that in a rat model of septic shock, melatonin decreased the concentration of the proinflammatory IL-10 and TNF-alpha, increasing simultaneously the concentration of anti-inflammatory IL-12 [107]. Melatonin may also be a crucial player in modulating the inflammatory response through the inhibition of the action of the inflammasome NLRP3 [108].

Promising results of studies performed on animal models and observations of a relation between low concentrations of melatonin and sepsis-related mortality [109] led to the first human trials. Alamili and colleagues induced endotoxemia in healthy human volunteers who were pre-treated with melatonin or a placebo. The researchers observed that melatonin (compared with placebo) led to a reduction in the concentration of some proinflammatory cytokines (e.g., interleukin Il-1beta) during the day, with no statistically significant effect during the night [110,111]. Galley et al. performed a study with a human whole blood model of sepsis, proving that melatonin reduced oxidative stress and mitochondrial failure as well as the production of proinflammatory cytokines [112]. Aisa-Alvares and co-workers performed a small, single-center clinical trial in septic patients, assessing the impact of various antioxidant molecules. Melatonin appeared to reduce the concentration of procalcitonin and lipid peroxidation [113]. Melatonin was also shown to reduce the need for mechanical ventilation and for the use of a vasopressor in another small (n = 40 patients), single-center clinical trial [114]. Mansilla-Roselló et al. published very recently the results of a clinical study comparing the 5-day therapy of septic patients with the i.v. infusion of 50 mg of melatonin versus a placebo. The authors observed that patients treated with melatonin had significantly shorter hospital stays, lower values of the SOFA score, lower concentrations of procalcitonin and more intense antioxidant activity [115].

The clinical usefulness of the antioxidant properties of melatonin was also shown in a recent clinical study focusing on septic patients, which compared the effectiveness of 50 mg of orally given melatonin with other antioxidants. It was shown that melatonin reduced the SOFA score and lipid peroxidation while increasing the total antioxidant capacity [116]. Details of all the above-mentioned clinical studies are presented in Table 1. Although the results are very promising, the limitations of the studies must be noted. All of them were single-center studies with a limited number of patients included. This does not allow the generalization of the conclusions. Multi-center studies with the randomization of large groups of patients are required.

Table 1. Clinical trials assessing usage of melatonin in sepsis.

Author of the Study (Year of Publication)	Number of Participants	Study Design, Dose of Melatonin Assessed	Assessed Outcome	Conclusions
Alamili et al. (2014) [110]	12 healthy volunteers with experimentally evoked endotoxaemia	Randomized, placebo-controlled, double-blinded cross-over trial, comparing infusion of 100 mg i.v. melatonin with placebo	Concentrations of pro- and anti-inflammatory markers in plasma of participants during night	Melatonin had no effect on concentratons of markers during the night
Alamili et al. (2014) [111]	12 healthy volunteers with experimentally evoked endotoxaemia	Randomized, placebo-controlled, double-blinded cross-over trial, comparing infusion of 100 mg i.v. melatonin with placebo	Concentrations of pro- and anti-inflammatory markers in plasma of participants during day	Melatonin reduced concentrations of pro-inflammatory biomarkers during the day
Aisa-Álvarez et. al. (2020) [113]	97 subjects divided into 5 subgroups	Randomized, controlled, triple-masked, and with parallel assignment clinical trial with a control group without treatment. Comparison of 4 molecules (vitamin C, vitamin E, n-acetylcysteine, melatonin—50 mg p.o. daily) and no antioxidant treatment	Concentrations of inflammation and oxidative stress markers in plasma	Reduction in procalcitonine and lipid peroxidation in subjects treated with melatonin, paralleled with reduction in sepsis severity measured with SOFA score
Taher et al. (2022) [114]	40 patients with septic shock	Prospective, two-arm, double-blind, randomized clinical trial; 50 mg melatonin given p.o. through 5 days was compared with placebo.	Change in SOFA score, need for mechanical ventilation, required dosage of vasopressor	Insignificant reduction in SOFA score, percentage of patients requiring ventilation and usage of vasopressors in melatonin-treated group
Mansilla-Roseló et al. (2023) [115]	29 patients (14 in placebo group, 15 in melatonin group)	Unicenter, randomized, placebo-controlled, double-blind trial; 60 mg of i.v. given melatonin was compared to placebo therapy	Change in SOFA score, concentrations of inflammatory markers, oxidative stress status	Treatment with melatonin led to decrease in SOFA score and in concentrations of inflammatory markers and improvement in oxidative stress status
Aisa-Álvarez et al. (2023) [116]	131 subjects divided into 5 subgroups	Randomized, controlled, triple-masked, and with parallel assignment clinical trial with a control group without treatment. Comparison of 4 molecules (vitamin C, vitamin E, n-acetylcysteine, melatonin—50 mg p.o. daily) and no antioxidant treatment	Change in SOFA score, change in concentrations of inflammatory markers and antioxidant activity	Therapy with melatonin was related to decrease in SOFA score, decrease in inflammatory markers and improvement in antioxidant activity

Human studies on melatonin in sepsis were performed with a variety of dosages, ranging from 50 mg p.o. daily to 100 mg i.v. daily. Establishing the proper dosage of a neurohormone that is released in a specific circadian rhythm is challenging task. Galley et al. undertook this and performed a clinical study aiming at assessing the clinical safety, tolerance and optimal dosing of melatonin. There were 10 patients with sepsis resulting from community-acquired pneumonia participating in the study. Five of them received 50 mg of melatonin p.o. daily and the other five were given 20 mg of the neurohormone daily. Clinical observation led to the conclusion that therapy with melatonin did not lead to any adverse event and was well tolerated. Twenty mg daily was found to be the optimal dose from a pharmacological point of view [117].

The implementation of melatonin in clinical practice must be performed cautiously as little is known about its potential interactions with other drugs, especially the ones used in critically ill subjects (e.g., sedatives, vasopressors, antibiotics or steroids). It is a consequence of the sporadic use of melatonin in clinical practice. On the other hand, the authors of clinical trials assessing melatonin in sepsis did not report any significant interactions with routine therapy.

4.2. Neuroprotective Potential of Melatonin in Sepsis-Associated Encephalopathy

Ji et al. performed an interesting experiment with the immediate or delayed treatment with melatonin for sepsis-related brain injury in a mouse model of sepsis. It was shown that while immediate therapy with melatonin led to an increase in survival with no positive neurobehavioral effect, the delayed therapy with melatonin caused a neurobehavioral improvement [118]. The potential (although speculative) explanation for this finding may be that melatonin given immediately protects the integrity of the BBB and thus prevents the development of further neuroinflammatory processes. Due to this, sepsis-related brain injury is diminished and this decreases mortality with no visible neurobehavioral effect. Meanwhile, melatonin given later may only have a neuroprotective effect in the already inflamed brain and this explains the neurobehavioral benefit. This observation is very important in a human and clinical context as it suggests that therapy with melatonin may be somewhat beneficial regardless of the moment at which it is started. The beneficial and diverse neurologic effects of therapy with melatonin are not surprising in the context of the multidirectional action of this neurohormone in sepsis. This multipotency of melatonin is also visible in terms of its neuroprotective effect in sepsis-associated encephalopathy. As was shown earlier in the paper, the first event leading to SAE is blood–brain barrier (BBB) leakage. Wang et al. have shown that pre-treatment with melatonin may sustain BBB integrity in lipopolysaccharide (LPS)-evoked sepsis in mice. The main reason for BBB damage in sepsis is the degradation of tight junction proteins (e.g., occluding or claudin-5), which was clearly stopped with melatonin in Wang and colleagues' experiment. This effect was achieved through the antioxidant action of melatonin, which inhibited the expression of enzymes involved in the production of free radicals (specifically, the gp91 phox subunit of NADPH oxidase) [119]. The same group of authors suggested that the supplementation of melatonin may protect BBB integrity in older patients suffering from sepsis [120].

As was mentioned earlier, melatonin exerts an anti-inflammatory effect, which also includes the suppression of neuroinflammation. Zhou et al. analyzed the effects of therapy with melatonin on neuroinflammation in an animal model of sepsis (lipopolysaccharide-injected rats). The authors reported a significant reduction in the production of proinflammatory mediators by microglial cells and the upregulation of the expression of anti-inflammatory factors in melatonin-treated animals. Moreover, a prominent shift towards the anti-inflammatory phenotype M2 of microglial cells was observed. These changes resulted in reduced white matter damage and demyelination in treated animals [121]. Melatonin reduces the LPS-evoked increase in the production of proinflammatory cytokines in the brain (e.g., IL-6, TNF-alpha) and simultaneously normalizes the expression and activity of redox signaling molecules (e.g., SOD2) [122]. Interestingly, the simultaneity of the anti-inflammatory and antioxidant action of melatonin in models of sepsis-associated encephalopathy is repeatedly described. Zhao et al. observed that in septic animals, therapy with melatonin led to a reduction in cerebral concentrations of proinflammatory cytokines (e.g., TNF-alpha, IN-1beta) and at the same time to a reduction in oxidative stress (confirmed by observations of the activation of SOD and diminished production of malondialdehyde) [123].

Apart from blood–brain barrier damage, neuroinflammation and oxidative stress, the septic brain is also injured due to multifocal ischemia resulting from decreased cerebral perfusion and inflammation-evoked thrombotic events. According to the results of a recently published, placebo-controlled, double-blind clinical trial, therapy with melatonin may intensify the neurologic improvement in patients after ischemic stroke [124]. This finding is not surprising as the neuroprotective action of melatonin in ischemic brain injury was proven in animal models. This neuroprotective effect is exerted through intensive antioxidant action (e.g., through an increase in the expression of glutathione peroxidase or through the attenuation of endoplasmic reticulum stress) [95,125] and the suppression of inflammatory processes (e.g., through promoting the anti-inflammatory phenotype of microglia and through a reduction in the expression of proinflammatory cytokines) [126,127]. It was also observed that in models of ischemic brain injury, melatonin could improve the integrity of the blood–brain barrier, similarly to what was observed in sepsis [128].

The result of the mechanisms described above is the suppression of apoptotic processes and reduction in brain injury during sepsis, which suggests potential for melatonin-driven better neurological outcomes of sepsis-associated encephalopathy.

The facts collected in the previous paragraphs suggest that melatonin may be a potent therapeutical agent in patient in sepsis and with SAE. This is strongly supported by the results of pre-clinical studies on animal models, which have found initial clinical confirmation in the human studies performed so far. The collected data allow us to plan more efficient protocols for clinical trials with melatonin in the upcoming future. Potential mechanisms through which melatonin may exert its effect in SAE are shown in Figure 2.

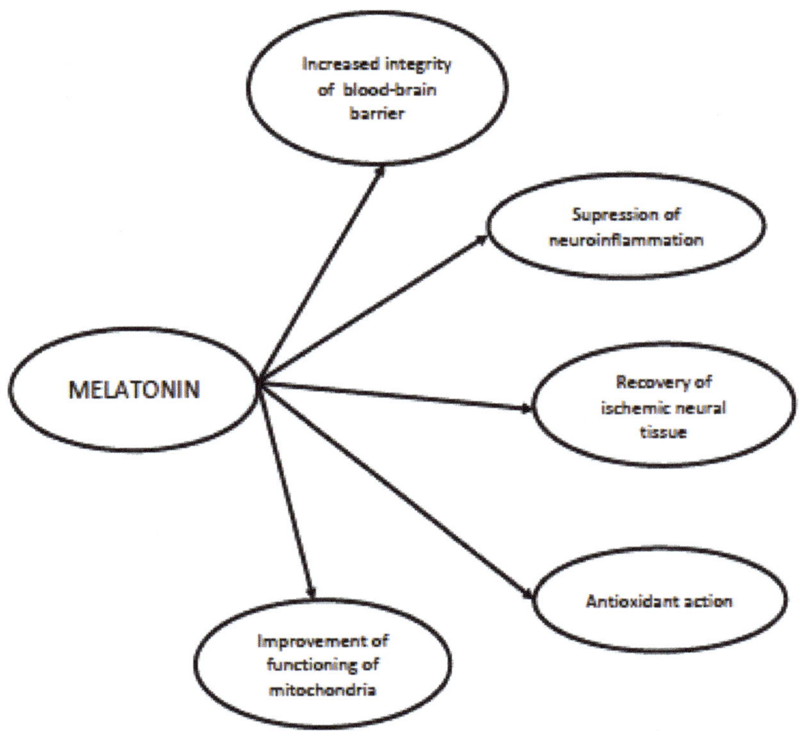

Figure 2. Potential therapeutical mechanisms of melatonin in sepsis and sepsis-associated encephalopathy.

5. Conclusions and Future Perspectives

Melatonin is a fascinating neurohormone with a surprisingly wide range of potential actions. Some of them, like antioxidant potential, anti-inflammatory properties, a beneficiary impact on blood–brain barrier integrity or an ability to restore mitochondrial equilibrium, make this molecule a natural ally in the fight against sepsis and its complications and consequences. It must be remembered that the brain is the most vulnerable organ to circulatory, inflammatory and metabolic disorders caused by uncontrolled inflammation during sepsis. Sepsis-associated encephalopathy, being a consequence of blood–brain barrier leakage, neuroinflammatory processes, the injury of neural tissue caused by free radicals and multifocal ischemia of the brain, increases the mortality of septic patients and leads to chronic cognitive decline in a significant proportion of sepsis survivors. There is a significant group of molecules that may be potentially neuroprotective in SAE [129]. The data collected in this review suggest that melatonin may potentially act as a preventive or therapeutic substance in septic patients. On the other hand, we must be aware that regardless of the large amount of data coming from animal studies, clinical information on the therapeutic effectiveness of melatonin in humans is scarce. This constitutes the most important challenge for the future—implementing actual knowledge of melatonin in clinical trials in sepsis.

Author Contributions: Conceptualization, M.S. and K.K.; methodology, K.K. and M.S.; formal analysis, M.S.; investigation, K.S.-R., J.S. and K.K.; writing—original draft preparation, M.S., K.S.-R., J.S. and K.K.; writing—review and editing, M.S.; supervision, M.S.; project administration. All authors have read and agreed to the published version of the manuscript.

Funding: This research received no external funding.

Institutional Review Board Statement: Not applicable.

Informed Consent Statement: Not applicable.

Data Availability Statement: Not applicable.

Conflicts of Interest: The authors declare no conflict of interest.

References

1. Singer, M.; Deutschman, C.S.; Seymour, C.W.; Shankar-Hari, M.; Annane, D.; Bauer, M.; Bellomo, R.; Bernard, G.R.; Chiche, J.-D.; Coopersmith, C.M.; et al. The Third International Consensus Definitions for Sepsis and Septic Shock (Sepsis-3). *JAMA* **2016**, *315*, 801–810. [CrossRef] [PubMed]
2. Seymour, C.W.; Liu, V.X.; Iwashyna, T.J.; Brunkhorst, F.M.; Rea, T.D.; Scherag, A.; Rubenfeld, G.; Kahn, J.M.; Shankar-Hari, M.; Singer, M.; et al. Assessment of Clinical Criteria for Sepsis: For the Third International Consensus Definitions for Sepsis and Septic Shock (Sepsis-3). *JAMA* **2016**, *315*, 762–774. [CrossRef] [PubMed]
3. Fleischmann-Struzek, C.; Mellhammar, L.; Rose, N.; Cassini, A.; Rudd, K.E.; Schlattmann, P.; Allegranzi, B.; Reinhart, K. Incidence and Mortality of Hospital- and ICU-Treated Sepsis: Results from an Updated and Expanded Systematic Review and Meta-Analysis. *Intensive Care Med.* **2020**, *46*, 1552–1562. [CrossRef] [PubMed]
4. Rhee, C.; Dantes, R.; Epstein, L.; Murphy, D.J.; Seymour, C.W.; Iwashyna, T.J.; Kadri, S.S.; Angus, D.C.; Danner, R.L.; Fiore, A.E.; et al. Incidence and Trends of Sepsis in US Hospitals Using Clinical vs Claims Data, 2009–2014. *JAMA* **2017**, *318*, 1241–1249. [CrossRef]
5. Fleischmann-Struzek, C.; Rose, N.; Freytag, A.; Spoden, M.; Prescott, H.C.; Schettler, A.; Wedekind, L.; Ditscheid, B.; Storch, J.; Born, S.; et al. Epidemiology and Costs of Postsepsis Morbidity, Nursing Care Dependency, and Mortality in Germany, 2013 to 2017. *JAMA Netw. Open* **2021**, *4*, e2134290. [CrossRef]
6. Chung, H.-Y.; Wickel, J.; Brunkhorst, F.M.; Geis, C. Sepsis-Associated Encephalopathy: From Delirium to Dementia? *J. Clin. Med.* **2020**, *9*, 703. [CrossRef]
7. Dantzer, R. Evolutionary Aspects of Infections: Inflammation and Sickness Behaviors. *Curr. Top. Behav. Neurosci.* **2022**, *61*, 1–14. [CrossRef]
8. Heming, N.; Mazeraud, A.; Verdonk, F.; Bozza, F.A.; Chrétien, F.; Sharshar, T. Neuroanatomy of Sepsis-Associated Encephalopathy. *Crit. Care* **2017**, *21*, 65. [CrossRef]
9. Wasyluk, W.; Wasyluk, M.; Zwolak, A. Sepsis as a Pan-Endocrine Illness-Endocrine Disorders in Septic Patients. *J. Clin. Med.* **2021**, *10*, 2075. [CrossRef]

10. Hosokawa, K.; Gaspard, N.; Su, F.; Oddo, M.; Vincent, J.-L.; Taccone, F.S. Clinical Neurophysiological Assessment of Sepsis-Associated Brain Dysfunction: A Systematic Review. *Crit. Care* **2014**, *18*, 674. [CrossRef]
11. Catarina, A.V.; Branchini, G.; Bettoni, L.; De Oliveira, J.R.; Nunes, F.B. Sepsis-Associated Encephalopathy: From Pathophysiology to Progress in Experimental Studies. *Mol. Neurobiol.* **2021**, *58*, 2770–2779. [CrossRef]
12. Bowman, E.M.L.; Cunningham, E.L.; Page, V.J.; McAuley, D.F. Phenotypes and Subphenotypes of Delirium: A Review of Current Categorisations and Suggestions for Progression. *Crit. Care* **2021**, *25*, 334. [CrossRef]
13. Mazeraud, A.; Righy, C.; Bouchereau, E.; Benghanem, S.; Bozza, F.A.; Sharshar, T. Septic-Associated Encephalopathy: A Comprehensive Review. *Neurother. J. Am. Soc. Exp. Neurother.* **2020**, *17*, 392–403. [CrossRef] [PubMed]
14. Polito, A.; Eischwald, F.; Maho, A.-L.; Polito, A.; Azabou, E.; Annane, D.; Chrétien, F.; Stevens, R.D.; Carlier, R.; Sharshar, T. Pattern of Brain Injury in the Acute Setting of Human Septic Shock. *Crit. Care* **2013**, *17*, R204. [CrossRef] [PubMed]
15. Young, G.B.; Bolton, C.F.; Archibald, Y.M.; Austin, T.W.; Wells, G.A. The Electroencephalogram in Sepsis-Associated Encephalopathy. *J. Clin. Neurophysiol.* **1992**, *9*, 145–152. [CrossRef] [PubMed]
16. Pantzaris, N.-D.; Platanaki, C.; Tsiotsios, K.; Koniari, I.; Velissaris, D. The Use of Electroencephalography in Patients with Sepsis: A Review of The Literature. *J. Transl. Intern. Med.* **2021**, *9*, 12–16. [CrossRef] [PubMed]
17. Foreman, B.; Mahulikar, A.; Tadi, P.; Claassen, J.; Szaflarski, J.; Halford, J.J.; Dean, B.C.; Kaplan, P.W.; Hirsch, L.J.; LaRoche, S.; et al. Generalized Periodic Discharges and "Triphasic Waves": A Blinded Evaluation of Inter-Rater Agreement and Clinical Significance. *Clin. Neurophysiol.* **2016**, *127*, 1073–1080. [CrossRef]
18. Orhun, G.; Tüzün, E.; Bilgiç, B.; Ergin Özcan, P.; Sencer, S.; Barburoğlu, M.; Esen, F. Brain Volume Changes in Patients with Acute Brain Dysfunction Due to Sepsis. *Neurocrit. Care* **2020**, *32*, 459–468. [CrossRef]
19. Masse, M.-H.; Richard, M.A.; D'Aragon, F.; St-Arnaud, C.; Mayette, M.; Adhikari, N.K.J.; Fraser, W.; Carpentier, A.; Palanchuck, S.; Gauthier, D.; et al. Early Evidence of Sepsis-Associated Hyperperfusion-A Study of Cerebral Blood Flow Measured With MRI Arterial Spin Labeling in Critically Ill Septic Patients and Control Subjects. *Crit. Care Med.* **2018**, *46*, e663–e669. [CrossRef]
20. Orhun, G.; Sencer, S.; Tüzün, E.; Bebek, N.; Ergin Özcan, P.; Barburoğlu, M.; Günver, M.G.; Esen, F. Posterior Reversible Encephalopathy in Sepsis-Associated Encephalopathy: Experience from a Single Center. *Neurocrit. Care* **2022**, *36*, 372–386. [CrossRef]
21. Young, G.B.; Bolton, C.F.; Austin, T.W.; Archibald, Y.M.; Gonder, J.; Wells, G.A. The Encephalopathy Associated with Septic Illness. *Clin. Investig. Med. Med. Clin. Exp.* **1990**, *13*, 297–304.
22. Chen, J.; Shi, X.; Diao, M.; Jin, G.; Zhu, Y.; Hu, W.; Xi, S. A Retrospective Study of Sepsis-Associated Encephalopathy: Epidemiology, Clinical Features and Adverse Outcomes. *BMC Emerg. Med.* **2020**, *20*, 77. [CrossRef]
23. Feng, Q.; Ai, Y.-H.; Gong, H.; Wu, L.; Ai, M.-L.; Deng, S.-Y.; Huang, L.; Peng, Q.-Y.; Zhang, L.-N. Characterization of Sepsis and Sepsis-Associated Encephalopathy. *J. Intensive Care Med.* **2019**, *34*, 938–945. [CrossRef] [PubMed]
24. Sonneville, R.; de Montmollin, E.; Poujade, J.; Garrouste-Orgeas, M.; Souweine, B.; Darmon, M.; Mariotte, E.; Argaud, L.; Barbier, F.; Goldgran-Toledano, D.; et al. Potentially Modifiable Factors Contributing to Sepsis-Associated Encephalopathy. *Intensive Care Med.* **2017**, *43*, 1075–1084. [CrossRef] [PubMed]
25. Iwashyna, T.J.; Ely, E.W.; Smith, D.M.; Langa, K.M. Long-Term Cognitive Impairment and Functional Disability among Survivors of Severe Sepsis. *JAMA* **2010**, *304*, 1787–1794. [CrossRef]
26. Iwashyna, T.J.; Cooke, C.R.; Wunsch, H.; Kahn, J.M. Population Burden of Long-Term Survivorship after Severe Sepsis in Older Americans. *J. Am. Geriatr. Soc.* **2012**, *60*, 1070–1077. [CrossRef] [PubMed]
27. Stallmach, A.; Kesselmeier, M.; Bauer, M.; Gramlich, J.; Finke, K.; Fischer, A.; Fleischmann-Struzek, C.; Heutelbeck, A.; Katzer, K.; Mutschke, S.; et al. Comparison of Fatigue, Cognitive Dysfunction and Psychological Disorders in Post-COVID Patients and Patients after Sepsis: Is There a Specific Constellation? *Infection* **2022**, *50*, 661–669. [CrossRef] [PubMed]
28. Wang, H.E.; Kabeto, M.M.; Gray, M.; Wadley, V.G.; Muntner, P.; Judd, S.E.; Safford, M.M.; Kempker, J.; Levine, D.A. Trajectory of Cognitive Decline After Sepsis. *Crit. Care Med.* **2021**, *49*, 1083–1094. [CrossRef] [PubMed]
29. Iuvone, P.M.; Tosini, G.; Pozdeyev, N.; Haque, R.; Klein, D.C.; Chaurasia, S.S. Circadian Clocks, Clock Networks, Arylalkylamine N-Acetyltransferase, and Melatonin in the Retina. *Prog. Retin. Eye Res.* **2005**, *24*, 433–456. [CrossRef]
30. Carrillo-Vico, A.; Calvo, J.R.; Abreu, P.; Lardone, P.J.; García-Mauriño, S.; Reiter, R.J.; Guerrero, J.M. Evidence of Melatonin Synthesis by Human Lymphocytes and Its Physiological Significance: Possible Role as Intracrine, Autocrine, and/or Paracrine Substance. *FASEB J.* **2004**, *18*, 537–539. [CrossRef]
31. Slominski, A.T.; Zmijewski, M.A.; Semak, I.; Kim, T.-K.; Janjetovic, Z.; Slominski, R.M.; Zmijewski, J.W. Melatonin, Mitochondria, and the Skin. *Cell. Mol. Life Sci. CMLS* **2017**, *74*, 3913–3925. [CrossRef]
32. Bertrand, P.P.; Polglaze, K.E.; Bertrand, R.L.; Sandow, S.L.; Pozo, M.J. Detection of Melatonin Production from the Intestinal Epithelium Using Electrochemical Methods. *Curr. Pharm. Des.* **2014**, *20*, 4802–4806. [CrossRef]
33. Shukla, M.; Vincent, B. Melatonin as a Harmonizing Factor of Circadian Rhythms, Neuronal Cell Cycle and Neurogenesis: Additional Arguments for Its Therapeutic Use in Alzheimer's Disease. *Curr. Neuropharmacol.* **2023**, *21*, 1273–1298. [CrossRef] [PubMed]
34. Tan, D.-X.; Reiter, R.J.; Zimmerman, S.; Hardeland, R. Melatonin: Both a Messenger of Darkness and a Participant in the Cellular Actions of Non-Visible Solar Radiation of Near Infrared Light. *Biology* **2023**, *12*, 89. [CrossRef]

35. Ramos, E.; Gil-Martín, E.; De Los Ríos, C.; Egea, J.; López-Muñoz, F.; Pita, R.; Juberías, A.; Torrado, J.J.; Serrano, D.R.; Reiter, R.J.; et al. Melatonin as Modulator for Sulfur and Nitrogen Mustard-Induced Inflammation, Oxidative Stress and DNA Damage: Molecular Therapeutics. *Antioxidants* **2023**, *12*, 397. [CrossRef] [PubMed]
36. Piekarska, K.; Bonowicz, K.; Grzanka, A.; Jaworski, Ł.M.; Reiter, R.J.; Slominski, A.T.; Steinbrink, K.; Kleszczyński, K.; Gagat, M. Melatonin and TGF-β-Mediated Release of Extracellular Vesicles. *Metabolites* **2023**, *13*, 575. [CrossRef] [PubMed]
37. Guan, Q.; Wang, Z.; Cao, J.; Dong, Y.; Chen, Y. Mechanisms of Melatonin in Obesity: A Review. *Int. J. Mol. Sci.* **2021**, *23*, 218. [CrossRef] [PubMed]
38. Cvikova, D.; Sutovska, H.; Babarikova, K.; Molcan, L. Hypotensive Effects of Melatonin in Rats: Focus on the Model, Measurement, Application, and Main Mechanisms. *Hypertens. Res.* **2022**, *45*, 1929–1944. [CrossRef]
39. Gao, Q.; Hernandes, M.S. Sepsis-Associated Encephalopathy and Blood-Brain Barrier Dysfunction. *Inflammation* **2021**, *44*, 2143–2150. [CrossRef]
40. Erikson, K.; Tuominen, H.; Vakkala, M.; Liisanantti, J.H.; Karttunen, T.; Syrjälä, H.; Ala-Kokko, T.I. Brain Tight Junction Protein Expression in Sepsis in an Autopsy Series. *Crit. Care* **2020**, *24*, 385. [CrossRef]
41. Torii, K.; Takagi, S.; Yoshimura, R.; Miyata, S. Microglial Proliferation Attenuates Sickness Responses in Adult Mice during Endotoxin-Induced Inflammation. *J. Neuroimmunol.* **2022**, *365*, 577832. [CrossRef] [PubMed]
42. van Munster, B.C.; Aronica, E.; Zwinderman, A.H.; Eikelenboom, P.; Cunningham, C.; Rooij, S.E.J.A. de Neuroinflammation in Delirium: A Postmortem Case-Control Study. *Rejuven. Res.* **2011**, *14*, 615–622. [CrossRef] [PubMed]
43. Zrzavy, T.; Höftberger, R.; Berger, T.; Rauschka, H.; Butovsky, O.; Weiner, H.; Lassmann, H. Pro-Inflammatory Activation of Microglia in the Brain of Patients with Sepsis. *Neuropathol. Appl. Neurobiol.* **2019**, *45*, 278–290. [CrossRef] [PubMed]
44. Sharshar, T.; Chevret, S.; Bourdain, F.; Raphaël, J.-C. Early Predictors of Mechanical Ventilation in Guillain-Barré Syndrome. *Crit. Care Med.* **2003**, *31*, 278–283. [CrossRef] [PubMed]
45. Zhao, Z.; Wang, Y.; Zhou, R.; Li, Y.; Gao, Y.; Tu, D.; Wilson, B.; Song, S.; Feng, J.; Hong, J.-S.; et al. A Novel Role of NLRP3-Generated IL-1β in the Acute-Chronic Transition of Peripheral Lipopolysaccharide-Elicited Neuroinflammation: Implications for Sepsis-Associated Neurodegeneration. *J. Neuroinflamm.* **2020**, *17*, 64. [CrossRef]
46. Ammar, M.A.; Ammar, A.A.; Wieruszewski, P.M.; Bissell, B.D.; Long, M.T.; Albert, L.; Khanna, A.K.; Sacha, G.L. Timing of Vasoactive Agents and Corticosteroid Initiation in Septic Shock. *Ann. Intensive Care* **2022**, *12*, 47. [CrossRef]
47. Smith, S.M.; Padayachee, S.; Modaresi, K.B.; Smithies, M.N.; Bihari, D.J. Cerebral Blood Flow Is Proportional to Cardiac Index in Patients with Septic Shock. *J. Crit. Care* **1998**, *13*, 104–109. [CrossRef]
48. Pfister, D.; Siegemund, M.; Dell-Kuster, S.; Smielewski, P.; Rüegg, S.; Strebel, S.P.; Marsch, S.C.U.; Pargger, H.; Steiner, L.A. Cerebral Perfusion in Sepsis-Associated Delirium. *Crit. Care* **2008**, *12*, R63. [CrossRef]
49. Caldas, J.; Quispe-Cornejo, A.A.; Crippa, I.A.; Subira, C.; Creteur, J.; Panerai, R.; Taccone, F.S. Cerebral Autoregulation Indices Are Not Interchangeable in Patients With Sepsis. *Front. Neurol.* **2022**, *13*, 760293. [CrossRef]
50. Okazaki, N.; Lankadeva, Y.R.; Peiris, R.M.; Birchall, I.E.; May, C.N. Rapid and Persistent Decrease in Brain Tissue Oxygenation in Ovine Gram-Negative Sepsis. *Am. J. Physiol. Regul. Integr. Comp. Physiol.* **2021**, *321*, R990–R996. [CrossRef]
51. Sharshar, T.; Annane, D.; de la Grandmaison, G.L.; Brouland, J.P.; Hopkinson, N.S.; Françoise, G. The Neuropathology of Septic Shock. *Brain Pathol.* **2004**, *14*, 21–33. [CrossRef] [PubMed]
52. Wang, W.; Xu, H.; Lin, H.; Molnar, M.; Ren, H. The Role of the Cholinergic Anti-Inflammatory Pathway in Septic Cardiomyopathy. *Int. Immunopharmacol.* **2021**, *90*, 107160. [CrossRef] [PubMed]
53. Erbaş, O.; Taşkıran, D. Sepsis-Induced Changes in Behavioral Stereotypy in Rats; Involvement of Tumor Necrosis Factor-Alpha, Oxidative Stress, and Dopamine Turnover. *J. Surg. Res.* **2014**, *186*, 262–268. [CrossRef] [PubMed]
54. Li, F.; Zhang, B.; Duan, S.; Qing, W.; Tan, L.; Chen, S.; Wang, Y.; Li, D.; Yang, J.; Tong, J.; et al. Small Dose of L-Dopa/Benserazide Hydrochloride Improved Sepsis-Induced Neuroinflammation and Long-Term Cognitive Dysfunction in Sepsis Mice. *Brain Res.* **2020**, *1737*, 146780. [CrossRef]
55. Serantes, R.; Arnalich, F.; Figueroa, M.; Salinas, M.; Andrés-Mateos, E.; Codoceo, R.; Renart, J.; Matute, C.; Cavada, C.; Cuadrado, A.; et al. Interleukin-1beta Enhances GABAA Receptor Cell-Surface Expression by a Phosphatidylinositol 3-Kinase/Akt Pathway: Relevance to Sepsis-Associated Encephalopathy. *J. Biol. Chem.* **2006**, *281*, 14632–14643. [CrossRef] [PubMed]
56. Gu, M.; Mei, X.-L.; Zhao, Y.-N. Sepsis and Cerebral Dysfunction: BBB Damage, Neuroinflammation, Oxidative Stress, Apoptosis and Autophagy as Key Mediators and the Potential Therapeutic Approaches. *Neurotox. Res.* **2021**, *39*, 489–503. [CrossRef]
57. Haileselassie, B.; Joshi, A.U.; Minhas, P.S.; Mukherjee, R.; Andreasson, K.I.; Mochly-Rosen, D. Mitochondrial Dysfunction Mediated through Dynamin-Related Protein 1 (Drp1) Propagates Impairment in Blood Brain Barrier in Septic Encephalopathy. *J. Neuroinflamm.* **2020**, *17*, 36. [CrossRef]
58. Huang, Z.-S.; Xie, D.-Q.; Xu, L.-J.; Huang, C.-S.; Zheng, M.; Chen, Y.-J.; Cao, Y. Tetramethylpyrazine Ameliorates Lipopolysaccharide-Induced Sepsis in Rats via Protecting Blood-Brain Barrier, Impairing Inflammation and Nitrous Oxide Systems. *Front. Pharmacol.* **2020**, *11*, 562084. [CrossRef]
59. Kawakami, M.; Hattori, M.; Ohashi, W.; Fujimori, T.; Hattori, K.; Takebe, M.; Tomita, K.; Yokoo, H.; Matsuda, N.; Yamazaki, M.; et al. Role of G Protein-Coupled Receptor Kinase 2 in Oxidative and Nitrosative Stress-Related Neurohistopathological Changes in a Mouse Model of Sepsis-Associated Encephalopathy. *J. Neurochem.* **2018**, *145*, 474–488. [CrossRef]
60. Wang, P.; Hu, Y.; Yao, D.; Li, Y. Omi/HtrA2 Regulates a Mitochondria-Dependent Apoptotic Pathway in a Murine Model of Septic Encephalopathy. *Cell. Physiol. Biochem.* **2018**, *49*, 2163–2173. [CrossRef]

61. Jiang, Y.; Zhang, K.; Yu, Y.; Wang, Y.; Lian, N.; Xie, K.; Yu, Y. Molecular Hydrogen Alleviates Brain Injury and Cognitive Impairment in a Chronic Sequelae Model of Murine Polymicrobial Sepsis. *Exp. Brain Res.* **2020**, *238*, 2897–2908. [CrossRef] [PubMed]
62. Boutin, J.A.; Jockers, R. Melatonin Controversies, an Update. *J. Pineal Res.* **2021**, *70*, e12702. [CrossRef] [PubMed]
63. Kennaway, D.J. Measuring Melatonin by Immunoassay. *J. Pineal Res.* **2020**, *69*, e12657. [CrossRef] [PubMed]
64. Maldonado, M.D.; Mora-Santos, M.; Naji, L.; Carrascosa-Salmoral, M.P.; Naranjo, M.C.; Calvo, J.R. Evidence of Melatonin Synthesis and Release by Mast Cells. Possible Modulatory Role on Inflammation. *Pharmacol. Res.* **2010**, *62*, 282–287. [CrossRef]
65. Cecon, E.; Oishi, A.; Jockers, R. Melatonin Receptors: Molecular Pharmacology and Signalling in the Context of System Bias. *Br. J. Pharmacol.* **2018**, *175*, 3263–3280. [CrossRef] [PubMed]
66. Janjetovic, Z.; Jarrett, S.G.; Lee, E.F.; Duprey, C.; Reiter, R.J.; Slominski, A.T. Melatonin and Its Metabolites Protect Human Melanocytes against UVB-Induced Damage: Involvement of NRF2-Mediated Pathways. *Sci. Rep.* **2017**, *7*, 1274. [CrossRef]
67. Tordjman, S.; Chokron, S.; Delorme, R.; Charrier, A.; Bellissant, E.; Jaafari, N.; Fougerou, C. Melatonin: Pharmacology, Functions and Therapeutic Benefits. *Curr. Neuropharmacol.* **2017**, *15*, 434–443. [CrossRef]
68. Lalanne, S.; Fougerou-Leurent, C.; Anderson, G.M.; Schroder, C.M.; Nir, T.; Chokron, S.; Delorme, R.; Claustrat, B.; Bellissant, E.; Kermarrec, S.; et al. Melatonin: From Pharmacokinetics to Clinical Use in Autism Spectrum Disorder. *Int. J. Mol. Sci.* **2021**, *22*, 1490. [CrossRef]
69. Erren, T.C.; Reiter, R.J. Melatonin: A Universal Time Messenger. *Neuro Endocrinol. Lett.* **2015**, *36*, 187–192.
70. Truong, K.K.; Lam, M.T.; Grandner, M.A.; Sassoon, C.S.; Malhotra, A. Timing Matters: Circadian Rhythm in Sepsis, Obstructive Lung Disease, Obstructive Sleep Apnea, and Cancer. *Ann. Am. Thorac. Soc.* **2016**, *13*, 1144–1154. [CrossRef]
71. Li, J.; Cai, S.; Liu, X.; Mei, J.; Pan, W.; Zhong, M.; Zhang, Y. Circadian Rhythm Disturbance and Delirium in ICU Patients: A Prospective Cohort Study. *BMC Anesthesiol.* **2023**, *23*, 203. [CrossRef] [PubMed]
72. Melone, M.-A.; Becker, T.C.; Wendt, L.H.; Ten Eyck, P.; Patel, S.B.; Poston, J.; Pohlman, A.S.; Pohlman, M.; Miller, A.; Nedeltcheva, A.; et al. Disruption of the Circadian Rhythm of Melatonin: A Biomarker of Critical Illness Severity. *Sleep Med.* **2023**, *110*, 60–67. [CrossRef] [PubMed]
73. Sørensen, K.M.; Olesen, C.; Meyhoff, C.S.; Andersen, L.P.K. The Use of Melatonin in Danish Intensive Care Departments-A Nationwide Observational Study. *Acta Anaesthesiol.* **2022**, *66*, 833–837. [CrossRef]
74. Juszczak, K.; Ziomber, A.; Machowska, A.; Furgała, A.; Dobrek, Ł.; Wyczółkowski, M.; Thor, P.J. The Ameliorating Effect of Exogenous Melatonin on Urinary Bladder Function in Hyperosmolar Bladder Overactivity and Its Influence on the Autonomic Nervous System Activity. *Acta Med.* **2011**, *54*, 63–68. [CrossRef]
75. Campos, L.A.; Bueno, C.; Barcelos, I.P.; Halpern, B.; Brito, L.C.; Amaral, F.G.; Baltatu, O.C.; Cipolla-Neto, J. Melatonin Therapy Improves Cardiac Autonomic Modulation in Pinealectomized Patients. *Front. Endocrinol.* **2020**, *11*, 239. [CrossRef] [PubMed]
76. Hsu, C.-T.; Tai, H.C.-H.; Chung, J.-Y.; Chen, J.-H.; Chen, W.-L. Depressed Sympathovagal Modulation Indicates Sepsis in Patients with Suspected Infection. *Medicine* **2020**, *99*, e18961. [CrossRef]
77. Kohoutova, M.; Horak, J.; Jarkovska, D.; Martinkova, V.; Tegl, V.; Nalos, L.; Vistejnova, L.; Benes, J.; Sviglerova, J.; Kuncova, J.; et al. Vagus Nerve Stimulation Attenuates Multiple Organ Dysfunction in Resuscitated Porcine Progressive Sepsis. *Crit. Care Med.* **2019**, *47*, e461–e469. [CrossRef]
78. Baker, J.; Kimpinski, K. Role of Melatonin in Blood Pressure Regulation: An Adjunct Anti-Hypertensive Agent. *Clin. Exp. Pharmacol. Physiol.* **2018**, *45*, 755–766. [CrossRef]
79. Cipolla-Neto, J.; Amaral, F.G.; Afeche, S.C.; Tan, D.X.; Reiter, R.J. Melatonin, Energy Metabolism, and Obesity: A Review. *J. Pineal Res.* **2014**, *56*, 371–381. [CrossRef]
80. Karamitri, A.; Jockers, R. Melatonin in Type 2 Diabetes Mellitus and Obesity. *Nat. Rev. Endocrinol.* **2019**, *15*, 105–125. [CrossRef]
81. Mannino, G.; Caradonna, F.; Cruciata, I.; Lauria, A.; Perrone, A.; Gentile, C. Melatonin Reduces Inflammatory Response in Human Intestinal Epithelial Cells Stimulated by Interleukin-1β. *J. Pineal Res.* **2019**, *67*, e12598. [CrossRef] [PubMed]
82. Liu, Z.; Gan, L.; Xu, Y.; Luo, D.; Ren, Q.; Wu, S.; Sun, C. Melatonin Alleviates Inflammasome-Induced Pyroptosis through Inhibiting NF-KB/GSDMD Signal in Mice Adipose Tissue. *J. Pineal Res.* **2017**, *63*, e12414. [CrossRef] [PubMed]
83. Galano, A.; Tan, D.-X.; Reiter, R.J. Melatonin: A Versatile Protector against Oxidative DNA Damage. *Molecules* **2018**, *23*, 530. [CrossRef]
84. Purushothaman, A.; Sheeja, A.A.; Janardanan, D. Hydroxyl Radical Scavenging Activity of Melatonin and Its Related Indolamines. *Free Radic. Res.* **2020**, *54*, 373–383. [CrossRef] [PubMed]
85. Matuszak, Z.; Reszka, K.; Chignell, C.F. Reaction of Melatonin and Related Indoles with Hydroxyl Radicals: EPR and Spin Trapping Investigations. *Free Radic. Biol. Med.* **1997**, *23*, 367–372. [CrossRef]
86. Galano, A. On the Direct Scavenging Activity of Melatonin towards Hydroxyl and a Series of Peroxyl Radicals. *Phys. Chem. Chem. Phys. PCCP* **2011**, *13*, 7178–7188. [CrossRef]
87. Chang, C.-C.; Huang, T.-Y.; Chen, H.-Y.; Huang, T.-C.; Lin, L.-C.; Chang, Y.-J.; Hsia, S.-M. Protective Effect of Melatonin against Oxidative Stress-Induced Apoptosis and Enhanced Autophagy in Human Retinal Pigment Epithelium Cells. *Oxid. Med. Cell. Longev.* **2018**, *2018*, 9015765. [CrossRef]
88. Aydogan, S.; Yerer, M.B.; Goktas, A. Melatonin and Nitric Oxide. *J. Endocrinol. Investig.* **2006**, *29*, 281–287. [CrossRef]
89. Limson, J.; Nyokong, T.; Daya, S. The Interaction of Melatonin and Its Precursors with Aluminium, Cadmium, Copper, Iron, Lead, and Zinc: An Adsorptive Voltammetric Study. *J. Pineal Res.* **1998**, *24*, 15–21. [CrossRef]

90. Romero, A.; Ramos, E.; de Los Ríos, C.; Egea, J.; Del Pino, J.; Reiter, R.J. A Review of Metal-Catalyzed Molecular Damage: Protection by Melatonin. *J. Pineal Res.* **2014**, *56*, 343–370. [CrossRef]
91. Gancitano, G.; Reiter, R.J. The Multiple Functions of Melatonin: Applications in the Military Setting. *Biomedicines* **2022**, *11*, 5. [CrossRef] [PubMed]
92. Colares, J.R.; Hartmann, R.M.; Schemitt, E.G.; Fonseca, S.R.B.; Brasil, M.S.; Picada, J.N.; Dias, A.S.; Bueno, A.F.; Marroni, C.A.; Marroni, N.P. Melatonin Prevents Oxidative Stress, Inflammatory Activity, and DNA Damage in Cirrhotic Rats. *World J. Gastroenterol.* **2022**, *28*, 348–364. [CrossRef] [PubMed]
93. Rehman, S.U.; Ikram, M.; Ullah, N.; Alam, S.I.; Park, H.Y.; Badshah, H.; Choe, K.; Kim, M.O. Neurological Enhancement Effects of Melatonin against Brain Injury-Induced Oxidative Stress, Neuroinflammation, and Neurodegeneration via AMPK/CREB Signaling. *Cells* **2019**, *8*, 760. [CrossRef] [PubMed]
94. Blanco, S.; Hernández, R.; Franchelli, G.; Ramos-Álvarez, M.M.; Peinado, M.Á. Melatonin Influences NO/NOS Pathway and Reduces Oxidative and Nitrosative Stress in a Model of Hypoxic-Ischemic Brain Damage. *Nitric Oxide Biol. Chem.* **2017**, *62*, 32–43. [CrossRef] [PubMed]
95. Gou, Z.; Su, X.; Hu, X.; Zhou, Y.; Huang, L.; Fan, Y.; Li, J.; Lu, L. Melatonin Improves Hypoxic-Ischemic Brain Damage through the Akt/Nrf2/Gpx4 Signaling Pathway. *Brain Res. Bull.* **2020**, *163*, 40–48. [CrossRef]
96. Barlow-Walden, L.R.; Reiter, R.J.; Abe, M.; Pablos, M.; Menendez-Pelaez, A.; Chen, L.D.; Poeggeler, B. Melatonin Stimulates Brain Glutathione Peroxidase Activity. *Neurochem. Int.* **1995**, *26*, 497–502. [CrossRef]
97. Slominski, A.T.; Semak, I.; Fischer, T.W.; Kim, T.-K.; Kleszczyński, K.; Hardeland, R.; Reiter, R.J. Metabolism of Melatonin in the Skin: Why Is It Important? *Exp. Dermatol.* **2017**, *26*, 563–568. [CrossRef]
98. Galano, A.; Reiter, R.J. Melatonin and Its Metabolites vs Oxidative Stress: From Individual Actions to Collective Protection. *J. Pineal Res.* **2018**, *65*, e12514. [CrossRef]
99. Zhen, G.; Liang, W.; Jia, H.; Zheng, X. Melatonin Relieves Sepsis-Induced Myocardial Injury via Regulating JAK2/STAT3 Signaling Pathway. *Minerva Med.* **2022**, *113*, 983–989. [CrossRef]
100. Rahim, I.; Sayed, R.K.; Fernández-Ortiz, M.; Aranda-Martínez, P.; Guerra-Librero, A.; Fernández-Martínez, J.; Rusanova, I.; Escames, G.; Djerdjouri, B.; Acuña-Castroviejo, D. Melatonin Alleviates Sepsis-Induced Heart Injury through Activating the Nrf2 Pathway and Inhibiting the NLRP3 Inflammasome. *Naunyn. Schmiedebergs Arch. Pharmacol.* **2021**, *394*, 261–277. [CrossRef]
101. Kang, J.-Y.; Xu, M.-M.; Sun, Y.; Ding, Z.-X.; Wei, Y.-Y.; Zhang, D.-W.; Wang, Y.-G.; Shen, J.-L.; Wu, H.-M.; Fei, G.-H. Melatonin Attenuates LPS-Induced Pyroptosis in Acute Lung Injury by Inhibiting NLRP3-GSDMD Pathway via Activating Nrf2/HO-1 Signaling Axis. *Int. Immunopharmacol.* **2022**, *109*, 108782. [CrossRef] [PubMed]
102. Shah, S.A.; Khan, M.; Jo, M.-H.; Jo, M.G.; Amin, F.U.; Kim, M.O. Melatonin Stimulates the SIRT1/Nrf2 Signaling Pathway Counteracting Lipopolysaccharide (LPS)-Induced Oxidative Stress to Rescue Postnatal Rat Brain. *CNS Neurosci. Ther.* **2017**, *23*, 33–44. [CrossRef] [PubMed]
103. Xu, S.; Li, L.; Wu, J.; An, S.; Fang, H.; Han, Y.; Huang, Q.; Chen, Z.; Zeng, Z. Melatonin Attenuates Sepsis-Induced Small-Intestine Injury by Upregulating SIRT3-Mediated Oxidative-Stress Inhibition, Mitochondrial Protection, and Autophagy Induction. *Front. Immunol.* **2021**, *12*, 625627. [CrossRef] [PubMed]
104. Colunga Biancatelli, R.M.L.; Berrill, M.; Mohammed, Y.H.; Marik, P.E. Melatonin for the Treatment of Sepsis: The Scientific Rationale. *J. Thorac. Dis.* **2020**, *12*, S54–S65. [CrossRef] [PubMed]
105. Lowes, D.A.; Webster, N.R.; Murphy, M.P.; Galley, H.F. Antioxidants That Protect Mitochondria Reduce Interleukin-6 and Oxidative Stress, Improve Mitochondrial Function, and Reduce Biochemical Markers of Organ Dysfunction in a Rat Model of Acute Sepsis. *Br. J. Anaesth.* **2013**, *110*, 472–480. [CrossRef]
106. Escames, G.; López, L.C.; Ortiz, F.; López, A.; García, J.A.; Ros, E.; Acuña-Castroviejo, D. Attenuation of Cardiac Mitochondrial Dysfunction by Melatonin in Septic Mice. *FEBS J.* **2007**, *274*, 2135–2147. [CrossRef]
107. Carrillo-Vico, A.; Lardone, P.J.; Naji, L.; Fernández-Santos, J.M.; Martín-Lacave, I.; Guerrero, J.M.; Calvo, J.R. Beneficial Pleiotropic Actions of Melatonin in an Experimental Model of Septic Shock in Mice: Regulation of pro-/Anti-Inflammatory Cytokine Network, Protection against Oxidative Damage and Anti-Apoptotic Effects. *J. Pineal Res.* **2005**, *39*, 400–408. [CrossRef]
108. Arioz, B.I.; Tarakcioglu, E.; Olcum, M.; Genc, S. The Role of Melatonin on NLRP3 Inflammasome Activation in Diseases. *Antioxidants* **2021**, *10*, 1020. [CrossRef]
109. Lorente, L.; Martín, M.M.; Abreu-González, P.; de la Cruz, T.; Ferreres, J.; Solé-Violán, J.; Labarta, L.; Díaz, C.; Jiménez, A.; Borreguero-León, J.M. Serum Melatonin Levels Are Associated with Mortality in Severe Septic Patients. *J. Crit. Care* **2015**, *30*, 860. [CrossRef]
110. Alamili, M.; Bendtzen, K.; Lykkesfeldt, J.; Rosenberg, J.; Gögenur, I. Effect of Melatonin on Human Nighttime Endotoxaemia: Randomized, Double-Blinded, Cross-over Study. *Vivo Athens Greece* **2014**, *28*, 1057–1063.
111. Alamili, M.; Bendtzen, K.; Lykkesfeldt, J.; Rosenberg, J.; Gögenur, I. Melatonin Suppresses Markers of Inflammation and Oxidative Damage in a Human Daytime Endotoxemia Model. *J. Crit. Care* **2014**, *29*, e9–e184. [CrossRef] [PubMed]
112. Galley, H.F.; Lowes, D.A.; Allen, L.; Cameron, G.; Aucott, L.S.; Webster, N.R. Melatonin as a Potential Therapy for Sepsis: A Phase I Dose Escalation Study and an Ex Vivo Whole Blood Model under Conditions of Sepsis. *J. Pineal Res.* **2014**, *56*, 427–438. [CrossRef] [PubMed]

113. Aisa-Alvarez, A.; Soto, M.E.; Guarner-Lans, V.; Camarena-Alejo, G.; Franco-Granillo, J.; Martínez-Rodríguez, E.A.; Gamboa Ávila, R.; Manzano Pech, L.; Pérez-Torres, I. Usefulness of Antioxidants as Adjuvant Therapy for Septic Shock: A Randomized Clinical Trial. *Med. Kaunas Lith.* **2020**, *56*, 619. [CrossRef]
114. Taher, A.; Shokoohmand, F.; Abdoli, E.; Mohammadi, Y.; Mehrpooya, M. A Pilot Study on the Melatonin Treatment in Patients with Early Septic Shock: Results of a Single-Center Randomized Controlled Trial. *Ir. J. Med. Sci.* **2022**, *191*, 1913–1924. [CrossRef]
115. Mansilla-Roselló, A.; Hernández-Magdalena, J.; Domínguez-Bastante, M.; Olmedo-Martín, C.; Comino-Pardo, A.; Escames, G.; Acuña-Castroviejo, D. A Phase II, Single-Center, Double-Blind, Randomized Placebo-Controlled Trial to Explore the Efficacy and Safety of Intravenous Melatonin in Surgical Patients with Severe Sepsis Admitted to the Intensive Care Unit. *J. Pineal Res.* **2023**, *74*, e12845. [CrossRef]
116. Aisa-Álvarez, A.; Pérez-Torres, I.; Guarner-Lans, V.; Manzano-Pech, L.; Cruz-Soto, R.; Márquez-Velasco, R.; Casarez-Alvarado, S.; Franco-Granillo, J.; Núñez-Martínez, M.E.; Soto, M.E. Randomized Clinical Trial of Antioxidant Therapy Patients with Septic Shock and Organ Dysfunction in the ICU: SOFA Score Reduction by Improvement of the Enzymatic and Non-Enzymatic Antioxidant System. *Cells* **2023**, *12*, 1330. [CrossRef] [PubMed]
117. Galley, H.F.; Allen, L.; Colin, P.J.; Galt, S.P.; Webster, N.R. Dose Assessment of Melatonin in Sepsis (DAMSEL2) Study: Pharmacokinetics of Two Doses of Oral Melatonin in Patients with Sepsis. *J. Pineal Res.* **2022**, *73*, e12830. [CrossRef] [PubMed]
118. Ji, M.-H.; Xia, D.-G.; Zhu, L.-Y.; Zhu, X.; Zhou, X.-Y.; Xia, J.-Y.; Yang, J.-J. Short- and Long-Term Protective Effects of Melatonin in a Mouse Model of Sepsis-Associated Encephalopathy. *Inflammation* **2018**, *41*, 515–529. [CrossRef]
119. Wang, X.; Xue, G.-X.; Liu, W.-C.; Shu, H.; Wang, M.; Sun, Y.; Liu, X.; Sun, Y.E.; Liu, C.-F.; Liu, J.; et al. Melatonin Alleviates Lipopolysaccharide-Compromised Integrity of Blood-Brain Barrier through Activating AMP-Activated Protein Kinase in Old Mice. *Aging Cell* **2017**, *16*, 414–421. [CrossRef]
120. Liu, W.-C.; Wang, X.; Zhang, X.; Chen, X.; Jin, X. Melatonin Supplementation, a Strategy to Prevent Neurological Diseases through Maintaining Integrity of Blood Brain Barrier in Old People. *Front. Aging Neurosci.* **2017**, *9*, 165. [CrossRef]
121. Zhou, Q.; Lin, L.; Li, H.; Wang, H.; Jiang, S.; Huang, P.; Lin, Q.; Chen, X.; Deng, Y. Melatonin Reduces Neuroinflammation and Improves Axonal Hypomyelination by Modulating M1/M2 Microglia Polarization via JAK2-STAT3-Telomerase Pathway in Postnatal Rats Exposed to Lipopolysaccharide. *Mol. Neurobiol.* **2021**, *58*, 6552–6576. [CrossRef]
122. Ali, T.; Rahman, S.U.; Hao, Q.; Li, W.; Liu, Z.; Ali Shah, F.; Murtaza, I.; Zhang, Z.; Yang, X.; Liu, G.; et al. Melatonin Prevents Neuroinflammation and Relieves Depression by Attenuating Autophagy Impairment through FOXO3a Regulation. *J. Pineal Res.* **2020**, *69*, e12667. [CrossRef]
123. Zhao, L.; An, R.; Yang, Y.; Yang, X.; Liu, H.; Yue, L.; Li, X.; Lin, Y.; Reiter, R.J.; Qu, Y. Melatonin Alleviates Brain Injury in Mice Subjected to Cecal Ligation and Puncture via Attenuating Inflammation, Apoptosis, and Oxidative Stress: The Role of SIRT1 Signaling. *J. Pineal Res.* **2015**, *59*, 230–239. [CrossRef] [PubMed]
124. Mehrpooya, M.; Mazdeh, M.; Rahmani, E.; Khazaie, M.; Ahmadimoghaddam, D. Melatonin Supplementation May Benefit Patients with Acute Ischemic Stroke Not Eligible for Reperfusion Therapies: Results of a Pilot Study. *J. Clin. Neurosci.* **2022**, *106*, 66–75. [CrossRef] [PubMed]
125. Lin, Y.W.; Chen, T.Y.; Hung, C.Y.; Tai, S.H.; Huang, S.Y.; Chang, C.C.; Hung, H.Y.; Lee, E.J. Melatonin Protects Brain against Ischemia/Reperfusion Injury by Attenuating Endoplasmic Reticulum Stress. *Int. J. Mol. Med.* **2018**, *42*, 182–192. [CrossRef]
126. Liu, Z.-J.; Ran, Y.-Y.; Qie, S.-Y.; Gong, W.-J.; Gao, F.-H.; Ding, Z.-T.; Xi, J.-N. Melatonin Protects against Ischemic Stroke by Modulating Microglia/Macrophage Polarization toward Anti-Inflammatory Phenotype through STAT3 Pathway. *CNS Neurosci. Ther.* **2019**, *25*, 1353–1362. [CrossRef]
127. Zhao, Y.; Wang, H.; Chen, W.; Chen, L.; Liu, D.; Wang, X.; Wang, X. Melatonin Attenuates White Matter Damage after Focal Brain Ischemia in Rats by Regulating the TLR4/NF-KB Pathway. *Brain Res. Bull.* **2019**, *150*, 168–178. [CrossRef] [PubMed]
128. Chen, S.; Sun, Y.; Li, F.; Zhang, X.; Hu, X.; Zhao, X.; Li, Y.; Li, H.; Zhang, J.; Liu, W.; et al. Modulation of A7nAchR by Melatonin Alleviates Ischemia and Reperfusion-Compromised Integrity of Blood-Brain Barrier Through Inhibiting HMGB1-Mediated Microglia Activation and CRTC1-Mediated Neuronal Loss. *Cell. Mol. Neurobiol.* **2022**, *42*, 2407–2422. [CrossRef] [PubMed]
129. Krzyzaniak, K.; Krion, R.; Szymczyk, A.; Stepniewska, E.; Sieminski, M. Exploring Neuroprotective Agents for Sepsis-Associated Encephalopathy: A Comprehensive Review. *Int. J. Mol. Sci.* **2023**, *24*, 10780. [CrossRef]

antioxidants

MDPI

Article

Agomelatine, a Melatonin-Derived Drug, as a New Strategy for the Treatment of Colorectal Cancer

Sara Moreno-SanJuan [1,2,†], Jose D. Puentes-Pardo [2,3,†], Jorge Casado [2], Julia Escudero-Feliu [2], Huda Khaldy [4], Javier Arnedo [5], Ángel Carazo [2,6,‡] and Josefa León [2,7,*,‡]

1 Cytometry and Microscopy Research Service, Biosanitary Research Institute of Granada (ibs.GRANADA), 18012 Granada, Spain
2 Biosanitary Research Institute of Granada (ibs.GRANADA), 18012 Granada, Spain
3 Department of Pharmacy, University of Granada, 18011 Granada, Spain
4 Fundamental Biology Service, Scientific Instrument Center, University of Granada, 18071 Granada, Spain
5 Department of Statistics and Operations Research, University of Granada, 18071 Granada, Spain
6 Clinical Management Unit of Microbiology, San Cecilio University Hospital, 18006 Granada, Spain
7 Clinical Management Unit of Digestive Disease, San Cecilio University Hospital, 18006 Granada, Spain
* Correspondence: pepileon@ugr.es; Tel.: +34-95-802-3709
† These authors contributed equally to this work.
‡ These authors contributed equally to this work.

Abstract: The potential use of agomelatine as an alternative treatment for colorectal cancer is evaluated in this work. The effect of agomelatine was studied in an in vitro model using two cell lines with different p53 statuses (HCT-116, wild-type p53, and HCT-116 p53 null) and an in vivo xenograft model. The inhibitory effects of agomelatine and melatonin were stronger in the cells harboring the wild-type p53, although in both cell lines, the effect of agomelatine was greater than that of the melatonin. In vivo, only agomelatine was able to reduce the volumes of tumors generated by the HCT-116-p53-null cells. Both treatments induced changes in the rhythmicity of the circadian-clock genes in vitro, albeit with some differences. Agomelatine and melatonin regulated the rhythmicity of Per1-3, Cry1, Sirt1, and Prx1 in the HCT-116 cells. In these cells, agomelatine also regulated Bmal1 and Nr1d2, while melatonin changed the rhythmicity of Clock. In the HCT-116-p53-null cells, agomelatine regulated Per1-3, Cry1, Clock, Nr1d2, Sirt1, and Prx1; however, melatonin only induced changes in Clock, Bmal1, and Sirt1. The differences found in the regulation of the clock genes may explain the greater oncostatic effect of agomelatine in CRC.

Keywords: agomelatine; melatonin; p53; circadian clock; colorectal cancer; SIRT1

check for updates

Citation: Moreno-SanJuan, S.; Puentes-Pardo, J.D.; Casado, J.; Escudero-Feliu, J.; Khaldy, H.; Arnedo, J.; Carazo, Á.; León, J. Agomelatine, a Melatonin-Derived Drug, as a New Strategy for the Treatment of Colorectal Cancer. *Antioxidants* **2023**, *12*, 926. https://doi.org/10.3390/antiox12040926

Academic Editors: Marina Garcia-Macia and Adrián Santos-Ledo

Received: 10 February 2023
Revised: 1 April 2023
Accepted: 6 April 2023
Published: 13 April 2023

1. Introduction

Colorectal cancer (CRC) is the third most common type of cancer and the second in terms of mortality worldwide [1]. The current treatments for CRC include 5-fluorouracil (5-FU) as a standard agent for chemotherapy, either as a single agent or in combination with other drugs, such as oxaliplatin or irinotecan [2]. However, these treatments have several limitations, such as severe side effects and the development of chemoresistance, which lead to unsatisfactory results [3]. These limitations have encouraged the development of the so-called targeted therapies, which are based on targeting concrete pathways according to each patient's characteristics. However, the clinical application of these therapies is far from being established because they do not provide an improvement in cost–benefit terms, they still possess adverse effects, they differ in efficacy between patients with similar characteristics, and resistance has been shown to arise [2,3]. Therefore, there is a necessity to develop new approaches to improve or replace the current treatments.

Circadian rhythms are cyclic biological processes with a duration of close to 24 h, which are implicated in the regulation of major physiological events, such as the sleep –wake cycle, reproduction, and inflammatory and immune responses, among others. At the molecular level, the circadian rhythm consists of transcription–translation feedback loops (TTFL). The transcription-factor brain and muscle ARNT-like (BMAL-1) dimerizes circadian locomotor cycles kaput (CLOCK) and stimulates the transcription of Period 1-3 (Per1-3), Cryptocrome 1-2 (Cry1-2), the nuclear receptor subfamily 1 group D member 2 (Nr1d2), and RAR-related orphan receptor alpha (RORα) genes. In turn, the PER1-3 and CRY1-2 proteins form complexes in the cytoplasm and, when their concentration reaches a certain level, they translocate to the nucleus, where they inhibit the action of BMAL-1 and CLOCK and, thus, the transcription of their own genes. The PER/CRY complex is degraded, allowing the cycle to be restored every 24 h [4].

The circadian clock is not only based on transcriptional mechanisms; post-translational modifications of core circadian genes have also been described. Sirtuin 1 (SIRT1) is a nicotinamide adenine dinucleotide (NAD+)-dependent class III histone deacetylase, which is involved in cellular redox control [5]. It regulates the circadian clock at the central-nervous-system level by activating BMAL1 and CLOCK transcription [6]. At the peripheral level, it deacetylates BMAL1, affecting its activity [7], and PER2, preventing its dimerization with CRY [8]. It can also bind to CLOCK/BMAL1 complexes, which alters the expression of other associated genes, such as PER, in a manner that is probably related to the variable concentration of its co-enzyme, NAD+ [9]. In addition, the existence of circadian rhythms independent of the canonical molecular clock has been demonstrated. Specifically, there is a circadian rhythm of oxidation-reduction of peroxiredoxin-1 (PRX1), an antioxidant protein that scavenges hydrogen peroxide. This cycle is interconnected with the TTFL cycle, and both are individually necessary for the maintenance of rhythms at the cellular level [10]. In fact, it has been demonstrated that SIRT1 can activate PRX1 transcription [11]. The human PRX1 gene is a target of nuclear factor erythroid 2 related factor-2 (Nrf2) [12], a transcription factor shared by several antioxidant enzymes [13]. Furthermore, SIRT1 increases Nrf2 activation and decreases its polyubiquitination by decreasing the expression of Kelch-like ECH-associated protein 1 (Keap1)/Cullin 3 (Cul3) and by increasing Nrf2's binding ability to anti-oxidant response element (ARE) [14].

Alterations in circadian rhythmicity have been linked to the onset and development of cancer [15]. Furthermore, the tolerability and efficacy of radiotherapy and chemotherapy also depend on circadian rhythms [16]. These experimental findings have raised interest in manipulating rhythms to prevent malignant transformation, to develop novel treatment strategies, and ultimately improve the outcomes of cancer patients [17]. In this sense, melatonin is an indolic compound secreted mainly by the pineal gland, although it can also be found in extrapineal tissues; it plays a key role in the control of circadian rhythms [18]. Melatonin exerts oncotastic properties in several cancer types. Regarding CRC, melatonin reduces tumour growth and proliferation and induced apoptosis in in vitro [19,20] and in vivo models [21]. Some of melatonin's functions are mediated by the melatonin receptors MT1 and MT2. In fact, in CRC patients showed decreased levels of MT1 and MT2, without changes in melatonin levels, compared to their paired mucosa, in association with gender and invasion [22]. In addition, MT1 and MT2 levels are negatively correlated with cancer-stem-cell markers (CSCs), considered as the origins and major contributors to cancer progression, metastasis, and therapy resistance [23]. Taken together, these reports suggest that the use of non-selective MT1/MT2 agonists may be an interesting new approach for the treatment of CRC.

Agomelatine is a naphthalene analogue of melatonin, which shows agonist activities toward MT1/MT2 receptors, as well as greater affinity with them than melatonin, and antagonist activity toward the serotonin 5-HT2c receptor, and its use as an antidepressant drug has been approved [24]. Different in vivo models have proven that agomelatine synchronizes circadian rhythms, probably through MT1/MT2-receptor interaction [25]. The activation of MT1/MT2 synergically works with the blockage of the 5-HT2c receptor, resulting in neurogenesis induction, synaptic remodeling, and glutamate signaling, which explains agomelatine's antidepressant role [24]. Curiously, serotonin has been associated with the progression of several cancer types, including CRC, and serotonin-receptor inhibitors may be used as therapy [26]. Along with the aforementioned influence of circadian rhythms and MT1/MT2 on cancer development and progression, this suggests that agomelatine could be used as a treatment for CRC. However, to date, no studies exploiting this alternative function have been reported.

An interconnected pathway of regulation exists between melatonin, circadian clocks, and SIRT1 [27]. Agomelatine also regulates SIRT1 in several models of disease [28,29]. In this study, we evaluate the use of agomelatine as an antitumor therapy in established colorectal cell lines in vitro and in vivo in nude mice. We also analyze the implication of the SIRT1/PRX1 pathway in the regulation of the rhythmicity of core clock genes by agomelatine.

2. Materials and Methods

2.1. Cell Culture and Reagents

The colorectal adenocarcinoma cell lines HCT-116 (p53 wild type) and HCT-116 p53 null (Horizon Discovery Ltd., Cambridge, UK) were grown under standard conditions (37 °C and 5% CO_2 in a humid atmosphere) using RPMI 1640 medium supplemented with 10% FBS and 1% penicillin-streptomycin (Gibco; Invitrogen, Carlsbad, CA, USA). For the experiments that required cell synchronization, cells were cultured in serum-free medium for 6 h prior to the initiation of the experiment [1]. Next, the cells were cultured in media with serum in the conditions required for the experiments carried out.

Melatonin, agomelatine, fluorouracil (5-FU), and the remaining reagents employed in this study were obtained from Sigma-Aldrich (Sigma Chemical Co., St. Louis, MO, USA).

2.2. Cell Viability Assay

Cell viability was analyzed with 3-(4,5-Dimethylthiazol-2-yl)-2,5-diphenyltetrazolium bromide (MTT) assay. Briefly, 4000 cells were seeded per well in a 96-well plate in a final volume of 100 μL. Twenty-four hours later, the cells were treated with different drugs and concentrations for 24, 48, and 72 h. Once the treatment was stopped, 10 μL of MTT (5 mg/mL) was added and the plate was incubated for 4 h at 37 °C and 5% CO_2. Subsequently, 100 μL of lysis buffer (20% SDS in 50% formamide, pH 4.7) was added, and the plate was kept at 37 °C and 5% CO2 overnight. Optical density was measured with a Triad Multimode Microplate reader (Cultek SL, Madrid, Spain) at 570 nm.

2.3. Sphere-Formation Assay

Three-dimensional spheroids were generated and cultured in the InSphero GravityPLUS^{TM} Hanging Drop System (PerkinElmer), according to the manufacturer's instructions. Briefly, cells were seeded at a density of 2500 cells per well in a 96-well GravityPLUS^{TM} Plate and placed in a 5% humidified CO_2 incubator at 37 °C. After 3 days in culture, cells formed visible spheroids that were transferred to the GravityTRAP^{TM} Plate and allowed to grow for 4 more days. Every 2 days, the medium was changed. Subsequently, spheroids were treated with different doses of melatonin and agomelatine for 3 more days. Spheroids were imaged using a 4× objective on the image module of an EnSight^{TM} plate reader (PerkinElmer, Waltham, MA, USA). Acquired Images were automatically analyzed by the Kaleido 2.0 software.

2.4. Clonogenic Assay

Cells were seeded in 6-well plates at a concentration of 10^3 cells/well and left to grow 72 h. Next, cells were treated for 72 h. Subsequently, the medium was changed, and cells were allowed to grow under standard conditions for 10 more days. To perform fixation and staining of colonies, the medium was removed, and the cells were incubated for 5 min with 0.5% crystal-violet-oxalate solution in 50% methanol. Colonies with more than 50 cells were counted.

2.5. Cell-Cycle Analysis

The percentage of cells in each cell-cycle phase was determined based on the cellular DNA content in at least 20,000 nuclei. Cells were seeded in 6-well plates and after treatments, they were harvested, washed with PBS, and fixed with 200 μL of 70% ice-cold ethanol at 4 °C for 30 min. Next, cells were washed with a solution of PBS containing 2% BSA and incubated in 500 μL of PI/RNase solution (Immunostep SL; Salamanca, Spain) at room temperature in darkness for 15 min. The percentages of cells in G0/G1, S, and G2/M phases were determined using a BD FACSAria IIIu flow cytometer (Becton Dickinson, BD Biosciences; Franklin Lakes, NJ, USA) from the Cytometry and Microscopy Research Service of the Biosanitary Research Institute of Granada. Experiments were performed at least three times and three samples per group were analyzed in each case.

2.6. Apoptosis Assay

The percentages of apoptotic cells in response to the different treatments were analyzed using a FITC Annexin V Apoptosis Detection Kit (BD Biosciences, Franklin Lakes, NJ, USA). Briefly, the cells were seeded in 6-well plates, and after treatments they were harvested and washed with PBS and concentrated at 1×10^6 cells/mL. About 10^5 cells (100 μL) were incubated with Annexin V-FITC and propidium iodide at room temperature and in darkness for 15 min. The samples were immediately analyzed using a BD FACSAria IIIu flow cytometer (Becton Dickinson, BD Biosciences; Franklin Lakes, NJ, USA) from the Cytometry and Microscopy Research Service of the Biosanitary Research Institute of Granada. The percentage of apoptosis was calculated by taking into account the sum of percentages of apoptotic cells (Annexin-FITC+/PI−) and late apoptotic cells (Annexin-FITC+/PI+).

2.7. Real-Time PCR

Total RNA was extracted using TRIzol reagent (Invitrogen, Carlsbad, CA, USA). The RNA was reverse-transcribed by RT-PCR into cDNA using the commercial kit, AccuScriptTM High Fidelity 1st Strand cDNA Synthesis Kit (Stratagene, Austin, TX, USA), according to the manufacturer's instructions.

About 5 μL of the cDNA was amplified with specific primers for Per1, Per2, Per3, Cry1, Clock, Bmal1, Nr1d2, SIRT1, PRX1, and UBC (Table S1). Furthermore, PCR reactions with SYBER-green were performed using the Mx3000P qPCR System (Stratagene, Austin, TX, USA). Relative expression was calculated using UBC as a reference gene. Standard curves for each gene were made by plotting C_t values versus log cDNA dilution.

2.8. Immunoblot

Proteins were isolated from the lysed cells in RIPA buffer supplemented with protease inhibitors) for 30 min at 4 °C. The amount of proteins was quantified through Bradford assay, and samples were loaded in equal amounts (50 μg) into 12% SDS-polyacrylamide gels. The proteins resolved were transferred into PVDF-transfer membranes using a Bio-Rad Trans-Blot Turbo Transfer System (Bio-Rad Laboratories, Inc., Hercules, CA, USA). The blots were probed with the appropriate antibodies for caspase-3 (Santa Cruz Biotechnology, dilution 1:200), p53 (Santa Cruz Biotechnology, dilution 1:200), and β-Actin (Santa Cruz Biotechnology, dilution 1:200). As secondary antibody, HRP-conjugated anti-mouse antibody (Santa Cruz Biotechnology, dilution 1:50.000) was used. Amersham ECL Select Western Blotting Detection Reagent (GE Healthcare, Chicago, Il, USA) was applied before luminography for protein detection in a ChemiDoc MP System (Bio-Rad Inc., Hercules, CA, USA).

2.9. In Vivo Anti-Tumor Xenograft Studies

Female athymic Balb/c (nu/nu) mice (Charles Rivers Laboratory, Wilmington, MA, USA) housed in a regular 12:12 light–dark (LD) cycle (lights on at 08:00 h) were used for the in vivo studies. The mice were maintained in quarantine for a week before the subcutaneous injection of 2.5×10^6 HCT-116 and HCT-116-p53-null cells suspended in 100 μL of RPMI 1640 medium into the right and left sides, respectively. Once the tumor size reached about 50–100 mm^3, the animals were randomly divided into 4 groups (n = 8) and intraperitoneally treated for 2 weeks. Groups were as follows: (1) control (vehicle, three times per week); (2) melatonin (5 mg/kg, three times per week); (3) agomelatine (5 mg/kg, three times [er week); (4) 5-fluorouracil (50 mg/kg, two times per week). The agomelatine and melatonin groups were treated two hours before lights off, whereas the 5-fluorouracil group was treated 2 h after lights on. The tumor growth was measured three times per week and calculated by the formula $V = (4\pi/3) \times (\text{width}/2)2 \times (\text{length}/2)$. Blood of each animal was extracted to analyze different biochemical parameters (glucose, urea, uric acid, AST, ALT, and amylase) by a Cobas c311 analyzer (Roche Diagnostics, Basel, Switzerland). Two weeks after the initiation of the treatments, the animals were sacrificed, and the tumors were extracted.

2.10. Statistical Analysis

All the experiments were performed at least in triplicate and data were expressed as mean ± SEM. To obtain circadian parameters, we determined the correct distribution (data interdependency and normal distribution) for each set of time-series data by lag plots/Q-test and normal probability plots/K-S test and, next, we calculated the acrophase and amplitude through the cosinor method using the TSA (Time Series Analysis—Cosinor 8.0 Lab View January 2020) software (http://www.euroestech.com/, accessed on 21 January 2020). Detection of rhythm was achieved by rejection of the zero-amplitude hypothesis with 95% certainty, as reflected by the *p* value. The amplitude (i.e., the difference between the peak or trough and the mean value of a cosine curve), acrophase (i.e., the phase angle of the peak of a cosine curve), and midline estimating statistic of rhythm (MESOR) (i.e., the average value of cosine curve fitted to the data) were compared where applicable using 48-h trial period, since, in some cases, the oscillation period of the genes studied was close to this time. Nevertheless, comparisons were also performed using a 24-h test period (Tables S2–S5), with similar results obtained at both periods. Data from fitted curves were transferred to GraphPad Prism (GraphPad, La Jolla, SD, USA). Rhythm characteristics (MESOR, amplitude, acrophase) for each variable were compared by a non-parametric test [30,31].

3. Results

3.1. Agomelatine Inibits the Growth of Human CRC Cells in a p53-Dependent Manner

Two established CRC cell lines, HCT-116 (wild-type p53) and HCT-116 p53 null, were used to evaluate the antiproliferative effects of agomelatine in vitro. Cells were cultured in the presence of increasing concentrations of agomelatine (0–1 mM) for 24, 48, and 72 h. Treatment of both cell lines with agomelatine led to a dose-dependent inhibitory effect on cell growth at all times studied (Figure S1a,b). Similarly, the effect of melatonin was dose-dependent in both cell lines, although in HCT-116-p53-null cells, the indoleamine reduced cell growth significantly after 48 h treatment and at the highest dose studied (1 mM) (Figure S1c,d). We next compared the effects of both drugs regarding the p53 status of the cells at 72 h. Treatment of both cells lines with agomelatine inhibited cell growth significantly from the lowest dose used (0.1 mM), although the effect was higher in the HCT-116 than in the HCT-116 p53 null (Figure 1a). As we expected, melatonin was more effective on the HCT-116 than on the HCT-116 p53 null. As shown in Figure 1b, melatonin significantly inhibited the HCT-116's growth from 0.25 mM, whereas in the HCT-116 p53 null, the effect was statistically significant from 0.5 mM. These results indicate that agomelatine is more potent than melatonin in inhibiting CRC growth in vitro.

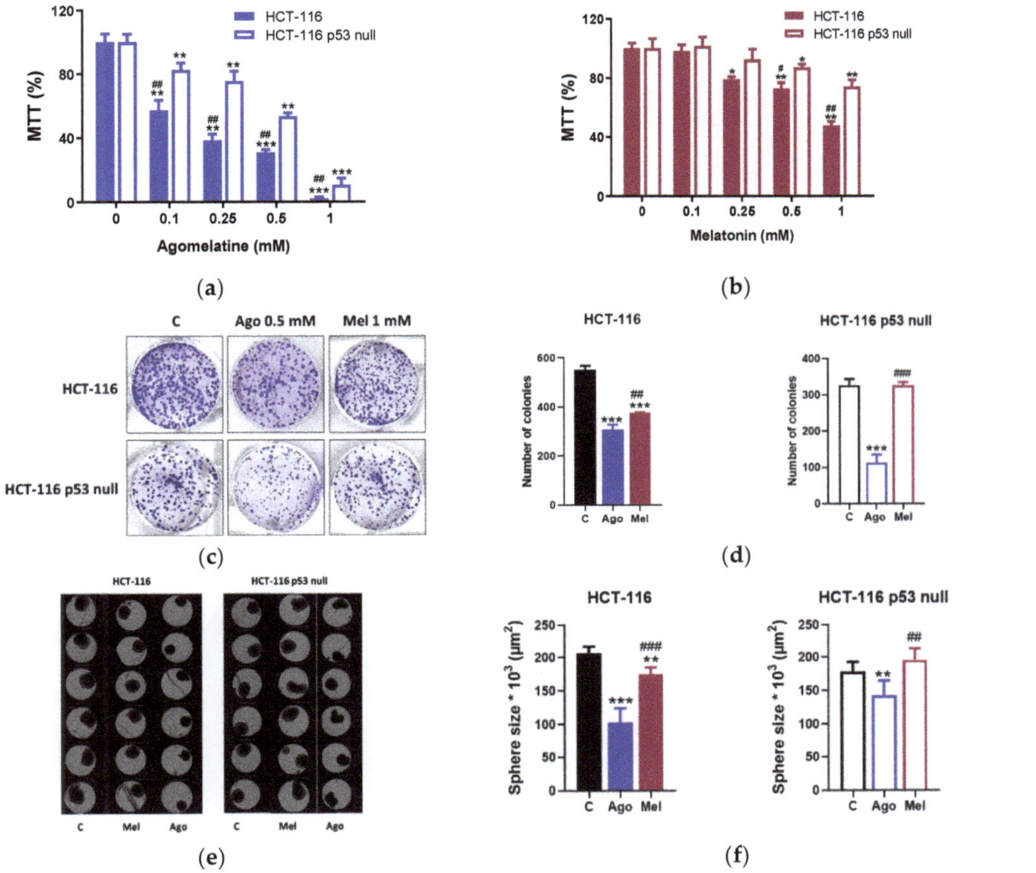

Figure 1. Inhibition of viability on HCT-116 and HCT-116-p53-null cells by (**a**) agomelatine and (**b**) melatonin. The results represent the mean ± SD of three experiments performed in quadruplicate. * $p < 0.05$ vs. C; ** $p < 0.01$ vs. C; *** $p < 0.001$ vs. C; ## $p < 0.01$ vs. the same concentration of agomelatine (Ago) or melatonin (Mel); # $p < 0.05$ vs. the same concentration of melatonin (Mel) or agomelatine (Ago). (**c**) Representative example of colony-formation assay and (**d**) graphical representation of three experiments in HCT-116 and HCT-116-p53-null cells after treatment with agomelatine (Ago) (0.5 mM) and melatonin (Mel) (1 mM) versus control (C) cells. Results are presented as means ± SD. *** Values of $p < 0.001$ versus C; ## $p < 0.01$ vs. agomelatine; ### $p < 0.001$ vs. agomelatine. In other experiments, (**e**) HTC-116 and HCT-116-p53-null cells were cultured in a 3D model and (**f**) the size of the spheroids was measured in control and after agomelatine (Ago) (0.5 mM) and melatonin (Mel) (1 mM) treatments during 72 h. ** Values of $p < 0.01$ vs. C; *** $p < 0.001$ vs. C; ## $p < 0.01$ vs. melatonin; ### $p < 0.001$ vs. melatonin.

We also conducted a clonogenic assay for up to 10 days after treatment to evaluate the long-term effects of agomelatine and melatonin on the cancer-cell survival (Figure 1c). Notably, the number of colonies in the HCT-116 cells reduced after the treatments with agomelatine (0.5 mM) and melatonin (1 mM), although the effect of agomelatine was significantly higher ($p < 0.01$) (Figure 1d). Only agomelatine (0.5 mM) was able to reduce the number of colonies in the HCT-116-p53-null cells (Figure 1d).

To corroborate these results, we also performed a 3D-culture model. To identify the sizes of the spheroids formed (Figure 1e), images were obtained and, subsequently, the perimeter was calculated (Figure 1f). According to the results described in the 2D model, both agomelatine (0.5 mM) and the melatonin (1 mM) reduced the sizes of the spheroids in the HCT-116, and agomelatine was more potent than melatonin ($p < 0.001$) at the doses used. In the HCT-116-p53-null cells, only agomelatine was able to reduce the sizes of the spheroids formed.

3.2. Agomelatine Induces Cell-Cycle Arrest and Caspase-Dependent Apoptosis in CRC Cells

The cell-cycle distribution after agomelatine and melatonin treatments was assessed by flow cytometry (Figures S2 and S3). The treatment with 0.5 mM of agomelatine for 72 h induced accumulation in the G2/M phase of the cycle in the HCT-116 (33.25 ± 1.91% of agomelatine-treated cells in the G2/M vs. 23.35 ± 1.42% of control cells in the same phase) and the HCT-116 p53 null (35.50 ± 1.13% of agomelatine-treated cells in the G2/M vs. 19.90 ± 1.42% of control cells in the same phase). Agomelatine treatment also resulted in a decrease in the number of cells in the G1 phase in the HCT-116 (33.25 ± 5.24% in the agomelatine-treated group vs. 69.47 ± 1.72% in the control cells) and HCT-116 p53 null (54.90 ± 2.07% in the agomelatine-treated group vs. 73.00 ± 2.68% in the control cells). The treatment of the cells with 0.25 mM agomelatine produced similar variations in cell-cycle distribution in the HCT-116-p53-null cells (Figure 2a,b).

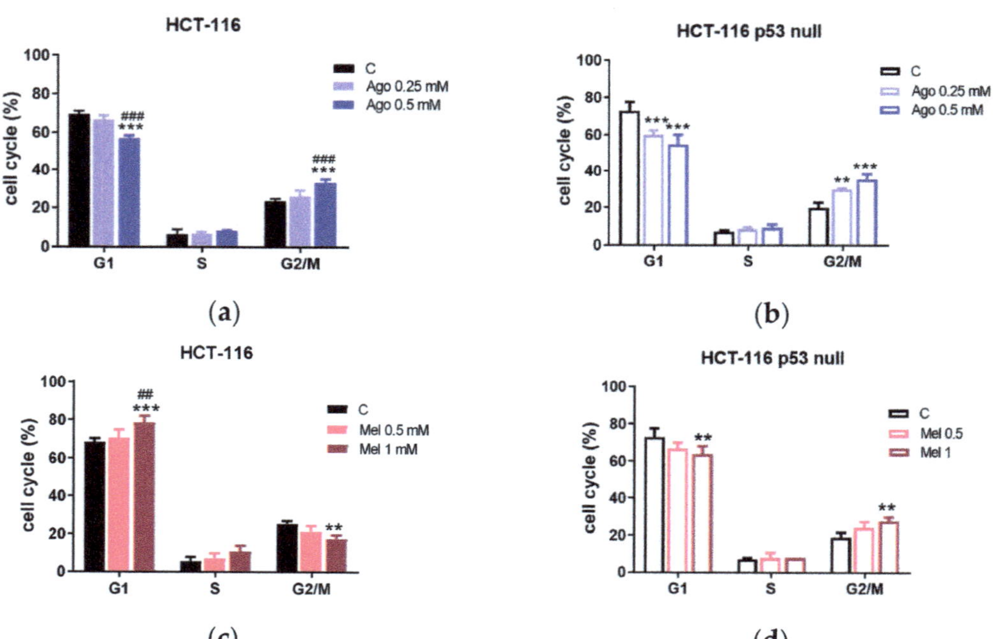

Figure 2. Percentage distribution in the different stages of the cell-cycle after treatment of HCT-116 and HCT-116-p53-null cells with agomelatine (Ago) (**a**,**b**) and melatonin (Mel) (**c**,**d**). Data represent the mean ± SD of three experiments performed in triplicate. ** Values of $p < 0.01$ vs. C; *** $p < 0.001$ vs. C. ## $p < 0.01$ vs. lower doses of treatment; ### $p < 0.001$ vs. lower doses of treatment.

The treatment with 1 mM melatonin for 72 h induced accumulation in the G1 phase of the cycle in the HCT-116 (78.70 ± 3.33% of melatonin-treated cells versus 68.5 ± 1.83% of control cells). This was associated with a decrease in the number of cells in the G2/M (17.70 ± 1.91% in the melatonin-treated group vs. 25.60 ± 1.42% in control cells). In the HCT-116-p53-null cells, we found an accumulation of cells in the G2/M phase (28.10 ± 1.91% of melatonin-treated cells versus 19.00 ± 2.73% of control cells) and a decrease in the G1 phase (64.00 ± 4.33% in the melatonin-treated group vs. 73.00 ± 4.59% in the control cells). The treatment with 0.5 mM melatonin did not produce any significant variations in the cell-cycle distribution in either of the two lines tested (Figure 2c,d).

In addition to the effects on the cell growth and cell cycle in vitro, we studied whether agomelatine and melatonin induced cell death through apoptosis using annexin V and propidium iodide and analyzing cell populations stained by flow cytometry (Figures S4 and S5).

The treatment with agomelatine induced a high percentage of apoptosis in both cell lines after 72 h at the doses studied (Figure 3a). Differences were found in the number of apoptotic cells in the two lines, although they were only observed at the lowest dose of agomelatine used (0.25 mM), and were greater in the HCT-116 cells (Figure 3a). When we used melatonin as a treatment for 72 h, we found increased apoptosis in the two cell lines tested only at 1 mM, while a smaller concentration of the indolamine (0.5 mM) did not produce cytotoxicity in either of the two lines analyzed. As expected, the percentage of cell death by apoptosis was higher in both cell lines treated with agomelatine.

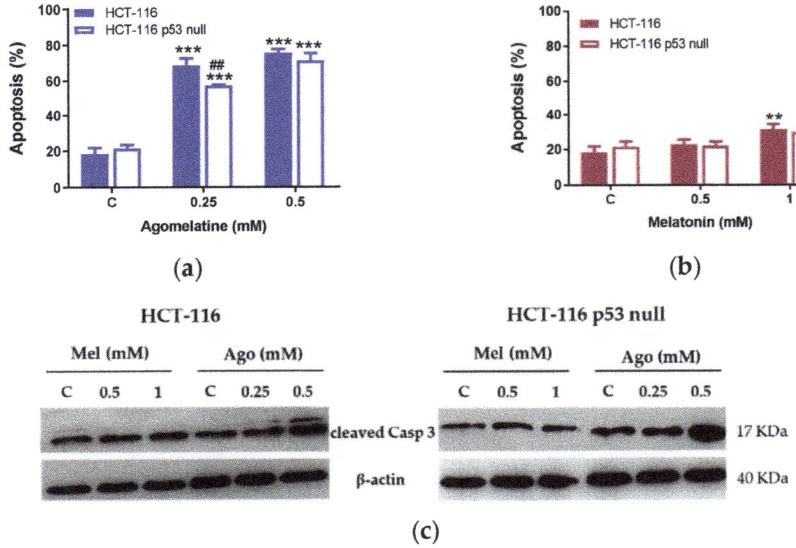

Figure 3. Percentage of apoptosis in HCT-116 and HCT-116-p53-null cells after agomelatine (**a**) and melatonin (**b**) treatments. Data represent the mean ± SD of three experiments performed in duplicate. * Values of $p < 0.05$ vs. C; ** $p < 0.01$ vs. C; *** $p < 0.001$ vs. C; ## $p < 0.01$ vs. HCT-116. (**c**) Expression of cleaved caspase-3 after treatment for 72 h with different doses of melatonin (Mel) and agomelatine (Ago) on HCT-116 and HCT-116 p53 null.

As shown in Figure 3c, the cell death induced by agomelatine was caspase-dependent, since this treatment induced increased the expression of active caspase-3 (cleaved caspase-3), at least at the higher doses used (0.5 mM), in the HCT-116 and HCT-116-null cells.

3.3. Regulation of Tumor Growth In Vivo by Agomelatine

To study their effect on tumor growth in vivo and to analyze possible side effects after treatment with both drugs, the HCT-116 and HCT-116-p53-null cell lines were injected into the right and left flanks of the immunosuppressed Balb/c nu/nu mice, respectively. Agomelatine and melatonin were injected at a dose of 5 mg/kg of body weight three times per week. A group of animals injected with 5-fluorouracil (5-FU) at a dose of 50 mg/kg twice per week was also used. This treatment was chosen to compare the effects of both agomelatine and melatonin, since it remains the first-line treatment for CRC, both alone (capecitabine) or in combination with other drugs (oxaliplatin and irinotecan, among others). The protein p53 serves as the major route for the anti-cancer effect of 5-FU and determines the cellular sensitivity to cytotoxic 5-FU [32] In fact, the absence of an active p53 drastically reduces its effectiveness [33].

At the doses used, all the drugs reduced the sizes of the tumors generated by the HCT-116 cells. In this case, agomelatine and 5-FU showed similar levels of potency ($p < 0.05$ vs. non-treated mice). Melatonin showed an almost statistically significant effect ($p = 0.076$) (Figure 4a). In the tumors generated by the HCT-116-p53-null cells, only agomelatine was able to reduce their size ($p < 0.05$), and its effect was significantly different from that of the 5-FU ($p < 0.01$) (Figure 4b). At the systemic level, agomelatine and melatonin led to an increase in blood glucose and alanine transaminase (ALT), whereas the treatment with 5-FU induced an increase in ALT levels (Figure S6).

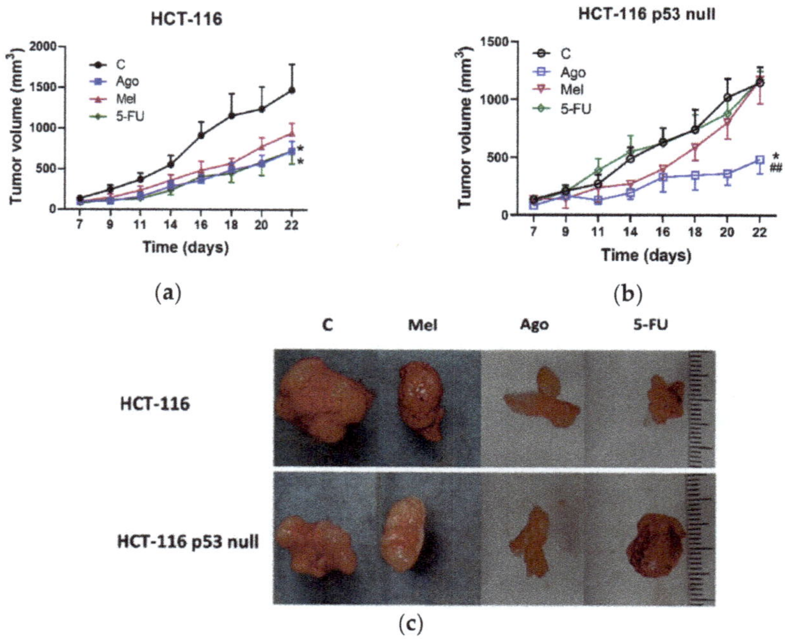

Figure 4. Tumor growth in the four groups of nude mice (controls and those treated with agomelatine, melatonin, or 5-FU, as described in Materials and Methods) in cell-line-derived xenografts in either (**a**) HCT-116 or (**b**) HCT-116 p53 null. Data represent mean ± SEM. * $p < 0.05$ vs. control mice; ## $p < 0.01$ vs. 5-FU group. (**c**) Representative images of tumors in control and treated mice in cell-line-derived xenografts. C: control; Mel: melatonin; Ago: agomelatine; 5-FU: 5-fluorouracil.

3.4. Agomelatine Regulates p53-Protein Levels in In Vitro and In Vivo Models of CRC

Melatonin and 5-FU regulate the expression of p53 protein levels and, as mentioned above, the presence of p53 is crucial for 5-FU's effectiveness [32–34]. As an analogue of melatonin, we analyzed whether agomelatine also regulated the levels of p53 in our in vitro and in vivo models. As shown in Figure 5a, agomelatine induced a small increase in the expression of the p53 protein after 72 h of treatment in the HCT-116 cells, although no effect was found after melatonin treatment at this point. Similar results were found in the in vivo model, in which the 5-FU displayed the highest effect (Figure 5b). As expected, the expression of the p53 protein was absent in the in vitro and in vivo models derived from the HCT-116-p53-null line (Figure 5).

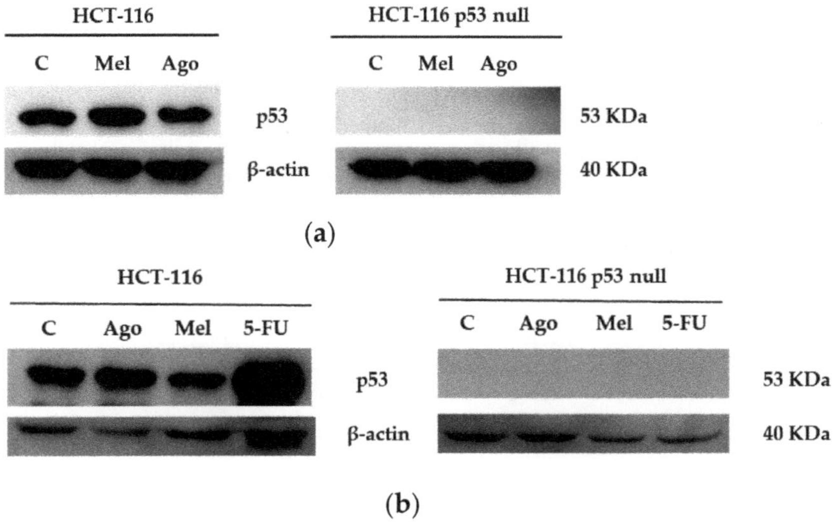

Figure 5. Expression levels of p53-protein in response to the treatments (**a**) with 0.5 mM agomelatine (Ago) and 1 mM melatonin (Mel) in the in vitro model, and (**b**) with agomelatine (Ago), melatonin (Mel), and 5-5-fluorouracil (FU) in tumors derived from HCT-116 and HCT-116-p53-null cell lines.

3.5. Agomelatine Regulates Circadian-Clock Genes' Rhythmicity in CRC Cell Lines In Vitro

Next, we analyzed the involvement of circadian-clock genes in the effects on the cell growth and viability induced by agomelatine and melatonin in both cell lines. Given that the effect of 0.5 mM agomelatine on cell growth was significant from 24 h in both cell lines, after cell-culture synchronization by serum shock for 6 h, the cells were cultured in the presence of 10% FBS and treated with vehicle, 0.5 mM agomelatine, or 1 mM melatonin for 48 h. The expressions of Per1-3, Cry1, Clock, Bmal1, and Nr1d2 were analyzed every 4 h (Figure 6).

We found a statistically significant rhythmicity in all the genes studied in the HCT-116 and HCT-116-p53-null cells (Tables 1 and 2). However, these cells showed some differences in the characteristics of their rhythmicity (Table 2). The acrophase of the Per2, Cry1, Bmal1, and Nr1d2 showed an advance in phase in the HCT-116-p53-null cells versus their isogenic HCT-116 cells. The MESOR levels of the Per1 and Per3 were lower in the HCT-116-p53-null cells than in the HCT-116 cells, but this parameter of Bmal1 was higher in the cell line HCT-116 p53 null. There were also differences in the amplitudes of some of the genes between these cells, although these were smaller than those found for the acrophase and the MESOR. The Per2 and Nr1d2 showed lower amplitudes in the HCT-116-p53-null cells than in the HCT-116, while this parameter was higher for the Bmal1.

The treatments with agomelatine and melatonin in the HCT-116 control cells maintained the rhythmicity of the genes studied, albeit with some differences (Table 1). Both drugs induced an advance in phase in the acrophase of the Per1, Per2, Per3, and Cry1; however, the effect of agomelatine was significantly higher in all these genes, except in the case of the Per3, in which both treatments had a similar effect. Only agomelatine induced a delay in the acrophase of the Nr1d2. No significant effects were found on the acrophase in the Clock or Bmal1. Agomelatine increased the MESOR in the Bmal1 and decreased it in the Nr1d2, while melatonin increased this parameter in the Clock. Both treatments increased the amplitude in the Per3. Agomelatine increased the amplitude in the Per2 and decreased it in the Nr1d2. Melatonin treatment decreased the amplitude in Per1 compared with the agomelatine treatment, while increased it in the Clock.

Similar to the results found in the HCT-116 cells, both treatments maintained the rhythmicity of the genes analyzed in the HCT-116-p53-null cells (Table 2). In these cells, only agomelatine was able to induce an advance in phase in the acrophase of the Per1, Per3, and Cry1 compared with the non-treated cells, whereas Per2 and Clock acrophase showed a delay in phase. Agomelatine also increased the MESOR in the Per1, Clock, and Nr1d2, while it decreased the MESOR in the Cry1. Melatonin also enhanced the MESOR in the Clock and reduced it in the Bmal1. Agomelatine increased the amplitude in the Per3 and Nr1d2. Melatonin did not change this parameter in any of the genes studied in this cell line.

We also analyzed the rhythmicity of the Sirt1 in the non-treated cells and after the agomelatine and melatonin treatments in both cell lines (Figure 7). The HCT-116 cells showed a delay in phase in the acrophase and lower MESOR levels than the HCT-116-p53-null cells (Table 3). Agomelatine and melatonin changed the acrophase of the genes in both types of cell, but while agomelatine induced an advance in phase of this parameter, melatonin delayed it. Melatonin also increased the amplitude of the SIRT1 rhythmicity.

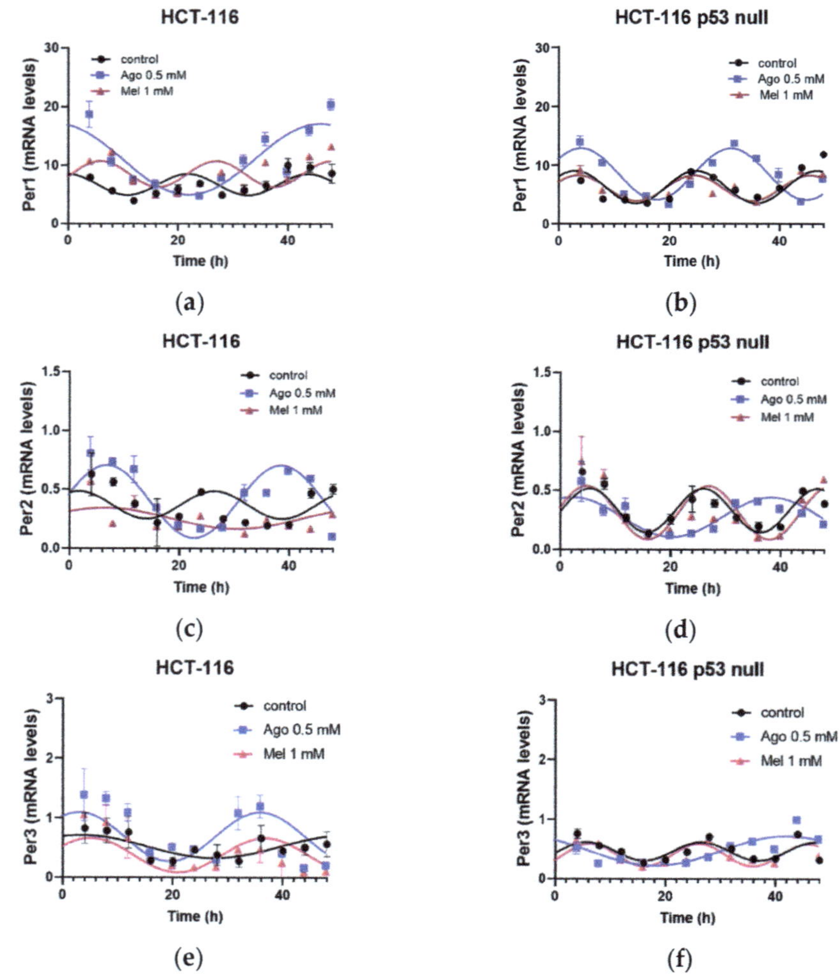

Figure 6. *Cont.*

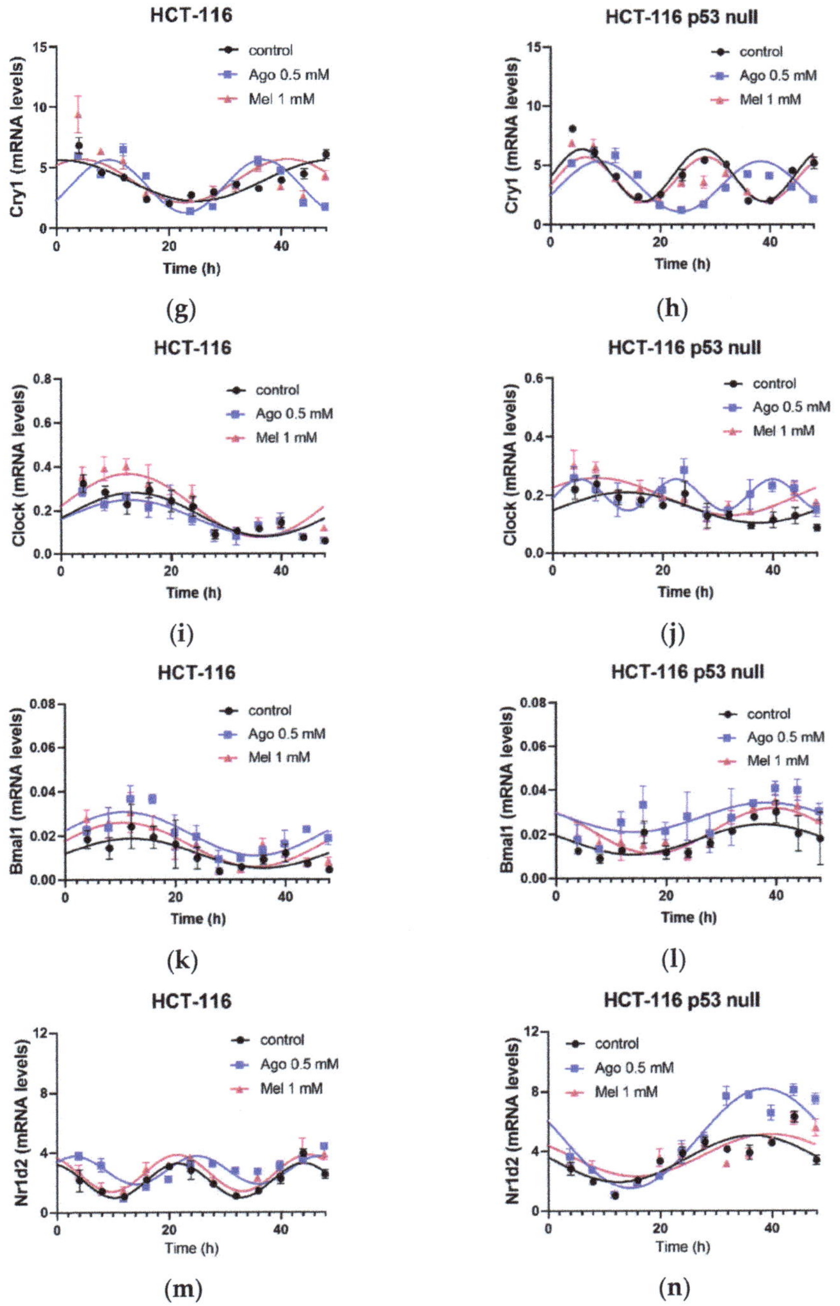

Figure 6. Characteristics of the rhythms obtained for Period 1 (Per1) (**a**,**b**), Period 2 (Per2) (**c**,**d**), Period 3 (Per3) (**e**,**f**), cryptocrome1 (Cry1) (**g**,**h**), circadian locomotor cycles kaput (Clock) (**i**,**j**), brain and muscle ARNT-like (Bmal1) (**k**,**l**) and nuclear receptor subfamily 1 group D member 2 (Nr1d2) (**m**,**n**) gene after agomelatine (Ago) and melatonin (Mel) treatments in the HCT-116 and HCT-116-p53-null cell lines. Curve fittings under the different conditions were performed using a 48-h test period.

Table 1. Results of the cosinor analysis of the clock genes' expression after agomelatine or melatonin treatment in the HCT-116 cell line.

Gene	Treatment	PR [1]	*p*-Value [2]	Amplitude (A.U.) [3]	Acrophase (h) [4]	MESOR (A.U.)
Per1	Control	29.41	0.03	5.17 ± 1.45	3.06 ± 1.91	9.77 ± 0.99
	Ago 0.5	42.85	0.003	6.06 ± 1.22	45.30 ± 1.53 ***	10.40 ± 0.86
	Mel 1	34.77	0.008	2.24 ± 0.53 #	32.83 ± 2.00 ***,###	8.25 ± 0.39
Per2	Control	31.66	0.006	0.076 ± 0.018	2.76 ± 1.19	0.28 ± 0.01
	Ago 0.5	42.02	0.001	0.140 ± 0.003 *	9.35 ± 0.08 ***	0.25 ± 0.02
	Mel 1	27.81	0.003	0.088 ± 0.024	6.58 ± 2.09 ***,###	0.26 ± 0.02
Per3	Control	25.36	0.008	0.33 ± 0.09	17.47 ± 1.15	0.72 ± 0.07
	Ago 0.5	60.62	0.0001	0.91 ± 0.14 *	6.06 ± 0.68 ***	0.94 ± 0.09
	Mel 1	31.33	0.002	0.96 ± 0.25 *	5.55 ± 1.44 ***	0.83 ± 0.18
Cry1	Control	63.01	<0.0001	1.62 ± 0.22	1.79 ± 1.02	3.81 ± 0.15
	Ago 0.5	73.64	<0.0001	2.19 ± 0.21	9.47 ± 0.48 ***	3.48 ± 0.16
	Mel 1	37.72	0.0004	1.79 ± 0.42	4.55 ± 1.31 **,###	3.90 ± 0.29
Clock	Control	57.99	<0.0001	0.101 ± 0.015	13.14 ± 1.13	0.18 ± 0.01
	Ago 0.5	56.69	0.001	0.111 ± 0.019	11.76 ± 2.97	0.16 ± 0.01
	Mel 1	56.75	<0.0001	0.189 ± 0.029 **,###	10.77 ± 1.16	0.26 ± 0.02 ***,###
Bmal1	Control	61.01	<0.0001	0.0088 ± 0.0012	12.99 ± 1.06	0.014 ± 0.001
	Ago 0.5	71.30	<0.0001	0.0138 ± 0.0015	10.47 ± 0.84	0.025 ± 0.001 ***
	Mel 1	38.63	0.0003	0.0092 ± 0.0020	10.85 ± 1.67	0.016 ± 0.001 ###
Nr1d2	Control	48.84	<0.0001	1.73 ± 0.32	12.19 ± 0.63	3.79 ± 0.22
	Ago 0.5	55.56	<0.0001	0.94 ± 0.15 **	3.32 ± 0.55 ***	2.86 ± 0.11 ***
	Mel 1	57.84	<0.0001	1.89 ± 0.28 ##	11.55 ± 0.56 ###	3.74 ± 0.20 ###

[1] PR: percentage of rhythm; [2] *p*-value: zero-amplitude test; [3] A.U.: arbitrary units; [4] h: hours. Comparisons of parameters under the different conditions were performed using a 48-h test period. * Values of $p < 0.05$ vs. control, ** $p < 0.01$ vs. control, *** $p < 0.001$ vs. control; # $p < 0.05$ vs. agomelatine, ## $p < 0.01$ vs. agomelatine, ### $p < 0.001$ vs. agomelatine.

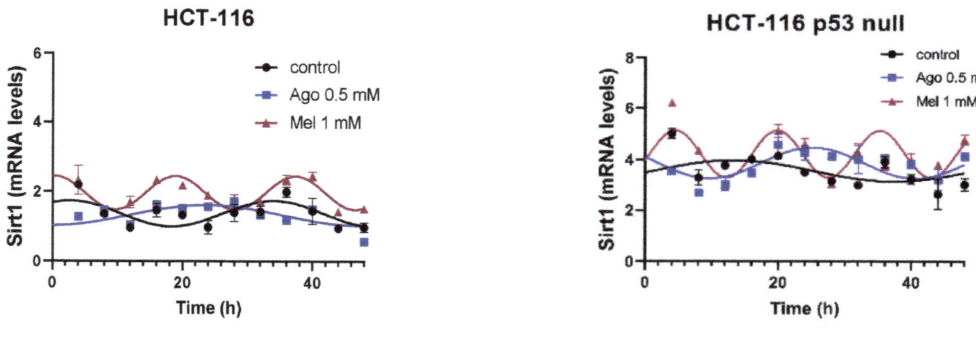

(a) (b)

Figure 7. Characteristics of each of the rhythms obtained for sirtuin 1 (SIRT1) gene after agomelatine (Ago) and melatonin (Mel) treatments in the HCT-116 (**a**) and HCT-116-p53-null (**b**) cell lines. Curve fittings under the different conditions were performed using a 48-h test period.

In order to find differences between the mechanisms of action of agomelatine and melatonin that could explain the differences in rhythmicity found between the genes studied in the two cell lines, we also analyzed the PRX1 in the non-treated cells and after the agomelatine and melatonin treatments (Figure 8). The PRX1 acrophase in the HCT-116 cells showed an advance in phase but lower MESOR than in the HCT-116-p53-null cells (Table 4). Agomelatine and melatonin induced changes in different parameters in the HCT-116 cells (Figure 8a). Melatonin induced an advance in phase in the acrophase and the

MESOR of the gene, while agomelatine delayed this parameter. By contrast, agomelatine induced an advance in phase in the acrophase and decreased the MESOR of the PRX1 in the HCT-116-p53-null cells (Figure 8b), while melatonin did not induce changes in any of the rhythmic parameters in this gene (Table 4).

Table 2. Results of the cosinor analysis of the clock genes' expression after agomelatine or melatonin treatment in the HCT-116-p53-null cell line.

Gene	Treatment	PR [1]	*p*-Value [2]	Amplitude (A.U.) [3]	Acrophase (h) [4]	MESOR (A.U.)
Per1	Control	47.47	0.0001	2.42 ± 0.45	3.01 ± 0.65	6.11 ± 0.33 †
	Ago 0.5	51.58	0.003	2.54 ± 1.22	4.80 ± 3.68 ***	8.36 ± 0.86 **
	Mel1	68.24	<0.0001	2.2 ± 0.26	2.61 ± 0.41 ###	6.13 ± 0.19 ##
Per2	Control	61.90	<0.0001	0.22 ± 0.03†	4.89 ± 0.48 †††	0.34 ± 0.02
	Ago 0.5	44.57	0.0001	0.19 ± 0.04	1.29 ± 1.16 ***	0.29 ± 0.03
	Mel 1	56.40	<0.0001	0.25 ± 0.04	5.44 ± 0.56 ###	0.32 ± 0.03
Per3	Control	40.65	0.0002	0.13 ± 0.03	6.02 ± 0.71	0.37 ± 0.02 ††
	Ago 0.5	65.87	<0.0001	0.21 ± 0.03 *	42.48 ± 0.95 ***	0.37 ± 0.02
	Mel 1	56.89	<0.0001	0.17 ± 0.02	6.41 ± 1.16 ###	0.34 ± 0.02
Cry1	Control	78.47	<0.0001	2.19 ± 0.20	5.73 ± 0.34 †††	4.16 ± 0.14
	Ago 0.5	79.43	<0.0001	2.13 ± 0.17	9.11 ± 0.42 *	3.25 ± 0.13 ***
	Mel 1	67.89	<0.0001	1.88 ± 0.22	6.51 ± 0.44	3.82 ± 0.16 ##
Clock	Control	41.13	0.0002	0.05 ± 0.01 †	11.99 ± 1.59	0.16 ± 0.01
	Ago 0.5	73.01	0.0001	0.04 ± 0.01	4.57 ± 0.76 ***	0.19 ± 0.01 **
	Mel 1	56.47	<0.0001	0.06 ± 0.01	8.86 ± 1.17 ##	0.19 ± 0.01 **
Bmal1	Control	58.43	<0.0001	0.018 ± 0.003 †††	35.12 ± 1.09 †††	0.025 ± 0.002 †††
	Ago 0.5	50.78	<0.0001	0.011 ± 0.002	39.95 ± 1.31	0.022 ± 0.001
	Mel 1	37.40	0.0004	0.010 ± 0.002	37.92 ± 1.56	0.020 ± 0.002 *
Nr1d2	Control	45.84	<0.0001	1.41 ± 0.27 ††	35.29 ± 1.44 †††	3.25 ± 0.19
	Ago 0.5	71.46	<0.0001	3.78 ± 0.41 ***	37.66 ± 0.84	5.54 ± 0.29 ***
	Mel 1	44.92	0.0001	2.08 ± 0.40 ##	36.94 ± 1.47	4.07 ± 0.28 ###

[1] PR: percentage of rhythm; [2] *p*-value: zero-amplitude test; [3] A.U.: arbitrary units; [4] h: hours. Comparisons of parameters at the different conditions were performed using a 48-h test period. * Values of $p < 0.05$ vs. control, ** $p < 0.01$ vs. control, *** $p < 0.001$ vs. control; ## $p < 0.01$ vs. agomelatine, ### $p < 0.001$ vs. agomelatine. † $p < 0.05$ vs. HCT-116, †† $p < 0.01$ vs. HCT-116, ††† $p < 0.001$ vs. HCT-116.

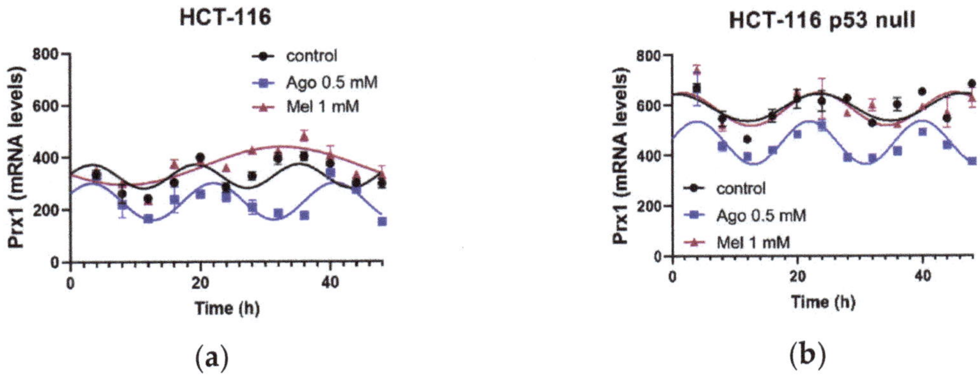

Figure 8. Circadian rhythms obtained for peroxiredoxin 1 (PRX1) gene after agomelatine (Ago) and melatonin (Mel) treatments in the HCT-116 (**a**) and HCT-116-p53-null (**b**) cell lines. Curve fittings under the different conditions were performed using a 48-h test period.

Table 3. Results of the cosinor analysis of Sirt1-gene expression after agomelatine or melatonin treatment in the HCT-116 and HCT-116-p53-null cell lines.

Cell Line	Treatment	PR [1]	*p*-Value [2]	Amplitude (A.U.) [3]	Acrophase (h) [4]	MESOR (A.U.)
HCT-116	Control	47.58	<0.0001	0.367 ± 0.071	2.80 ± 0.92	1.37 ± 0.15
	Ago 0.5	47.05	<0.0001	0.456 ± 0.084	22.31 ± 1.41 ***	1.23 ± 0.16
	Mel 1	80.65	<0.0001	0.483 ± 0.042	0.52 ± 0.24 ***,###	1.96 ± 0.13
HCT-116 p53 null	Control	34.8	0.0009	0.392 ± 0.093	16.42 ± 1.82 †††	3.48 ± 0.17 †††
	Ago 0.5	35.92	0.0006	0.518 ± 0.114	25.48 ± 1.12 ***	3.91 ± 0.18
	Mel 1	62.29	<0.0001	0.766 ± 0.102 **	4.59 ± 0.34 ***,###	4.14 ± 0.17

[1] PR: percentage of rhythm; [2] *p*-value: zero-amplitude test; [3] A.U.: arbitrary units; [4] h: hours. Comparisons of parameters at the different conditions were performed using a 48-h test period. ** $p < 0.01$ vs. control, *** $p < 0.001$ vs. control; ### $p < 0.001$ vs. agomelatine. ††† $p < 0.001$ vs. HCT-116.

Table 4. Results of the cosinor analysis of PEX1-gene expression after agomelatine or melatonin treatment in the HCT-116 and HCT-116-p53-null cell lines.

Cell Line	Treatment	PR [1]	*p*-Value [2]	Amplitude (A.U) [3]	Acrophase (h) [4]	MESOR (A.U.)
HCT-116	Control	38.04	0.0004	45.7 ± 10.2	3.38 ± 0.56	329 ± 7
	Ago 0.5	74.50	<0.0001	70.4 ± 7.4	3.27 ± 0.31	232 ± 5 ***
	Mel 1	67.01	<0.0001	71.6 ± 8.7	32.54 ± 0.93 ***,###	368 ± 6 ***,###
HCT-116 p53 null	Control	38.90	0.0003	54 ± 12	0.63 ± 0.79 ††	591 ± 8 †††
	Ago 0.5	64.23	<0.0001	81 ± 10	3.83 ± 0.39 ***	447 ± 8 ***
	Mel 1	46.83	<0.0001	65 ± 12	1.71 ± 0.62 ##	583 ± 9 ###

[1] PR: percentage of rhythm; [2] *p*-value: zero-amplitude test; [3] A.U.: arbitrary units; [4] h: hours. Comparisons of parameters at the different conditions were performed using a 48-h test period *** $p < 0.001$ vs. control; ## $p < 0.01$ vs. agomelatine, ### $p < 0.001$ vs. agomelatine. †† $p < 0.01$ vs. HCT-116, ††† $p < 0.001$ vs. HCT-116.

4. Discussion

This study describes, for the first time, the potential use of agomelatine as a treatment for CRC. Agomelatine inhibited cell growth in our in vitro and in vivo models of this disease and induced cell-cycle arrest and caspase-dependent apoptotic cell death. We propose that agomelatine exerted these effects through the regulation of the rhythmicity of several core Clock genes. The effect of agomelatine was independent of the p53, although this treatment decreased the cell growth more efficiently in the in vitro models of cells harboring the wild-type p53.

Similarly, melatonin induced cell-growth inhibition in vitro and in vivo and cell-cycle arrest. However, it was not able to increase apoptosis, as expected. These effects were only observed in the presence of the wild-type p53, since p53 activation is critical for the oncostatic effect of melatonin, encouraging p53 accumulation at the cellular level [34]. However, the expression of the protein increases transiently, displaying a maximum at three hours after the treatment of with melatonin cells [34]. This could explain the lack of effect found in our study, in which the effect on the p53-protein levels was assessed after 72 h of melatonin treatment. The inhibition of tumor growth after the treatment with melatonin was related to the inhibition of cell progression from the G0/G1 phase to the S phase. This effect was dependent on the p53 status, since melatonin induces the activation of p53 and p21. The observed increase in the percentage of apoptosis after the treatment with the indoleamine may have been due to the blockade of the Akt/MDM2 pathway, an effect that was previously observed in gastric cancer cells [35]. Increased apoptosis and cell-cycle arrest should trigger the activation of the caspase-3 pathway, as described for hepatocarcinoma cells [36]. Considering these observations, in our model, the fact that we did not observe an increase in caspase-3 expression after melatonin treatment may have been due to the small increase in apoptosis we obtained.

Intracellular melatonin signaling through its receptors can also be affected by the status of p53. It has been reported that the expression of the membrane (MT1 and MT2) receptors of melatonin decreases in patients with colon cancer [22], but mainly in those with mutations in the p53 gene [23], although melatonin levels and the expression of nuclear (RORα) receptors do not change [22]. On the other hand, several works previously showed the implication of melatonin receptors in its oncostatic actions in this disease [37,38]. This could indicate that non-selective MT1/MT2 agonists with a higher affinity with these receptors than melatonin itself could be of interest in the treatment of CRC [22]. In line with this observation, we found that agomelatine was more potent than melatonin, at least in the in vitro and in vivo models in which the tumor-suppressor gene p53 was in its wild form. In this case, agomelatine seemed to induce a slight increase in the p53 protein levels at the time at which it was analyzed. In addition, agomelatine was more potent than the indoleamine in inhibiting cell growth in the in vitro models with the non-active p53. Interestingly, unlike melatonin, as described above, agomelatine inhibited tumor growth in vivo in the cases with the non-functional p53, indicating that mechanisms other than p53 regulation may be involved in its mechanism of action, at least under these conditions. Agomelatine has been shown to be more effective than melatonin when used as a treatment for insomnia, depression, or obesity-associated comorbidities [39–41]. Agomelatine is a non-selective agonist of MT1 and MT2 receptors, with a higher affinity with these receptors than melatonin itself [42]. In addition, agomelatine has a longer half-life and better oral absorption [43]. On the other hand, agomelatine is a serotonin-HT-2c- and -HT-2b-receptor antagonist. The involvement of serotonin in tumor growth, differentiation, and gene expression has long been known. In fact, its mitogenic effect has been demonstrated in various types of cancer, including prostate, bladder, breast, and colon cancer [44]. In the specific case of colon cancer, selective serotonin reuptake inhibitors (SSRIs), which increase its concentration in nerve terminals both centrally and peripherally, have been shown to have preventive effects [45].

In our in vitro model, we found the resynchronization of circadian rhythmicity by agomelatine and melatonin in vitro. Both drugs regulated Per1, Per2, Per3, Cry1, and Clock in cells harboring the wild-type p53. The change induced in the characteristics of the rhythms analyzed (i.e., amplitude, acrophase, and/or MESOR) was always higher after the treatment with agomelatine. In these cells, agomelatine also regulated Bmal1 and Nr1d2. The deletion of p53 induced a loss of effect in both drugs, although agomelatine was also found to be more potent than melatonin. In this type of cell, we also found a decrease in the number of circadian genes regulated by both drugs, indicating the influence of the circadian clock on the antiproliferative effect of these treatments. Previous reports showed that melatonin regulates the circadian clock after the activation of MT1 and MT2 receptors at the central level [39] and in cancer [46], whereas agomelatine acts at the central level in a complementary and possibly synergistic manner on the MT1 and MT2 receptors of melatonin and the HT2c receptor of serotonin to resynchronize circadian rhythms [47,48], which could explain the greater effect found after the treatment with agomelatine.

The SIRT1 is a NAD+ dependent class III histone deacetylase, which is involved in the control of cellular redox homeostasis [6]. It regulates the circadian clock by activating the transcription of BMAL1 and Clock [6], and the intracellular levels of SIRT1 and NAD+ oscillate with a circadian pattern [7]. Although the exact role of SIRT1 is not well understood, it has been implicated in tumor growth and resistance to therapy in several types of cancer, including CRC [49,50]. An important relationship between melatonin, the circadian clock, and SIRT1 has been described, although it remains unclear whether SIRT1 is a key mediator of circadian-clock regulation by melatonin or whether melatonin acts through circadian genes to regulate SIRT1 [51]. Peroxiredoxins (PRX) are a family of conserved enzymes involved in the regulation of peroxide levels. The oxidation-reduction states of PRX1 proteins exhibit self-sustained oscillation in the absence of TTFL mechanisms. In addition, the knockdown of PRX1 proteins affects circadian rhythms in nucleated cells, indicating that the oxidation-reduction cycle of PRX1 is interconnected with the main circadian-clock cycle [52]. Furthermore, Tp53 and SIRT1 are mediators of these connections [52], which are important regulators of circadian-clock-gene expression [8,53,54]. In fact, in the present study, we found that agomelatine and melatonin can regulate the rhythmicity of SIRT1, regardless of p53, although they differ in the characteristics of the rhythmicity affected. Interestingly, only agomelatine was able to regulate the oscillations of the PRX1 in both the wild-type p53 and p53 null cells, whereas melatonin only affected it in the wild-type p53 cells. Further research is warranted to determine the mechanism underlying these differences. The antioxidant ability of melatonin is extended to the regulation of the expression and activity of PRX1 [55], although there are no available data regarding the role of p53 in this action of the indoleamine.

The antitumor efficacy of 5-FU is due to its ability to induce cell-cycle arrest and p53-dependent apoptosis [56]. Similarly to melatonin, several lines of research demonstrated that this chemotherapeutic agent has a reduced ability to inhibit cell growth in CRC with mutated or inactive p53 [57]. In this study, we used the treatment with 5-FU as a control for the in vivo model carrying p53 null tumors, and only agomelatine was able to reduce tumor growth in these animals. Although the side effects found at the systemic level were similar after the three treatments, the 5-FU induced the greatest increase in AST activity. At the metabolic level, melatonin treatment produced an increase in glucose and alanine-aminotransferase levels, probably due to the role of melatonin in processes such as glycogenesis or glycolysis [36].

5. Conclusions

Recent studies have shown that some anti-depressant treatments also exert an anti-tumor effect, which is due to the modification of the tumor environment or the alteration of the immune response. This is the case in colon-cancer , in which the use of fluoxetine inhibits the transcriptional activity of NF-kappa B, and the proliferation of cells [58]. In addition, in general, the use of these drugs does not present a large number of side effects in the organism [59]. Therefore, taking into account the fact that treatment with antidepressant drugs after a cancer diagnosis is quite frequent, if it is possible to perform both treatments with the same drug, as is the case with agomelatine, conferring a double benefit. In addition, the presence of a non-functional p53 has been implicated in 5-FU resistance in CRC patients [57]. This barrier could be overcome with treatment with agomelatine, since it has similar effectiveness regardless of p53 status, at least in the xenograft model used in this study.

Supplementary Materials: The following supporting information can be downloaded at: https://www.mdpi.com/article/10.3390/antiox12040926/s1. Figure S1: Growth inhibition of HCT-116 and HCT-116-p53-null cells after treatment with (a) (c) agomelatine (0–1 mM) and (b) (d) melatonin (0–1 mM) for 24, 48, and 72 h; Figure S2: Raw data of cell-cycle analysis by flow cytometry in HCT-116 cell line after treatment with agomelatine or melatonin; Figure S3: Raw data of cell-cycle analysis by flow cytometry in HCT-116-p53-null cell line after treatment with agomelatine or melatonin; Figure S4: Raw data of apoptosis analysis by flow cytometry in HCT-116 cell line after treatment with agomelatine or melatonin; Figure S5: Raw data of apoptosis analysis by flow cytometry in HCT-116 cell line after treatment with agomelatine or melatonin; Figure S6: Levels of glucose (a), alanine aminotransferase (ALT) (b), amylase (c), urea (d), aspartame aminotransferase (e), and uric acid (f) in sera of nude mice (controls and those treated with agomelatine, melatonin, or 5-FU, as described in Materials and Methods); Table S1: Primers used in the real-time PCR assay; Table S2: Results at 24 h of the cosinor analysis of the Clock genes' expression after agomelatine or melatonin treatment in the HCT-116 cell line; Table S3: Results at 24 h of the cosinor analysis of the Clock genes' expression after agomelatine or melatonin treatment in the HCT-116-p53-null cell line; Table S4: Results at 24 h of the cosinor analysis of SIRT1 gene's expression after agomelatine or melatonin treatment in the HCT-116 and HCT-116-p53-null cell lines; Table S5: Results at 24 h of the cosinor analysis of PRX1 gene expression after agomelatine or melatonin treatment in the HCT-116 and HCT-116-p53-null cell line; Table S6: Median Ct values obtained in the circadian rhythmicity analysis for each gene under all conditions included in the study in HCT-116 cells; Table S7: Median Ct values obtained in the circadian rhythmicity analysis for each gene under all conditions included in the study in HCT-116-p53-null cells.

Author Contributions: Conceptualization: J.L.; methodology: S.M.-S., J.D.P.-P., J.C., J.E.-F. and H.K.; investigation: J.L.; data curation: J.D.P.-P., S.M.-S., J.C. and J.A.; writing—original draft preparation: S.M.-S., J.D.P.-P. and J.L.; writing—review and editing: J.L. and Á.C.; supervision: J.L.; project administration: J.L.; funding acquisition: J.L. and Á.C. All authors have read and agreed to the published version of the manuscript.

Funding: This work was supported by a research grant from the Instituto de Salud Carlos III- FEDER (PI18/01947) and the Consejería de Salud from the Junta de Andalucía (PI-067/2013). J.L. was supported by the Nicolás Monardes Program in the Andalusian Health Service (C-0033-2015). J.D.P.-P. is funded by a FPU2019 fellowship (FPU190/02269) from the Ministerio de Universidades (Spain).

Institutional Review Board Statement: Not applicable.

Informed Consent Statement: Not applicable.

Data Availability Statement: Data is contained within the article and Supplementary Materials.

Conflicts of Interest: The authors declare no conflict of interest.

Abbreviations

5-FU	5-fluorouracil
Ago	agomelatine
ALT	alanine transaminase
ARE	anti-oxidant-response element
BMAL	brain and muscle ARNT-like
Clock	circadian locomotor cycles kaput
CRC	colorectal cancer
Cry	cryptocrome
CSCs	cancer stem cells
Cul3	cullin 3
Keap1	kelch-like ECH-associated protein 1
Mel	melatonin
MESOR	midline estimating statistic of rhythm
MT	melatonin receptor
NAD	nicotinamide adenine dinucleotide
Nr1d2	nuclear receptor subfamily 1 group D member 2
Nrf2	nuclear factor erythroid 2 related factor-2
Per	period
PR	percentage of rhythm
Prx-1	peroxiredoxin-1
RORα	RAR-related orphan receptor alpha
SIRT1	sirtuin 1
TTFL	transcription–translation feedback loops.

References

1. Bray, F.; Ferlay, J.; Soerjomataram, I.; Siegel, R.L.; Torre, L.A.; Jemal, A. Global Cancer Statistics 2018: GLOBOCAN Estimates of Incidence and Mortality Worldwide for 36 Cancers in 185 Countries. *CA A Cancer J. Clin.* **2018**, *68*, 394–424. [CrossRef] [PubMed]
2. Xie, Y.-H.; Chen, Y.-X.; Fang, J.-Y. Comprehensive Review of Targeted Therapy for Colorectal Cancer. *Signal Transduct. Target. Ther.* **2020**, *5*, 22. [CrossRef]
3. Vodenkova, S.; Buchler, T.; Cervena, K.; Veskrnova, V.; Vodicka, P.; Vymetalkova, V. 5-Fluorouracil and Other Fluoropyrimidines in Colorectal Cancer: Past, Present and Future. *Pharmacol. Ther.* **2020**, *206*, 107447. [CrossRef] [PubMed]
4. Robinson, I.; Reddy, A.B. Molecular Mechanisms of the Circadian Clockwork in Mammals. *FEBS Lett.* **2014**, *588*, 2477–2483. [CrossRef]
5. Tang, B.L. Sirt1 and the Mitochondria. *Mol. Cells* **2016**, *39*, 87–95. [CrossRef] [PubMed]
6. Kleszcz, R.; Paluszczak, J.; Baer-Dubowska, W. Targeting Aberrant Cancer Metabolism—The Role of Sirtuins. *Pharmacol. Rep.* **2015**, *67*, 1068–1080. [CrossRef] [PubMed]
7. Nakahata, Y.; Sahar, S.; Astarita, G.; Kaluzova, M.; Sassone-Corsi, P. Circadian Control of the NAD+ Salvage Pathway by CLOCK-SIRT1. *Science* **2009**, *324*, 654–657. [CrossRef] [PubMed]

8. Asher, G.; Gatfield, D.; Stratmann, M.; Reinke, H.; Dibner, C.; Kreppel, F.; Mostoslavsky, R.; Alt, F.W.; Schibler, U. SIRT1 Regulates Circadian Clock Gene Expression through PER2 Deacetylation. *Cell* **2008**, *134*, 317–328. [CrossRef] [PubMed]
9. Bellet, M.M.; Orozco-Solis, R.; Sahar, S.; Eckel-Mahan, K.; Sassone-Corsi, P. The Time of Metabolism: NAD+, SIRT1, and the Circadian Clock. *Cold Spring Harb. Symp. Quant. Biol.* **2011**, *76*, 31–38. [CrossRef]
10. Edgar, R.S.; Green, E.W.; Zhao, Y.; van Ooijen, G.; Olmedo, M.; Qin, X.; Xu, Y.; Pan, M.; Valekunja, U.K.; Feeney, K.A.; et al. Peroxiredoxins Are Conserved Markers of Circadian Rhythms. *Nature* **2012**, *485*, 459–464. [CrossRef]
11. Yang, Y.; Li, X.; Peng, L.; An, L.; Sun, N.; Hu, X.; Zhou, P.; Xu, Y.; Li, P.; Chen, J. Tanshindiol C Inhibits Oxidized Low-Density Lipoprotein Induced Macrophage Foam Cell Formation via a Peroxiredoxin 1 Dependent Pathway. *Biochim. Biophys. Acta Mol. Basis Dis.* **2018**, *1864*, 882–890. [CrossRef] [PubMed]
12. Kim, Y.-J.; Ahn, J.-Y.; Liang, P.; Ip, C.; Zhang, Y.; Park, Y.-M. Human Prx1 Gene Is a Target of Nrf2 and Is Up-Regulated by Hypoxia/Reoxygenation: Implication to Tumor Biology. *Cancer Res.* **2007**, *67*, 546–554. [CrossRef]
13. Tonelli, C.; Chio, I.I.C.; Tuveson, D.A. Transcriptional Regulation by Nrf2. *Antioxid. Redox. Signal.* **2018**, *29*, 1727–1745. [CrossRef] [PubMed]
14. Patel, S.; Khan, H.; Majumdar, A. Crosstalk between Sirtuins and Nrf2: SIRT1 Activators as Emerging Treatment for Diabetic Neuropathy. *Metab. Brain Dis.* **2022**, *37*, 2181–2195. [CrossRef] [PubMed]
15. Kelleher, F.C.; Rao, A.; Maguire, A. Circadian Molecular Clocks and Cancer. *Cancer Lett.* **2014**, *342*, 9–18. [CrossRef]
16. Van der Jeught, K.; Xu, H.-C.; Li, Y.-J.; Lu, X.-B.; Ji, G. Drug Resistance and New Therapies in Colorectal Cancer. *World J. Gastroenterol.* **2018**, *24*, 3834–3848. [CrossRef]
17. Lee, Y. Roles of Circadian Clocks in Cancer Pathogenesis and Treatment. *Exp. Mol. Med.* **2021**, *53*, 1529–1538. [CrossRef]
18. García-Costela, M.; Escudero-Feliú, J.; Puentes-Pardo, J.D.; San Juán, S.M.; Morales-Santana, S.; Ríos-Arrabal, S.; Carazo, Á.; León, J. Circadian Genes as Therapeutic Targets in Pancreatic Cancer. *Front. Endocrinol.* **2020**, *11*, 638. [CrossRef]
19. León, J.; Casado, J.; Ruiz, S.M.J.; Zurita, M.S.; González-Puga, C.; Rejón, J.D.; Gila, A.; de Rueda, P.M.; Pavón, E.J.; Reiter, R.J.; et al. Melatonin Reduces Endothelin-1 Expression and Secretion in Colon Cancer Cells through the Inactivation of FoxO-1 and NF-κβ. *J. Pineal Res.* **2014**, *56*, 415–426. [CrossRef]
20. Wei, J.-Y.; Li, W.-M.; Zhou, L.-L.; Lu, Q.-N.; He, W. Melatonin Induces Apoptosis of Colorectal Cancer Cells through HDAC4 Nuclear Import Mediated by CaMKII Inactivation. *J. Pineal Res.* **2015**, *58*, 429–438. [CrossRef]
21. Kannen, V.; Marini, T.; Zanette, D.L.; Frajacomo, F.T.; Silva, G.E.B.; Silva, W.A.; Garcia, S.B. The Melatonin Action on Stromal Stem Cells within Pericryptal Area in Colon Cancer Model under Constant Light. *Biochem. Biophys. Res. Commun.* **2011**, *405*, 593–598. [CrossRef] [PubMed]
22. León, J.; Casado, J.; Carazo, Á.; Sanjuán, L.; Maté, A.; Rueda, P.M.; de Cueva, P.; de la Quiles, R.; Ruíz, S.; Ruíz-Extremera, Á.; et al. Gender-Related Invasion Differences Associated with MRNA Expression Levels of Melatonin Membrane Receptors in Colorectal Cancer. *Mol. Carcinog.* **2012**, *51*, 608–618. [CrossRef] [PubMed]
23. Casado, J.; Iñigo-Chaves, A.; Jiménez-Ruiz, S.M.; Ríos-Arrabal, S.; Carazo-Gallego, Á.; González-Puga, C.; Núñez, M.I.; Ruíz-Extremera, Á.; Salmerón, J.; León, J. AA-NAT, MT1 and MT2 Correlates with Cancer Stem-Like Cell Markers in Colorectal Cancer: Study of the Influence of Stage and P53 Status of Tumors. *Int. J. Mol. Sci.* **2017**, *18*, 1251. [CrossRef] [PubMed]
24. Guardiola-Lemaitre, B.; De Bodinat, C.; Delagrange, P.; Millan, M.J.; Munoz, C.; Mocaër, E. Agomelatine: Mechanism of Action and Pharmacological Profile in Relation to Antidepressant Properties. *Br. J. Pharm.* **2014**, *171*, 3604–3619. [CrossRef] [PubMed]
25. Mairesse, J.; Silletti, V.; Laloux, C.; Zuena, A.R.; Giovine, A.; Consolazione, M.; Van Camp, G.; Malagodi, M.; Gaetani, S.; Cianci, S.; et al. Chronic Agomelatine Treatment Corrects the Abnormalities in the Circadian Rhythm of Motor Activity and Sleep/Wake Cycle Induced by Prenatal Restraint Stress in Adult Rats. *Int. J. Neuropsychopharmacol.* **2013**, *16*, 323–338. [CrossRef]
26. Kannen, V.; Bader, M.; Sakita, J.Y.; Uyemura, S.A.; Squire, J.A. The Dual Role of Serotonin in Colorectal Cancer. *Trends Endocrinol. Metab.* **2020**, *31*, 611–625. [CrossRef]
27. Soni, S.K.; Basu, P.; Singaravel, M.; Sharma, R.; Pandi-Perumal, S.R.; Cardinali, D.P.; Reiter, R.J. Sirtuins and the Circadian Clock Interplay in Cardioprotection: Focus on Sirtuin 1. *Cell. Mol. Life Sci.* **2021**, *78*, 2503–2515. [CrossRef]
28. Mahmoud, A.M.; Abd El-Ghafar, O.A.M.; Alzoghaibi, M.A.; Hassanein, E.H.M. Agomelatine Prevents Gentamicin Nephrotoxicity by Attenuating Oxidative Stress and TLR-4 Signaling, and Upregulating PPARγ and SIRT1. *Life Sci.* **2021**, *278*, 119600. [CrossRef]
29. Mahmoud, N.M.; Elshazly, S.M.; Hassan, A.A.; Soliman, E. Agomelatine Improves Streptozotocin-Induced Diabetic Nephropathy through Melatonin Receptors/SIRT1 Signaling Pathway. *Int. Immunopharmacol.* **2023**, *115*, 109646. [CrossRef]
30. Refinetti, R. Non-Parametric Procedures for the Determination of Phase Markers of Circadian Rhythms. *Int. J. Biomed. Comput.* **1992**, *30*, 49–56. [CrossRef]
31. Orozco-Solis, R.; Ramadori, G.; Coppari, R.; Sassone-Corsi, P. SIRT1 Relays Nutritional Inputs to the Circadian Clock through the Sf1 Neurons of the Ventromedial Hypothalamus. *Endocrinology* **2015**, *156*, 2174–2184. [CrossRef] [PubMed]
32. Longley, D.B.; Harkin, D.P.; Johnston, P.G. 5-Fluorouracil: Mechanisms of Action and Clinical Strategies. *Nat. Rev. Cancer* **2003**, *3*, 330–338. [CrossRef] [PubMed]
33. Violette, S.; Poulain, L.; Dussaulx, E.; Pepin, D.; Faussat, A.-M.; Chambaz, J.; Lacorte, J.-M.; Staedel, C.; Lesuffleur, T. Resistance of Colon Cancer Cells to Long-Term 5-Fluorouracil Exposure Is Correlated to the Relative Level of Bcl-2 and Bcl-X(L) in Addition to Bax and P53 Status. *Int. J. Cancer* **2002**, *98*, 498–504. [CrossRef] [PubMed]

34. Santoro, R.; Marani, M.; Blandino, G.; Muti, P.; Strano, S. Melatonin Triggers P53Ser Phosphorylation and Prevents DNA Damage Accumulation. *Oncogene* **2012**, *31*, 2931–2942. [CrossRef] [PubMed]
35. Song, J.; Ma, S.-J.; Luo, J.-H.; Zhang, H.; Wang, R.-X.; Liu, H.; Li, L.; Zhang, Z.-G.; Zhou, R.-X. Melatonin Induces the Apoptosis and Inhibits the Proliferation of Human Gastric Cancer Cells via Blockade of the AKT/MDM2 Pathway. *Oncol. Rep.* **2018**, *39*, 1975–1983. [CrossRef]
36. Samec, M.; Liskova, A.; Koklesova, L.; Zhai, K.; Varghese, E.; Samuel, S.M.; Šudomová, M.; Lucansky, V.; Kassayova, M.; Pec, M.; et al. Metabolic Anti-Cancer Effects of Melatonin: Clinically Relevant Prospects. *Cancers* **2021**, *13*, 3018. [CrossRef]
37. Winczyk, K.; Fuss-Chmielewska, J.; Lawnicka, H.; Pawlikowski, M.; Karasek, M. Luzindole but Not 4-Phenyl-2- Propionamidotetralin (4P-PDOT) Diminishes the Inhibitory Effect of Melatonin on Murine Colon 38 Cancer Growth In Vitro. *Neuro Endocrinol. Lett.* **2009**, *30*, 657–662.
38. Karasek, M.; Carrillo-Vico, A.; Guerrero, J.M.; Winczyk, K.; Pawlikowski, M. Expression of Melatonin MT(1) and MT(2) Receptors, and ROR Alpha(1) Receptor in Transplantable Murine Colon 38 Cancer. *Neuro Endocrinol. Lett.* **2002**, *23*, 55–60.
39. Liu, J.; Clough, S.J.; Hutchinson, A.J.; Adamah-Biassi, E.B.; Popovska-Gorevski, M.; Dubocovich, M.L. MT1 and MT2 Melatonin Receptors: A Therapeutic Perspective. *Annu. Rev. Pharm. Toxicol.* **2016**, *56*, 361–383. [CrossRef]
40. Diez-Echave, P.; Vezza, T.; Algieri, F.; Ruiz-Malagón, A.J.; Hidalgo-García, L.; García, F.; Morón, R.; Sánchez, M.; Toral, M.; Romero, M.; et al. The Melatonergic Agonist Agomelatine Ameliorates High Fat Diet-Induced Obesity in Mice through the Modulation of the Gut Microbiome. *Biomed. Pharm.* **2022**, *153*, 113445. [CrossRef]
41. Kim, H.K.; Yang, K.I. Melatonin and Melatonergic Drugs in Sleep Disorders. *Transl. Clin. Pharm.* **2022**, *30*, 163–171. [CrossRef] [PubMed]
42. Liu, Z.; Zou, D.; Yang, X.; Xue, X.; Zuo, L.; Zhou, Q.; Hu, R.; Wang, Y. Melatonin Inhibits Colon Cancer RKO Cell Migration by Downregulating Rho-associated Protein Kinase Expression via the P38/MAPK Signaling Pathway. *Mol. Med. Rep.* **2017**, *16*, 9383–9392. [CrossRef] [PubMed]
43. Kast, R.E. Agomelatine or Ramelteon as Treatment Adjuncts in Glioblastoma and Other M1- or M2-Expressing Cancers. *Contemp. Oncol. Współczesna Onkol.* **2015**, *2*, 157–162. [CrossRef] [PubMed]
44. Siddiqui, E.J.; Shabbir, M.A.; Mikhailidis, D.P.; Mumtaz, F.H.; Thompson, C.S. The Effect of Serotonin and Serotonin Antagonists on Bladder Cancer Cell Proliferation. *BJU Int.* **2006**, *97*, 634–639. [CrossRef]
45. Pantaleo, M.A.; Palassini, E.; Labianca, R.; Biasco, G. Targeted Therapy in Colorectal Cancer: Do We Know Enough? *Dig. Liver Dis.* **2006**, *38*, 71–77. [CrossRef]
46. Hill, S.M.; Belancio, V.P.; Dauchy, R.T.; Xiang, S.; Brimer, S.; Mao, L.; Hauch, A.; Lundberg, P.W.; Summers, W.; Yuan, L.; et al. Melatonin: An Inhibitor of Breast Cancer. *Endocr. Relat. Cancer* **2015**, *22*, R183–R204. [CrossRef]
47. de Bodinat, C.; Guardiola-Lemaitre, B.; Mocaër, E.; Renard, P.; Muñoz, C.; Millan, M.J. Agomelatine, the First Melatonergic Antidepressant: Discovery, Characterization and Development. *Nat. Rev. Drug Discov.* **2010**, *9*, 628–642. [CrossRef]
48. Tardito, D.; Molteni, R.; Popoli, M.; Racagni, G. Synergistic Mechanisms Involved in the Antidepressant Effects of Agomelatine. *Eur. Neuropsychopharmacol.* **2012**, *22*, S482–S486. [CrossRef]
49. Jung-Hynes, B.; Reiter, R.J.; Ahmad, N. Sirtuins, Melatonin and Circadian Rhythms: Building a Bridge between Aging and Cancer. *J. Pineal Res.* **2010**, *48*, 9–19. [CrossRef]
50. Ríos-Arrabal, S.; Muñoz-Gámez, J.A.; Jiménez-Ruíz, S.M.; Casado-Ruíz, J.; Artacho-Cordón, F.; León-López, J. *Circadian Regulation of Colon Cancer Stem Cells: Implications for Therapy*; IntechOpen: London, UK, 2016; ISBN 978-953-51-2545-7.
51. Mayo, J.C.; Sainz, R.M.; González Menéndez, P.; Cepas, V.; Tan, D.-X.; Reiter, R.J. Melatonin and Sirtuins: A "Not-so Unexpected" Relationship. *J. Pineal Res.* **2017**, *62*, e12391. [CrossRef]
52. Ray, S.; Valekunja, U.K.; Stangherlin, A.; Howell, S.A.; Snijders, A.P.; Damodaran, G.; Reddy, A.B. Circadian Rhythms in the Absence of the Clock Gene Bmal1. *Science* **2020**, *367*, 800–806. [CrossRef] [PubMed]
53. Miki, T.; Matsumoto, T.; Zhao, Z.; Lee, C.C. P53 Regulates Period2 Expression and the Circadian Clock. *Nat. Commun.* **2013**, *4*, 2444. [CrossRef] [PubMed]
54. Nakahata, Y.; Kaluzova, M.; Grimaldi, B.; Sahar, S.; Hirayama, J.; Chen, D.; Guarente, L.P.; Sassone-Corsi, P. The NAD+-Dependent Deacetylase SIRT1 Modulates CLOCK-Mediated Chromatin Remodeling and Circadian Control. *Cell* **2008**, *134*, 329–340. [CrossRef] [PubMed]
55. Torres, F.F.; Bernardo, V.S.; de Paula, C.P.; da Silva, J.P.M.D.O.; de Almeida, E.A.; da Cunha, A.F.; da Silva, D.G.H. Influence of Melatonin Treatment on Cellular Mechanisms of Redox Adaptation in K562 Erythroleukemic Cells. *Genes* **2022**, *13*, 2337. [CrossRef]
56. Sun, X.-X.; Dai, M.-S.; Lu, H. 5-Fluorouracil Activation of P53 Involves an MDM2-Ribosomal Protein Interaction. *J. Biol. Chem.* **2007**, *282*, 8052–8059. [CrossRef]
57. Bunz, F.; Hwang, P.M.; Torrance, C.; Waldman, T.; Zhang, Y.; Dillehay, L.; Williams, J.; Lengauer, C.; Kinzler, K.W.; Vogelstein, B. Disruption of P53 in Human Cancer Cells Alters the Responses to Therapeutic Agents. *J. Clin. Investig.* **1999**, *104*, 263–269. [CrossRef]

58. Stopper, H.; Garcia, S.B.; Waaga-Gasser, A.M.; Kannen, V. Antidepressant Fluoxetine and Its Potential against Colon Tumors. *World J. Gastrointest. Oncol.* **2014**, *6*, 11–21. [CrossRef]
59. Zheng, Y.; Chang, X.; Huang, Y.; He, D. The Application of Antidepressant Drugs in Cancer Treatment. *Biomed. Pharm.* **2023**, *157*, 113985. [CrossRef]

antioxidants MDPI

Review

Melatonin as Modulator for Sulfur and Nitrogen Mustard-Induced Inflammation, Oxidative Stress and DNA Damage: Molecular Therapeutics

Eva Ramos [1], Emilio Gil-Martín [2], Cristóbal De Los Ríos [3,4], Javier Egea [5], Francisco López-Muñoz [6,7], René Pita [8], Antonio Juberías [9], Juan J. Torrado [10], Dolores R. Serrano [10], Russel J. Reiter [11] and Alejandro Romero [1,*]

[1] Department of Pharmacology and Toxicology, Faculty of Veterinary Medicine, Complutense University of Madrid, 28040 Madrid, Spain
[2] Department of Biochemistry, Genetics and Immunology, Faculty of Biology, University of Vigo, 36310 Vigo, Spain
[3] Health Research Institute, Hospital Universitario de la Princesa, 28006 Madrid, Spain
[4] Departamento de Ciencias Básicas de la Salud, Universidad Rey Juan Carlos, 28922 Alcorcón, Spain
[5] Molecular Neuroinflammation and Neuronal Plasticity Research Laboratory, Hospital Universitario Santa Cristina, Instituto de Investigación Sanitaria-Hospital Universitario de la Princesa, 28006 Madrid, Spain
[6] Faculty of Health, Camilo José Cela University of Madrid (UCJC), 28692 Madrid, Spain
[7] Neuropsychopharmacology Unit, Hospital 12 de Octubre Research Institute, 28041 Madrid, Spain
[8] Chemical Defense Department, Chemical, Biological, Radiological, and Nuclear Defense School, Hoyo de Manzanares, 28240 Madrid, Spain
[9] Dirección de Sanidad Ejército del Aire, Cuartel General Ejército del Aire, 28008 Madrid, Spain
[10] Department of Pharmaceutics and Food Technology, Complutense University of Madrid, 28040 Madrid, Spain
[11] Department of Cell Systems and Anatomy, UT Health, San Antonio, TX 78229, USA
* Correspondence: manarome@ucm.es; Tel.: +34-913943970

Citation: Ramos, E.; Gil-Martín, E.; De Los Ríos, C.; Egea, J.; López-Muñoz, F.; Pita, R.; Juberías, A.; Torrado, J.J.; Serrano, D.R.; Reiter, R.J.; et al. Melatonin as Modulator for Sulfur and Nitrogen Mustard-Induced Inflammation, Oxidative Stress and DNA Damage: Molecular Therapeutics. *Antioxidants* **2023**, *12*, 397. https://doi.org/10.3390/antiox12020397

Academic Editors: Marina Garcia-Macia and Adrián Santos-Ledo

Received: 10 December 2022
Revised: 27 January 2023
Accepted: 4 February 2023
Published: 6 February 2023

Abstract: Sulfur and nitrogen mustards, bis(2-chloroethyl)sulfide and tertiary bis(2-chloroethyl) amines, respectively, are vesicant warfare agents with alkylating activity. Moreover, oxidative/nitrosative stress, inflammatory response induction, metalloproteinases activation, DNA damage or calcium disruption are some of the toxicological mechanisms of sulfur and nitrogen mustard-induced injury that affects the cell integrity and function. In this review, we not only propose melatonin as a therapeutic option in order to counteract and modulate several pathways involved in physiopathological mechanisms activated after exposure to mustards, but also for the first time, we predict whether metabolites of melatonin, cyclic-3-hydroxymelatonin, N1-acetyl-N2-formyl-5-methoxykynuramine, and N1-acetyl-5-methoxykynuramine could be capable of exerting a scavenger action and neutralize the toxic damage induced by these blister agents. NLRP3 inflammasome is activated in response to a wide variety of infectious stimuli or cellular stressors, however, although the precise mechanisms leading to activation are not known, mustards are postulated as activators. In this regard, melatonin, through its anti-inflammatory action and NLRP3 inflammasome modulation could exert a protective effect in the pathophysiology and management of sulfur and nitrogen mustard-induced injury. The ability of melatonin to attenuate sulfur and nitrogen mustard-induced toxicity and its high safety profile make melatonin a suitable molecule to be a part of medical countermeasures against blister agents poisoning in the near future.

Keywords: sulfur and nitrogen mustard; melatonin; oxidative stress; melatonin metabolites; NLRP3 inflammasome; DNA damage

1. Introduction

Sulfur mustards, especially bis(2-chloroethyl) sulfide, known as yperite or mustard gas, and nitrogen mustards, with similar properties to those of their sulfur analogues, being bis(2-chloroethyl)methylamine (HN2) the best known, are the most important blister agents, but not because they are considered a chemical weapon, but because of their use in cancer treatment. These substances are included in the lists of substances subject to CWC verification inspections [1]. In reference to the molecular mechanisms of toxicity induced by blister agents, the following have been reported: the formation of DNA double-strand breaks [2], alkylation of cellular macromolecules [3], activation of Poly ADP ribose polymerase-1 (PARP-1) [4], oxidative stress and generation of reactive oxygen and nitrogen species (RONS) [5], dysregulation of intracellular Ca^{+2} [6], inflammation [7], proteolytic activation [8], and epigenetic modifications [9] have been reported. Unfortunately, immediate total decontamination after mustards exposure is difficult to achieve, and there are not any completely effective and/or efficient antidotes and treatments do not exist. In this complex scenario, the use of a broad-spectrum agent, with a high safety profile and the ability to act on multiple intracellular signaling pathways favoring cell survival, would be a good therapeutic strategy to counteract vesicant agents-induced damage. In this regard, we proposed that melatonin, due to its wide range of biological actions, serving as an indirect antioxidant and free radical scavenger, anti-inflammatory and immunomodulatory agent, or as an epigenetic modulator, and taking into account its low toxicity and high efficacy in reducing oxidative damage and improving human health, should be considered as a strong candidate against exposure and toxicity caused by the most widely used blister agents.

Melatonin and its metabolites, such as [cyclic-3-hydroxymelatonin (c3OHM), N1-acetyl-N2-formyl-5-methoxykynuramine (AFMK), and N1-acetyl-5-methoxykynuramine (AMK)] exert cell protection against oxidative stress, scavenging and inhibiting the generation of free radicals. Therefore, in this review, we propose mechanisms by which some of these metabolites could mitigate the toxic events elicited by blister agents. Consequently, this protection would further enhance the therapeutic profile of melatonin.

Another important target activated in response to vesicants is the NLRP3 inflammasome [10], which is associated with a large number of diseases, including inflammatory diseases, metabolic pathologies and carcinogenesis. In this context, melatonin inhibits the NLRP3 inflammasome activity [11] suggesting a new strategy for protecting against blister agents-induced cell death.

Current understanding of the toxicology associated with exposure to vesicant agents is insufficient to explain, in mechanistic terms, their long-term pathology. In this sense, the effects of variable severity that can occur in the long term after acute intoxication by blister agents, makes us think, once again, on the use of melatonin to modulate epigenetic mechanisms in the clinical treatment of exposed patients.

The purpose of this review article is to highlight the protective role of melatonin and its metabolites in counteracting sulfur and nitrogen mustard-induced damage.

2. Molecular Toxicity Mechanisms of Sulfur and Nitrogen Mustards

The toxicological mechanism of action of vesicants (blistering agents), including the cytotoxic vesicating sulfur and nitrogen mustards, is related to their high reactivity with proteins, DNA, and other cellular components. However, oxidative stress induction (glutathione depletion, lipid peroxidation and reactive oxygen and nitrogen species generation), the activation of poly (ADP-ribose) polymerases (PARPs) (NAD+ depletion and decrease of ATP production), mitochondrial disruption, changes in matrix metalloproteinase-9 (MMP-9) expression, intracellular Ca^{+2} overload, epigenetic mechanisms (long-term toxicity), and programmed cell death, are significant events involved in the molecular toxicity of these vesicating agents [12].

The disturbances between reactive oxygen and nitrogen species (RONS) generation and antioxidant defense mechanisms balance, after acute toxicity of sulfur and nitrogen mustards, lead to oxidative stress [13], which is a crucial event of the pathological process causing oxidation of macromolecules, including proteins, nucleic acids and lipids. Currently, all efforts in drug development are directed towards the attenuation of DNA alkylating capacity. However, the generation of RONS and establishment of a scenario of oxidative stress after vesicants exposure, is a more immediate event (hours) than the formation of guanine adducts in contaminated individuals (weeks) [14]. Therefore, in a first approach to reduce sulfur and nitrogen mustards-induced toxicity, the main target would be to counteract oxidative stress.

Production and accumulation of mitochondrial RONS induced by blister agents lead to mitochondrial disruption, mtDNA damage and inhibition of mitochondrial oxidative phosphorylation (OXPHOS) complex inhibition [15]. Recently, Meng et al. [16] have evidenced the presence of a sulfur mustard in the mitochondria of living cells, which indicates that mustards can exert toxicity in cytoplasmic organelles, particularly, dysfunction of mitochondrial dynamics [17]. This increases RONS producing-related enzymes, such as aldehyde oxidase-1 (AOX1), dual oxidases (DUOXs), inducible nitric oxide synthase (iNOS) and thyroid peroxidase (TPO) by reducing both activities of the cytosolic antioxidant enzymes superoxide dismutase (SOD), catalase (CAT) and intercellular glutathione (GSH) [18]. Furthermore, the thioredoxin system and NADPH, critical elements in the protection of all living cells, are necessary for controlling the antioxidant defense system against oxidative stress and protein folding through thiol redox control. In this regard, the nitrogen mustard-derivative mechlorethamine (HN2) inhibits the thioredoxin system inducing overload of RONS and, subsequently, oxidative stress and cellular damage [19]. In addition, RONS can also induce lipid peroxidation (LPO), which leads to LPO–DNA adducts in targeted tissues exposed to mustards [20]. Moreover, an increase in the production of superoxide anions ($O_2^{\bullet-}$) from mitochondria associated with an uncoupler in the mitochondrial respiratory chain has been observed after mustards exposure [21]. These events compromise the activity of antioxidant defense systems and, therefore, cellular survival.

The most significant effect of mustards in a low dose of exposure is DNA alkylation and cross-linking. PARP-1, in response to DNA damage, uses NAD+ as a substrate to catalyze the covalent binding of poly-ADP-ribose (PAR) and some other nuclear proteins involved in DNA repair. Subsequently, overactivation of PARP-1 and depletion of NAD+-dependent nuclear enzymes by the complex mustards–DNA adducts, without recruitment of the repair system, triggers cell death via necrosis/apoptosis and autophagy [22]. PARP-1 activation is involved in DNA repair; however, overactivation by mustards-induced DNA damage coupled with fast consumption of the NAD+ leads to ATP depletion and cell death [4]. In this respect, intracellular ATP levels act as a cellular sensor to switch the apoptotic or necrotic pathways in response to mustard injury, which may affect the microfilament architecture. In this line, the cytoskeleton organization disruption by exposure to bifunctional alkylating agents sulfur [23] and nitrogen mustards [24], interfere with the mechanisms required for homeostasis maintenance and actin filament cell morphology. It has been proposed that after mustard exposure, PARP-1 is activated and mediates $ONOO^-$ induced necrosis; under conditions of low level cell damage, PARP-1 allows DNA repair and cell recovery [25]. This may lead to the delayed toxicity of mustards since cells are able to divide but not be free of damage.

Matrix metalloproteinases (MMPs) are a family of calcium-dependent zinc-binding proteolytic enzymes essential for the remodeling of extracellular matrix (ECM). Gelatinase MMP-9, which promotes proinflammatory pathways and degrades ECM constituents, is up-regulated in lung tissue following sulfur mustard exposure and aggravates the pathogenesis during the progression of a disease [8]. More recently, sulfur mustard was reported to induce corneal structural damage through changes in gene expression of MMPs [26]. Therefore, extracellular matrix proteins and matrix metalloproteinases could be a goal of strategy as a direct therapeutic target against vesicant injury in the ocular tissue. In this context, topical exposure to nitrogen mustard significantly up-regulates MMP9 via MAPK/Akt-AP1 pathway, increasing vesicant-induced skin damage [27].

Intracellular calcium $[Ca^{2+}]_i$ overload is another molecular toxicity mechanism by which vesicants produce cell damage at the acute phase [5]. Ca^{+2} is an essential second messenger for cell homeostasis maintenance and its disturbance provokes triggering cellular pathways that contribute to cytotoxicity. Thus, intracellular $[Ca^{2+}]_i$ overload after nitrogen mustard exposure triggers autophagy through the TRPV1-Ca^{2+}-CaMKKβ-AMPK-ULK1 signaling pathway [28]. Likewise, sulfur mustard is able to increase $[Ca^{2+}]_i$ mediated by transient receptor potential ankyrin 1 (TRPA1) ion channel/ heat shock 70 kDa protein 6 (HSPA6)-induction [29], which leads to oxidative stress and stimulates cell death.

Long-term toxicity of mustards affects the quality of life of patients leading to neurobehavioral impairment, cognitive disorders, and severe depression, among others [30]. Furthermore, mustards induce epigenetic modifications without altering the primary DNA nucleotide sequence in cells and tissues exerted by modulators such as histone acetyltransferases (HATs), histone deacetylases (HDACs) and DNA methyltransferases (DNMTs), which play crucial roles in histone acetylation and deacetylation, modulation of chromatin and DNA expression, and cytosine methylation, respectively. Regarding this issue, sulfur mustard induces epigenetic disturbances through DNA methylation and acetylation both in endothelial cells and in vivo skin samples [31]. Dysregulation of HATs and HDACs was also observed after 24 h at low and high dose of sulfur mustard exposure [32]. Regarding the post-transcriptional epigenetic modifications, increasing in serum levels, non-coding microRNAs (miRNAs) were found in humans [33,34] and in rats [35] after sulfur and nitrogen mustard exposure. In this sense, we believe that further clarification of epigenetic mechanisms of mustards may be useful in the development of new therapeutic options.

3. Protective Cellular and Molecular Mechanisms of Melatonin against Sulfur and Nitrogen Mustard-Induced Oxidative Stress

As mentioned above, oxidative stress (OS) events have been proposed to play a major role after sulfur and nitrogen mustard exposure. Indeed, after mustard exposure, the induced oxidative environment (ROS, NO and $ONOO^-$), including GSH depletion, triggers its detrimental effects cascade [25]. Consequently, a molecule with an effective antioxidant activity would be of therapeutic interest to counteract mustard effects against this key toxic event [36]. Noteworthy, conventional antioxidants cannot remove $ONOO^-$. In this regard, melatonin emerges as a promising candidate for a medical countermeasure, with unique features and pleiotropic activities including its well-known antioxidant properties and RONS scavenging potential. Melatonin exerts these effects through several mechanisms [37,38]. Antioxidant therapy is intimately related to inflammation modulation, as NOS inhibitors effectively counteract OS and OS seems to trigger an inflammatory cascade [39].

As a consequence of the described protective effects of melatonin, several authors have studied its antioxidant potency versus mustards, its capacity to detoxify free radicals, as well as other antioxidant properties in rats (Table 1).

The exposure to mustards reduces glutathione peroxidase (GPx) activity which leads to an OS environment. A pre-administration of melatonin in nitrogen mustard-exposed animals protected from GPx loss in a dose-dependent manner [40]. This is quite relevant because while mustards did not diminish significantly SOD levels, other OS markers were altered, i.e., MDA or GPx, which denotes that an OS environment is present in exposed animals. Remarkably, melatonin enhanced SOD activity despite it not previously diminished by mustards [40]. Melatonin showed significant protection for increased oxidized glutathione (GSSG) levels in addition to counteracting the decrease of GSH after nitrogen mustard administration [41]. These results reveal that melatonin's capacity to neutralize RONS is extensive and complex and, in particular, can directly and immediately combat the short-term damage caused by acute exposure to mustards, as previously reported [14,18].

After nitrogen mustard exposure, RONS overload induces LPO, which is estimated by measuring Malondialdehyde (MDA) or Thiobarbituric acid reactive substance (TBARS) levels, as LPO markers. A study determined that mustard-exposed animals developed strong oxidative stress increasing MDA levels, and melatonin administration significantly diminished MDA levels [40,41], which contributed to membrane protection from LPO. Whereas Pohanka et al. [42] observed that, despite sulfur mustard not seeming to significantly induce changes in plasma levels of LPO compared to the control, when melatonin was previously administered to animals, the LPO marker TBARS was approximately three folds lower, indicating the significative antioxidant activity exerted by melatonin, reducing LPO (Table 1).

Nitrosative stress markers are just as relevant as oxidative markers. In this regard, there is a large body of evidence that melatonin has an important role in counteracting an increased activity of reactive nitrogen species (RNS) producing related enzymes, as well as selectively inhibiting iNOS [43–45]. There is a strong nitrosative stress induction after nitrogen mustard exposure, the expression of iNOS increases generating nitric oxide (NO), which is an unstable nitrogen radical related to $ONOO^-$ formation by reacting with superoxide anion ($O_2^{\bullet -}$). Both NO and $ONOO^-$ induce cytotoxicity, modifying membrane lipids, proteins, and DNA covalently [40,44,46,47]; the protection of melatonin at this stage will be extensively discussed in Section 5. Nevertheless, the administration of melatonin seems to reduce nitrosative stress markers (Table 1), measured as the urinary excretion of NO metabolites, nitrite–nitrate (NOx), which is in accordance with the observed inhibition of iNOS activity suppression in treated animals [40,47,48]. Similarly, mice administered intraperitoneally with a single toxic dose of alkylating cyclophosphamide (CP; 200 mg/kg) developed intense oxidative stress in lung homogenates (reducing GSH levels and SOD and CAT antioxidant enzymes), while pre-treatment for seven consecutive days with melatonin (2.5–20 mg/kg) quenched lipid peroxidation and restored normal oxidative parameters [49].

Table 1. In vitro and in vivo studies about the protective actions of melatonin against sulfur and nitrogen mustard-induced damage.

Model	Blister Agent (Dose)	Melatonin Dose	Results	Reference
Wistar Rats	Mechlorethamine (0.5 mg/kg)	20 and 40 mg/kg (pretreatment)	↓ LPO (MDA)	[40]
Wistar Rats	Mechlorethamine (0.5 mg/kg)	40 mg/kg (pretreatment)	↑ SOD activity	[40]
Wistar Rats	Mechlorethamine (0.5 mg/kg)	20 and 40 mg/kg (pretreatment)	↑ GPx	[40]
Wistar Rats	Mechlorethamine (0.5 mg/kg)	40 mg/kg (pretreatment)	↓ NOx	[40]
Wistar Rats	Sulfur mustard (20 and 80 mg/kg)	25 and 50 mg/kg (pretreatment)	↓ LPO (TBARS)	[42]
Wistar Rats	Sulfur mustard (20 and 80 mg/kg)	25 and 50 mg/kg (pretreatment)	↑antioxidant power (FRAP)	[42]
Sprague-Dawley rats	Mechlorethamine (3.5 mg/kg)	100 mg/kg/12 h (6 doses post-exposure)	↓ NOx	[47,48]
Swiss mice	Mechlorethamine (10 mg/kg)	250 mg/kg	↓ GSSG	[41]
Swiss mice	Mechlorethamine (5 and 10 mg/kg)	250 mg/kg	↓ LPO (MDA)	[41]
Swiss mice	Mechlorethamine (5 mg/kg)	250 mg/kg	↑ GSH	[41]

Malondialdehyde (MDA), Superoxide dismutase (SOD), Glutathione peroxidase (GPX), Lipid peroxidation (LPO), Nitrosative stress markers nitrite-nitrate (NOx), Thiobarbituric acid reactive substance (TBARS), Ferric-reducing antioxidant power (FRAP), reduced and oxidized glutathione (GSSG), glutathione (GSH),↑ increase, ↓ decrease.

The studies mentioned in this section treated animals with doses of melatonin from 20 mg/kg to 250 mg/kg (Table 1). Undoubtedly, further investigations are needed to establish the effective dose for each of the wide variety of possible treatments as a preventive measure, and for acute or delayed mustard toxicity.

When melatonin antioxidant power has been studied (Table 1), with the ferric-reducing antioxidant power (FRAP) assay in plasma, a significant situation has been found. While untreated melatonin rats did not present changes, the FRAP for Sulfur mustard + melatonin animals were significantly improved, and this increment was significantly higher compared to only melatonin-treated rats. This leads the authors to think that melatonin's antioxidant power involves other factors and does not only respond to concentration [42].

Therefore, the described free radical scavenging capacity of melatonin would facilitate the neutralization of sulfur and nitrogen mustards-induced oxidative damage. In this regard, we have previously proposed three mechanisms to achieve melatonin's protective action: hydrogen-atom transfer (HAT), single electron transfer (SET) and radical adduct formation (RAF) [50,51]. Moreover, melatonin metabolites, c3OHM, AFMK and AMK are also capable of exerting a scavenger action [52]. Therefore, and considering the ability of melatonin metabolites to quench free radicals and thus protect against oxidative damage, we herein propose the possible mechanisms by which some of them could mitigate the toxic events elicited by blister agents such as the nitrogen mustard-2 (HN2) (Figure 1).

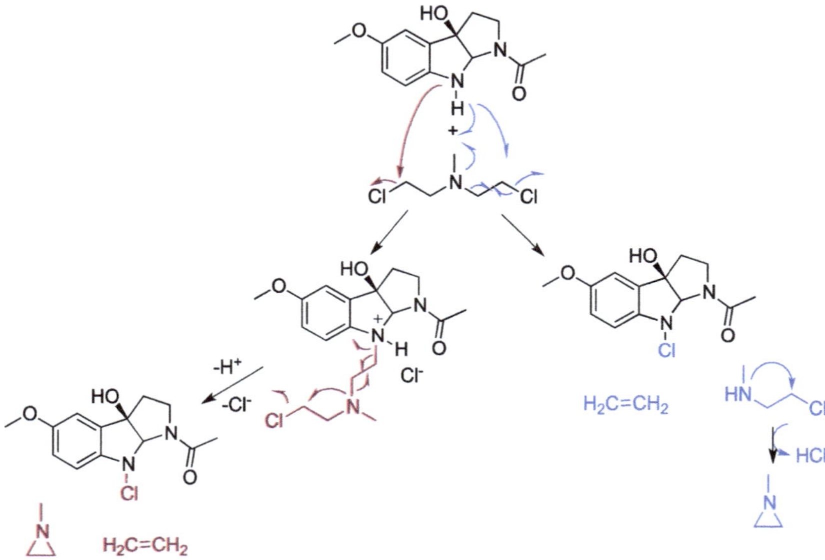

Figure 1. Two proposed mechanisms that may explain the quenching ability of c3OHM on HN2.

The melatonin metabolite c3OHM may trap HN2 by nucleophilic substitution, releasing a chloride ion. Otherwise, this chloride ion could leave previously due to an intramolecular nucleophilic attack of the tertiary amine, being then the target for the nucleophilic substitution of the N-methylaziridinium cation. In both cases, the pending alkylating agent may decompose to form ethylene and the less toxic N-methylaziridine, while the chlorine radical is captured by the c3OHM (purple mechanism, Figure 1). Alternatively, this metabolite of melatonin would furnish radical substitution over HN2, taking chlorine radical and provoking the rapid decomposition of this blistering agent to form similar products (Figure 1, blue mechanism). On the other hand, the AFMK metabolite of melatonin does not seem to have a chemical nature to quench this blistering agent. Conversely, the metabolite AMK, despite its soft nucleophilic nature, would develop radical substitution over HN2 to bind chlorine in that step, and thus, inducing the decomposition of this blistering agent in a similar fashion to the c3OHM (Figure 2).

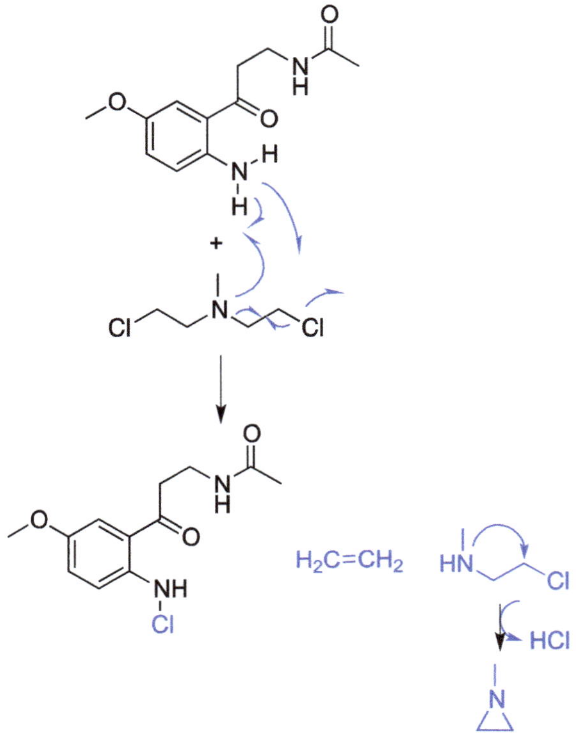

Figure 2. Proposed mechanism of HN2 quenching by the melatonin metabolite AMK.

Similar behavior may be displayed by melatonin metabolites quenching other blistering agents, such as sulfur mustards, which have similar toxic actions. For these mustards, a relevant toxicity mechanism is the formation of alkylthiiranium (namely cyclic alkylsulfonium), which rapidly reacts with water producing thiodiglycol. Consequently, it is essential that these active metabolites of melatonin, c3OHM or AMK, dissipate the sulfur mustard before the alkylthiiranium generation avoiding, thus, this toxic mechanism (Figure 3).

Subsequently, and taking into account the review by Tan et al. [53], the capacity of melatonin to counteract the oxidative environment induced by mustards does not only lie in melatonin as the parent compound, but also in the endless fashion due to the action of many melatonin metabolites, i.e., the antioxidant cascade. Together with its unique features, its metabolite activities may reinforce the potency of melatonin reverting a mustard's toxic actions compared to other classical antioxidants.

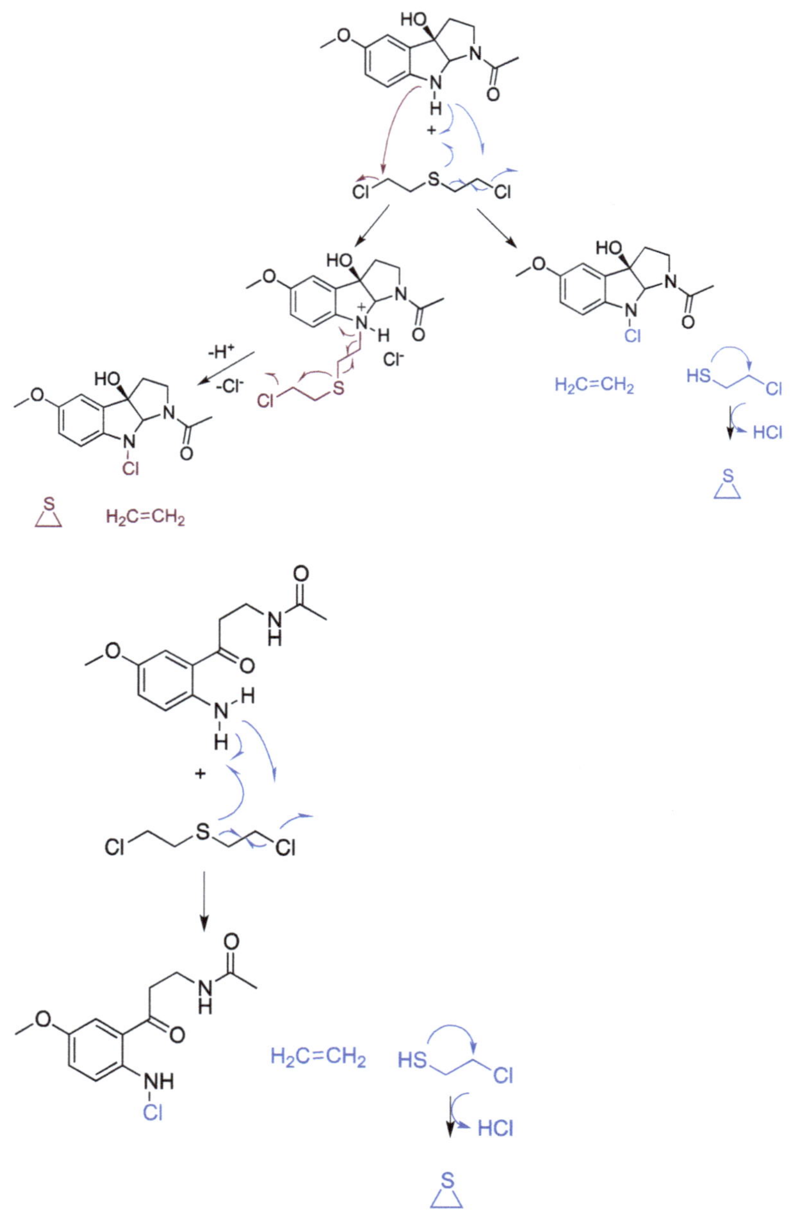

Figure 3. Proposed scavenging actions and mechanisms of breakdown of sulfur mustard by melatonin metabolites.

4. Sulfur and Nitrogen Mustard-Induced Inflammation: Therapeutic Regulation of the NLRP3 Inflammasome by Melatonin

As previously mentioned, sulfur and nitrogen mustards are bifunctional lipophilic alkylating agents that rapidly penetrate tissues and cells and react with sulfhydryl, carboxyl, and aliphatic amino groups to form stable adducts [54]. This causes oxidative and nitrosative stress, impairs cell function, causes DNA damage, and triggers cell death, apoptosis and autophagy [55,56]. In experimental models and humans, mustard-induced acute and long-term injury is associated with RONS, which are closely associated with mustard-induced inflammation. DNA-damaged cells undergo activation of various signaling pathways, such as PARP, p53 and NF-κB, which play an important role in DNA damage repair, cell cycle arrest and/or inflammation [57]. It has been shown that exposure to mustards increases p53 levels and p53 phosphorylation, mediating early apoptosis and inflammation [58,59]. Apoptotic cell death itself can activate innate immune responses and increase inflammation [60]. Indeed, cell death induced by mustard exposure produces the release of various pro-inflammatory mediators, such as TNF-α, IL-1α or IL-1β, and stimulates the activation of resident macrophages and mast cells, hence activating an immune response. These inflammatory cells also release inflammatory mediators that activate neutrophil extravasation and accumulation at the injured area [61]. After tissue infiltration, neutrophils generate chemotactic signals to attract monocytes and macrophages to the injured area. Depending on the stage of injury, these cells may exhibit a proinflammatory phenotype or aid in wound healing. Tissue repair involves the phagocytosis of apoptotic neutrophils, dead cells and debris at the site of injury and the release of various growth factors that may help promote cell proliferation and extracellular matrix synthesis [62]. However, during prolonged tissue stress, these cells may serve as a source of inflammatory mediators and cytokines that may support further neutrophil infiltration into the injury site [62]. Exposure of rats to vesicants results in marked increases in iNOS, COX-2 and TNFα positive-macrophages in the lung [63]. Moreover, mustards induce the increase of pro-inflammatory cytokines such as IL-6, IL-8, IL-12, and the fibrogenic cytokine transforming growth factor (TGF) β [64].

NOD-like receptor family pyrin domain containing 3 (NLRP3) inflammasome is the most widely characterized inflammasome that activates caspase-1 and induces IL-1β release. NLRP3 inflammasome is an important step in innate immune responses. Recently, the NLRP3 inflammasome was shown to play an important role in vesicant-induced cutaneous inflammation [10]. The authors describe that nitrogen mustard activates NLRP3 inflammasome through the SIRT3–SOD2–mtROS signaling pathway. The NLPR3 inflammasome has recently been implicated in the immune pathogenesis of several diseases including cardiac, gastrointestinal, pulmonary, metabolic and neurodegenerative diseases. In this regard, inhibition of NLRP3 inflammasome activation has been proposed as a promising novel therapeutic target for inflammation-related disorders. Therefore, mtROS-dependent activation of the NLRP3 inflammasome may be an excellent target to counteract vesicant exposure-induced damage.

Melatonin pleiotropy effects (free-radical scavenger, antioxidant, cytoprotective, anti-inflammatory, oncostatic, anti-aging, immunomodulatory), includes inhibition of NLRP3 inflammasome. ROS is a main trigger of NLRP3 inflammasome activation [65]. Melatonin reduces ROS production in various in vitro models through its antioxidant effect modifying several antioxidant proteins or its ability to scavenge free radicals. In this sense, melatonin reduces NLRP3 inflammasome activation through Nrf2-mediated ROS scavenging [66,67]. Moreover, melatonin diminishes the levels of TXNIP, suppressing ROS production and NLRP3 activity [68]. NF-κB is a key regulator of the first phase (priming phase) of NLRP3 inflammasome activation. Melatonin prevents NLRP3 inflammasome activation by inhibiting NF-κB signaling induced by LPS administration [69]. It is well-known that autophagy is a negative regulator for NLRP3 activation. Recently, our research group showed that LPS inhibits autophagy and increases NLRP3 protein levels and NLRP3 inflammasome activation. Melatonin reversed LPS-induced cognitive decline, decreased NLRP3 levels and promoted autophagic flux in mice [67]. Additionally, melatonin increased of LC3-II/LC3-I, Atg 5 expression and suppressed NLRP3 inflammasome activation in a model of subarachnoid hemorrhage [70]. Furthermore, the regulatory function of melatonin in NLRP3 inflammasome activation in part occurs through post-transcriptional mechanisms, by altering the expression of miRNAs and long noncoding RNAs [71]. In recent years, different groups have proven the modulatory effect of melatonin on the inflammatory response. Hence, on the one hand the NLRP3 inflammasome plays an important role in vesicant-induced inflammation and, on the other hand, melatonin exerts a strong regulatory effect on NLRP3 inflammasome activation. In view of these evidences, this indoleamine may serve as a good therapeutic candidate against vesicant-induced inflammation.

5. Role of Melatonin in Counteracting Sulfur and Nitrogen Mustard-Induced DNA Damage

The poisoning by mustards and the subsequent casualties and incapacitating toxicity produced are dire since none of their principal acute effects, including skin blistering, eye disease and acute respiratory distress [72], has a palliative agent [73]. Furthermore, the prolongation of sequelae over time in a pathological network of systemic involvement [74] puts us in the position of finding antidote-based pleiotropic remedies that address the multiplicity of actions gathered by vesicants. In this regard, the spectrum of signaling deployed by melatonin and its metabolites [75,76], as well as their wide distribution in nuclei and organelles [77,78], are consistent with the countermeasures demanded by mustard-induced dysfunctionalities [79]. Specifically, this indoleamine appears to directly remove DNA damage from physical and chemical aggressors [76]. The melatonin cascade also enables extremely efficient electron transfer [80] to scavenge most RONS [25,53,81,82] and the $ONOO^-$ [83,84]. In addition, genomic regulation exerted by melatonin modulates pro-oxidant and antioxidant mechanisms and mitochondrial metabolism [85,86], therefore enhancing cellular responses against DNA alteration and chemical damage [51,87–89]. This multidimensional response of melatonin towards the repair of modified sites, the elimination of macromolecule damaging agents and the activation of antioxidative defense gives it proficiency to neutralize mustard pathology.

Along with lung, eye and skin damage [90], sleep disorders are very common in mustard victims [91] and, in this regard, a cross-sectional study conducted with 100 sulfur mustard-injured Iranians from one-third of victims still suffering from poor sleep, respiratory malfunction and other delayed illness from the Iran–Iraq war (1980–1988), showed reduced nocturnal serum melatonin [92]. Moreover, a randomized, double-blind and placebo-controlled trial study in another cohort of 30 Iranian veterans showed the efficiency of melatonin in improving sleep quality [93]. Consequently, the current state of knowledge nurtured from clinical evidence and the functional biology of this indoleamine suggests that replacement therapy with melatonin is a plausible alternative to manage mustard toxicity.

Early pharmacological cytotoxicity of mustard exposure involves chromosomal DNA-damage by readily direct alkylation of purines (in addition to RNA and other macromolecules) and subsequent cross-linking and single or double DNA-strand breaks [22]. In the case that alkylating positions are found on the same strand, the so-called "limpet junction" hinders the access of vital DNA processing enzymes [94]. Otherwise, interstrand DNA cross-linking prevents the uncoiling and strand separation during replication and transcription and therefore, poses serious side effects when cells activate the genome or undergo division [94]. Either way, extensive DNA and/or protein damage exceeds the cell's clearance capacity, and the resulting processes activate repair mechanisms, oxidative stress and inflammation pathways [22,95,96]. Eventually, this response to genotoxicity can alert the checkpoint machinery to arrest cell cycle progression and activate apoptosis or necrosis [97–99]. In this last respect, cellular destruction accounted after extensive chemical damage and DNA fragmentation drives the clinical expression of mustard toxicity. So, skin samples of old male SKH-1 hairless mice treated with 3.2 mg/mouse nitrogen mustard for 30 min resulted in direct DNA damage, as suggested by both increased phosphorylation at Ser 139 of the histone variant H2A.X (called γH2A.X), which signals double-stranded breaks in DNA, and elevation of the oxidative damage marker 8-oxo-7,8-dihydro-2′-deoxyguanosine [100]. Oxidative stress, inflammation, dead epidermis and keratinization disorders were also observed [27,100]. Noteworthy, genome and expressed RNA comprehensive analyses have revealed that melatonin and its two indolic metabolites modulate multiple signaling pathways in cultured human epidermal keratinocytes, including cell differentiation and other relevant activities (antioxidative, antiaging, antiproliferative, etc) for cell homeostasis [101]. Similar to skin-induced injury, human corneal epithelial cells derived from the outermost and highly vesicant-susceptible ocular layer, once exposed to 100 μM nitrogen mustard for 12 to 48 h showed, among other expressions of cell failure, a dose-dependent increase of γH2A.X, proving the damage inflicted to DNA [102]. In view of the pathological expression described, it should be mentioned that at micromolar concentration, melatonin-receptor signaling activated the p38-dependent phosphorylation of p53 and histone H2AX at Ser139, thus gathering DNA-repairing cascades to prevent accumulation of DNA lesions in normal and cancer cells [103,104]. With regard to lung disease, the human bronchial epithelium line 16HBE exposed up to 24 h to yperite (0.25 and 0.5 mM) or its structural and functional analog mechlorethamine (0.1 and 0.25 mM) suffered from both cell detachment and DNA fragmentation in a time/concentration-dependent manner [105]. Regarding the remediation of this harmful, 100 μM melatonin reduced cell detachment by about half, as well as mustard's interference with cell proliferation, viability and apoptosis by one third [105].

Other evidence supporting the role of melatonin in maintaining molecular machinery and restoring cellular biosystems, is the reported ability to reduce the incidence of spontaneous and chemically induced carcinogenesis, which also invokes upregulation and activation of DNA repair mechanisms [106,107]. It is important to note in this context that DNA maintenance and tumor suppressors appear to be under circadian control [108–110], and also, alterations of melatonin regulators affect circadian genes, cancer etiology and patient survival [111]. Indeed, the reduction in breast cancer incidence among blind women has been linked to improved endogenous melatonin and rhythmicity synchronization compared to light-perceptive counterparts [112]. In keeping with melatonin's capacity to counteract potentially cancerous lesions, the comet assay and wide-genome gene expression screening have shown that pre-treatment with the indoleamine (1 nM) stimulated a functional response in breast and colon cancer cells to repair DNA strand breaks caused by the carcinogen methyl methanesulfonate [113]. Likewise, melatonin (100 and 400 μM) inhibited the induction of sister chromatid exchange by the antineoplastic and potent alkylating nitrogen mustard-derived melphalan [114]. In the same way, low (0.2 mg/kg) and high (0.4 mg/kg) doses of melatonin reduced the accumulation of DNA adducts instilled by the carcinogen safrole [115,116]. Likewise, the radioprotective potential of melatonin has shown to preclude double-stranded breaks in peripheral lymphocytes caused by ^{131}I [117] and overall DNA damage in spleen/cerebral [118], hematopoietic [119] and intestinal cells [120] of irradiated mice. Within the scope of genome protection, melatonin also assists reproductive technology to enhance cell homeostasis and functionality. At micromolar or millimolar concentrations, melatonin durably protected the DNA integrity and biochemical functionality of ram [121,122] and rabbit [123] sperm during cryopreservation, as well as reduced the DNA fragmentation and chromatin dispersion of sperm from asthenoteratozoospermic men [124]. Likewise, by inhibiting phospho-histone H2A.X, melatonin prevented double chain breaks in mouse oocytes during prophase arrest [125,126] and in porcine embryos coming from somatic cell nuclear transfer [127]. In short, these and other published data [128] clearly show the anti-mutational and genome-protective capacities that equip melatonin with aptitudes to mitigate the genotoxicity triggered by mustard intoxication.

In addition, to directly damaging the physical structure of macromolecules, strong sulfur and nitrogen mustard poisoning triggers other acute cytotoxicity mediated by the complex signaling pathways of DNA damage repair, cell death, oxidative stress, apoptosis and inflammation [27,59,129,130]. Indeed, high redox imbalance and nitrooxidative stress, in parallel to the decline of antioxidant mechanisms, are crucial in hurting nitric bases and the deoxyribose backbone of DNA [76], as well as releasing pro-inflammatory factors that affect signaling and collapse membranes [40,131]. However, the specific species responsible for the oxidative pathology induced by mustards have not been identified [132], although there is compelling evidence that reactive nitrogen species play a pivotal role in the respiratory damage induced by sulfur mustard [82]. In this regard, it is known that sulfur and nitrogen mustard cause depletion of glutathione (decreases GSH/GSSG ratio) and antioxidant enzymes, resulting in RONS overproduction [41,133,134]. Subsequently, nitric oxide and superoxide combine into the highly reactive nitrosating $ONOO^-$ [135], which cannot be cleared by conventional antioxidants and originates a great part of mustard toxicity [39,72,136–138]. Peroxynitrite reacts covalently with all major macromolecules and thiols and stresses cellular biosystems to compromise viability and cause apoptosis or primarily necrosis [139–141]. Thus, for example, peroxynitrite disrupts membranes through the oxidation of their structural lipids and inactivates the intermediate metabolism due to the inactivating oxidation of enzymes and ion channels essential for DNA repair and energy production [39,142,143]. Moreover, at very low doses, it attacks DNA (with a small preference for guanine and 2′-deoxyguanosine to form 8-nitroguanine and 4,5-dihydro-5-hydroxy-4-(nitrosooxy)-2′-deoxyguanosine adducts, respectively) and produces single strand breaks [144,145].

Despite the relevance of peroxynitrite toxicity, to fully understand mustard pathology [146], it is mandatory to take into account the oxidative damage coming from the hydroxyl radical, which is the most common affecting DNA and also displaying preference towards guanosines [147]. Likewise, female mice given percutaneously with sublethal nitrogen mustard exhibited exacerbation of oxidative stress markers in the liver, and DNA damage, which were antagonized at variable levels by oral melatonin (250 mg/kg), better at low (5 mg/kg) than at high (10 mg/kg) concentration of nitrogen mustard [41]. On the other hand, pinealectomized rats receiving melatonin at darkness (1 mg/kg in the drinking water) for 15 days and a terminal intraperitoneal dose of cyclophosphamide (CP) (20 or 50 mg/kg) showed strongly reduced spontaneous chromosomal abnormalities and oxidative lesions of DNA in their bone marrow cells, as well as upregulated DNA excision repair [148]. Furthermore, Sprague-Dawley rats were treated via transdermal with 3.5 mg/kg of the nitrogen mustard-related compound mechlorethamine (MEC) and 30 min later intraperitoneally with melatonin (100 mg/kg), subsequently repeated five times every 12 h, normalized the increase of TNF-α, IL-1β and iNOS produced by the MEC dosage (Table 1) [47]. Histologically, melatonin significantly reduced the alveolar epithelial damage, inflammation and interalveolar septal thickening observed in MEC-treated animals. An identical experimental schedule allowed melatonin to protect against mustard-induced kidney toxicity [48]. By activating the same defense responses, melatonin has also shown significant evidence of being able to protect DNA in other damaging contexts. Thus, the indoleamine provided genetic protection to primary cortical neurons from a rat model of intracerebral hemorrhage [149], human melanocytes [150] and skin fibroblasts [151] damaged by UVB, as well as murine ovaries exposed to genotoxicity and loss of fertility by cisplatin [152]. Notably, melatonin also alleviated oxidative damage of lipids and DNA in primate liver [153] and peripheral blood lymphocytes of healthy, non-smoking males [154]. Additionally, in vitro and in vivo settings have illuminated the broad potential of melatonin to address the genotoxicity of stressors such as arecoline [155], lead ([156]), formaldehyde [157], the endocrine-disrupting bisphenol A [158] or engineered titanium dioxide nanoparticles [159].

Remarkably, the detoxifying function of melatonin from oxidizing reactive species has recently been expanded with the suggestion that indoleamine and six of its metabolites are capable of directly reversing oxidative alterations in DNA by transferring electrons to oxidized guanine sites and hydrogen atoms to the sugar moieties of 2-deoxyguanosines [76]. As the authors themselves express, this hypothesis requires experimental research to be verified and thus enlarges the potential of melatonin from prophylaxis and/or attenuation to the direct reversal of oxidative diseases, including mustard poisoning. In view of this and all the other observations compiled earlier, it can be concluded that the anti-genotoxic and macromolecular integrity capacities of melatonin, together with the reinforcement of antioxidant and anti-inflammatory responses, outline a promising horizon for the indoleamine in therapies against morbidity and mortality associated with mustard intoxication. However, in the effort to unravel new clinical management strategies, more research should focus on the myriad actions of melatonin and its proficiency to mitigate the side-effects of mustards.

6. Conclusions

In view of the knowledge summarized herein, we have tried to provide an overview of the main molecular and cellular mechanisms involved in sulfur and nitrogen mustard toxicity, which, however, are still being investigated. In this complex scenario, various cytoprotective agents have been suggested to counteract the toxic effects of vesicants, although none of them protect against all pathophysiological manifestations and have so far provided only partial protection. Because it is endowed with multiple functions and a good safety profile, melatonin has a high relevance in the protection against numerous chemical compounds, among them, the blistering agents. Moreover, melatonin also has a differentiating element with respect to other cytoprotective agents; its own metabolites, c3OHM and AMK, exert cytoprotective actions that potentiate the therapeutic capacity reducing the oxidative damage of vesicants. For this reason, it is necessary to further research the role of this indolamine as a preventive/prophylactic agent in preclinical models, as well as in controlled translational trials. In this context, clinicians are paying increasing attention to melatonin, which could potentially have important applications to ameliorate the alkylating agents-induced side effects. But, for this to happen, randomized clinical studies to translate the therapeutic potential of melatonin to clinical practice are needed.

Author Contributions: Conceptualization, A.R.; formal analysis, C.D.L.R.; investigation, A.R., E.R., E.G.-M. and R.P.; resources, A.R., E.R., E.G.-M., J.J.T., D.R.S., J.E. and R.P.; writing—original draft preparation, A.R., E.R., E.G.-M., C.D.L.R. and J.E.; writing—review and editing, A.R., E.R., E.G.-M., C.D.L.R., R.J.R., J.E. and F.L.-M.; supervision, A.R.; project administration, A.R. and A.J.; funding acquisition, A.R. All authors have read and agreed to the published version of the manuscript.

Funding: This research was funded by the Spanish DGAM General Directorate of Armament and Material/SDG PLATIN for the MELVES project 10032/20/0085/00 (2020/SP03390102/00000359) and UCJC (QUIMELTER 2020-21 and ONCOMEL 2021-23 projects).

Institutional Review Board Statement: Not applicable.

Informed Consent Statement: Not applicable.

Data Availability Statement: Data is contained within the article.

Acknowledgments: Authors thank the Spanish Ministry of Defense and UCJC for its continued support.

Conflicts of Interest: The authors declare no conflict of interest.

References

1. UN Security Council. *Seventh Report of the Organisation for the Prohibition of Chemical Weapons-United Nations Joint Investigative Mechanism*; S/2017/904; UN Security Council: New York, NY, USA, 2017.
2. Zhao, Z.; Yan, X.; Li, L.; Shu, Y.; He, J.; Wang, L.; Huang, Q.; Xie, J.; Zhao, J.; Peng, S. Proliferating Stem Cells are Acutely Affected by DNA Damage Induced by Sulfur Mustard. *DNA Cell Biol.* **2022**, *41*, 716–726. [CrossRef] [PubMed]
3. Jan, Y.-H.; Heck, D.E.; Laskin, D.L.; Laskin, J.D. DNA damage signaling in the cellular responses to mustard vesicants. *Toxicol. Lett.* **2020**, *326*, 78–82. [CrossRef] [PubMed]
4. Liu, F.; Jiang, N.; Xiao, Z.-Y.; Cheng, J.-P.; Mei, Y.-Z.; Zheng, P.; Wang, L.; Zhang, X.-R.; Zhou, X.-B.; Zhou, W.-X.; et al. Effects of poly (ADP-ribose) polymerase-1 (PARP-1) inhibition on sulfur mustard-induced cutaneous injuries in vitro and in vivo. *PeerJ* **2016**, *4*, e1890. [CrossRef] [PubMed]
5. Borna, H.; Noe, S.H.H.Q.; Harchegani, A.B.; Talatappe, N.R.; Ghatrehsamani, M.; Ghanei, M.; Shahriary, A. A review on proteomics analysis to reveal biological pathways and predictive proteins in sulfur mustard exposed patients: Roles of inflammation and oxidative stress. *Inhal. Toxicol.* **2019**, *31*, 3–11. [CrossRef]
6. Panahi, Y.; Abdolghaffari, A.H.; Sahebkar, A. A review on symptoms, treatments protocols, and proteomic profile in sulfur mustard-exposed victims. *J. Cell. Biochem.* **2018**, *119*, 197–206. [CrossRef]
7. Murray, A.; Banota, T.; Guo, G.L.; Smith, L.C.; Meshanni, J.A.; Lee, J.; Kong, B.; Abramova, E.V.; Goedken, M.; Gow, A.J.; et al. Farnesoid X receptor regulates lung macrophage activation and injury following nitrogen mustard exposure. *Toxicol. Appl. Pharmacol.* **2022**, *454*, 116208. [CrossRef]

8. Malaviya, R.; Abramova, E.V.; Rancourt, R.C.; Sunil, V.R.; Napierala, M.; Weinstock, D.; Croutch, C.R.; Roseman, J.; Tuttle, R.; Peters, E.; et al. Progressive Lung Injury, Inflammation, and Fibrosis in Rats Following Inhalation of Sulfur Mustard. *Toxicol. Sci.* **2020**, *178*, 358–374. [CrossRef]
9. Li, K.; Li, Z.; Wu, J.; Gong, Y.; Guo, L.; Xie, J. In Vitro Evaluation of DNA Damage Effect Markers toward Five Nitrogen Mustards Based on Liquid Chromatography–Tandem Mass Spectrometry. *Chem. Res. Toxicol.* **2022**, *35*, 99–110. [CrossRef]
10. Dong, X.; He, Y.; Ye, F.; Zhao, Y.; Cheng, J.; Xiao, J.; Yu, W.; Zhao, J.; Sai, Y.; Dan, G.; et al. Vitamin D3 ameliorates nitrogen mustard-induced cutaneous inflammation by inactivating the NLRP3 inflammasome through the SIRT3–SOD2–mtROS signaling pathway. *Clin. Transl. Med.* **2021**, *11*, e312. [CrossRef]
11. Zhang, Y.; Li, X.; Grailer, J.J.; Wang, N.; Wang, M.; Yao, J.; Zhong, R.; Gao, G.F.; Ward, P.A.; Tan, D.-X.; et al. Melatonin alleviates acute lung injury through inhibiting the NLRP3 inflammasome. *J. Pineal Res.* **2016**, *60*, 405–414. [CrossRef]
12. Nair, A.; Yadav, P.; Behl, A.; Sharma, R.K.; Kulshrestha, S.; Butola, B.S.; Sharma, N. Toxic blister agents: Chemistry, mode of their action and effective treatment strategies. Chem. *Biol. Interact.* **2021**, *350*, 109654. [CrossRef] [PubMed]
13. Harchegani, A.B.; Khor, A.; Tahmasbpour, E.; Ghatrehsamani, M.; Kaboutaraki, H.B.; Shahriary, A. Role of oxidative stress and antioxidant therapy in acute and chronic phases of sulfur mustard injuries: A review. *Cutan. Ocul. Toxicol.* **2019**, *38*, 9–17. [CrossRef] [PubMed]
14. Mangerich, A.; Debiak, M.; Birtel, M.; Ponath, V.; Balszuweit, F.; Lex, K.; Martello, R.; Burckhardt-Boer, W.; Strobelt, R.; Siegert, M.; et al. Sulfur and nitrogen mustards induce characteristic poly(ADP-ribosyl)ation responses in HaCaT keratinocytes with distinctive cellular consequences. *Toxicol. Lett.* **2016**, *244*, 56–71. [CrossRef] [PubMed]
15. Sabnam, S.; Rizwan, H.; Pal, S.; Pal, A. CEES-induced ROS accumulation regulates mitochondrial complications and inflammatory response in keratinocytes. Chem. *Biol. Interact.* **2020**, *321*, 109031. [CrossRef]
16. Meng, W.; Sun, M.; Xu, Q.; Cen, J.; Cao, Y.; Li, Z.; Xiao, K. Development of a Series of Fluorescent Probes for the Early Diagnostic Imaging of Sulfur Mustard Poisoning. *ACS Sens.* **2019**, *4*, 2794–2801. [CrossRef] [PubMed]
17. Kia, A.; Nadi, M.; Hajhasan, V.; Salimian, J. Alterations in Mitochondrial and Inflammasome Homeostasis by 2-Chloroethyl Ethyl Sulfide and their Mitigation by Curcumin: An In vitro Study. *Iran. J. Allergy Asthma Immunol.* **2021**, *20*, 614–622. [CrossRef]
18. Jost, P.; Svobodova, H.; Stetina, R. Induction and repair of DNA cross-links induced by sulfur mustard in the A-549 cell line followed by a comet assay. *Chem. Interact.* **2015**, *237*, 31–37. [CrossRef]
19. Jan, Y.-H.; Heck, D.E.; Malaviya, R.; Casillas, R.P.; Laskin, D.L.; Laskin, J.D. Cross-Linking of Thioredoxin Reductase by the Sulfur Mustard Analogue Mechlorethamine (Methylbis(2-chloroethyl)amine) in Human Lung Epithelial Cells and Rat Lung: Selective Inhibition of Disulfide Reduction but Not Redox Cycling. *Chem. Res. Toxicol.* **2014**, *27*, 61–75. [CrossRef]
20. Laskin, J.D.; Black, A.T.; Jan, Y.-H.; Sinko, P.J.; Heindel, N.D.; Sunil, V.; Heck, D.E.; Laskin, D.L. Oxidants and antioxidants in sulfur mustard-induced injury. *Ann. N. Y. Acad. Sci.* **2010**, *1203*, 92–100. [CrossRef]
21. Harchegani, A.B.; Tahmasbpour, E.; Borna, H.; Imamy, A.; Ghanei, M.; Shahriary, A. Free Radical Production and Oxidative Stress in Lung Tissue of Patients Exposed to Sulfur Mustard: An Overview of Cellular and Molecular Mechanisms. *Chem. Res. Toxicol.* **2018**, *31*, 211–222. [CrossRef]
22. Kehe, K.; Balszuweit, F.; Steinritz, D.; Thiermann, H. Molecular toxicology of sulfur mustard-induced cutaneous inflammation and blistering. *Toxicology* **2009**, *263*, 12–19. [CrossRef] [PubMed]
23. Werrlein, R.J.; Madren-Whalley, J.S. Multiphoton microscopy: An optical approach to understanding and resolving sulfur mustard lesions. *J. Biomed. Opt.* **2003**, *8*, 396–409. [CrossRef] [PubMed]
24. Giuliani, I.; Boivieux-Ulrich, E.; Houcine, O.; Guennou, C.; Marano, F. Toxic effects of mechlorethamine on mammalian respiratory mucociliary epithelium in primary culture. *Cell Biol. Toxicol.* **1994**, *10*, 231–246. [CrossRef] [PubMed]
25. Korkmaz, A.; Kunak, Z.I.; Paredes, S.D.; Yaren, H.; Tan, D.-X.; Reiter, R.J. The use of melatonin to combat mustard toxicity. *REVIEW. Neuro Endocrinol. Lett.* **2008**, *29*, 614–619. [PubMed]
26. Joseph, L.B.; Gordon, M.K.; Zhou, P.; Hahn, R.A.; Lababidi, H.; Croutch, C.R.; Sinko, P.J.; Heck, D.E.; Laskin, D.L.; Laskin, J.D. Sulfur mustard corneal injury is associated with alterations in the epithelial basement membrane and stromal extracellular matrix. *Exp. Mol. Pathol.* **2022**, *128*, 104807. [CrossRef]
27. Kumar, D.; Tewari-Singh, N.; Agarwal, C.; Jain, A.K.; Inturi, S.; Kant, R.; White, C.W.; Agarwal, R. Nitrogen mustard exposure of murine skin induces DNA damage, oxidative stress and activation of MAPK/Akt-AP1 pathway leading to induction of inflammatory and proteolytic mediators. *Toxicol. Lett.* **2015**, *235*, 161–171. [CrossRef]
28. Chen, M.; Dong, X.; Deng, H.; Ye, F.; Zhao, Y.; Cheng, J.; Dan, G.; Zhao, J.; Sai, Y.; Bian, X.; et al. Targeting TRPV1-mediated autophagy attenuates nitrogen mustard-induced dermal toxicity. *Signal Transduct. Target. Ther.* **2021**, *6*, 29. [CrossRef]
29. Lüling, R.; John, H.; Gudermann, T.; Thiermann, H.; Mückter, H.; Popp, T.; Steinritz, D. Transient Receptor Potential Channel A1 (TRPA1) Regulates Sulfur Mustard-Induced Expression of Heat Shock 70 kDa Protein 6 (HSPA6) In Vitro. *Cells* **2018**, *7*, 126. [CrossRef]
30. Jiang, N.; Zhang, P.; Shen, W.; Zhang, Y.; Zhou, W. Clinical and experimental research progress on neurotoxicity of sulfur mustard and its possible mechanisms. *Toxicology* **2022**, *483*, 153372. [CrossRef]
31. Simons, T.; Steinritz, D.; Bölck, B.; Schmidt, A.; Popp, T.; Thiermann, H.; Gudermann, T.; Bloch, W.; Kehe, K. Sulfur mustard-induced epigenetic modifications over time—A pilot study. *Toxicol. Lett.* **2018**, *293*, 45–50. [CrossRef]

32. Steinritz, D.; Schmidt, A.; Balszuweit, F.; Thiermann, H.; Simons, T.; Striepling, E.; Bölck, B.; Bloch, W. Epigenetic modulations in early endothelial cells and DNA hypermethylation in human skin after sulfur mustard exposure. *Toxicol. Lett.* **2016**, *244*, 95–102. [CrossRef] [PubMed]
33. Gharbi, S.; Khateri, S.; Soroush, M.R.; Shamsara, M.; Naeli, P.; Najafi, A.; Korsching, E.; Mowla, S.J. MicroRNA expression in serum samples of sulfur mustard veterans as a diagnostic gateway to improve care. *PLoS ONE* **2018**, *13*, e0194530. [CrossRef] [PubMed]
34. Salimi, S.; Noorbakhsh, F.; Faghihzadeh, S.; Ghaffarpour, S.; Ghazanfari, T. Expression of miR-15b-5p, miR-21-5p, and SMAD7 in Lung Tissue of Sulfur Mustard-exposed Individuals with Long-term Pulmonary Complications. *Iran. J. Allergy Asthma Immunol.* **2019**, *18*, 332–339. [CrossRef] [PubMed]
35. Venosa, A.; Smith, L.C.; Gow, A.J.; Zarbl, H.; Laskin, J.D.; Laskin, D.L. Macrophage activation in the lung during the progression of nitrogen mustard induced injury is associated with histone modifications and altered miRNA expression. *Toxicol. Appl. Pharmacol.* **2021**, *423*, 115569. [CrossRef] [PubMed]
36. Anderson, D.R.; Taylor, S.L.; Fetterer, D.P.; Holmes, W.W. Evaluation of protease inhibitors and an antioxidant for treatment of sulfur mustard-induced toxic lung injury. *Toxicology* **2009**, *263*, 41–46. [CrossRef]
37. Reiter, R.J.; Tan, D.-X.; Manchester, C.; Burillo, S.L.; Juan, M.S.; Mayo, C. Melatonin: Detoxification of Oxygen And Nitrogen-Based Toxic Reactants. *Adv. Exp. Med. Biol.* **2003**, *527*, 539–548. [CrossRef]
38. Reiter, R.; Paredes, S.; Korkmaz, A.; Jou, M.-J.; Tan, D.-X. Melatonin combats molecular terrorism at the mitochondrial level. *Interdiscip. Toxicol.* **2008**, *1*, 137–149. [CrossRef]
39. Korkmaz, A.; Yaren, H.; Topal, T.; Oter, S. Molecular targets against mustard toxicity: Implication of cell surface receptors, peroxynitrite production, and PARP activation. *Arch. Toxicol.* **2006**, *80*, 662–670. [CrossRef]
40. Ucar, M.; Korkmaz, A.; Reiter, R.J.; Yaren, H.; Öter, S.; Kurt, B.; Topal, T. Melatonin alleviates lung damage induced by the chemical warfare agent nitrogen mustard. *Toxicol. Lett.* **2007**, *173*, 124–131. [CrossRef]
41. Sharma, M.; Vijayaraghavan, R.; Gautam, A. DRDE-07 and its analogues as promising cytoprotectants to nitrogen mustard (HN-2)—An alkylating anticancer and chemical warfare agent. *Toxicol. Lett.* **2009**, *188*, 243–250. [CrossRef]
42. Pohanka, M.; Sobotka, J.; Jilkova, M.; Stetina, R. Oxidative stress after sulfur mustard intoxication and its reduction by melatonin: Efficacy of antioxidant therapy during serious intoxication. *Drug Chem. Toxicol.* **2011**, *34*, 85–91. [CrossRef] [PubMed]
43. Escames, G.; López, L.C.; Ortiz, F.; Ros, E.; Acuña-Castroviejo, D. Age-dependent lipopolysaccharide-induced iNOS expression and multiorgan failure in rats: Effects of melatonin treatment. *Exp. Gerontol.* **2006**, *41*, 1165–1173. [CrossRef] [PubMed]
44. García, J.A.; Ortiz, F.; Miana, J.; Doerrier, C.; Fernández-Ortiz, M.; Rusanova, I.; Escames, G.; García, J.J.; Acuña-Castroviejo, D. Contribution of inducible and neuronal nitric oxide synthases to mitochondrial damage and melatonin rescue in LPS-treated mice. *J. Physiol. Biochem.* **2017**, *73*, 235–244. [CrossRef] [PubMed]
45. Cuzzocrea, S.; Mazzon, E.; Serraino, I.; Lepore, V.; Terranova, M.L.; Ciccolo, A.; Caputi, A.P. Melatonin reduces dinitrobenzene sulfonic acid-induced colitis. *J. Pineal Res.* **2001**, *30*, 1–12. [CrossRef] [PubMed]
46. Sunil, V.R.; Patel, K.J.; Shen, J.; Reimer, D.; Gow, A.J.; Laskin, J.D.; Laskin, D.L. Functional and inflammatory alterations in the lung following exposure of rats to nitrogen mustard. *Toxicol. Appl. Pharmacol.* **2011**, *250*, 10–18. [CrossRef] [PubMed]
47. Macit, E.; Yaren, H.; Aydin, I.; Kunak, Z.I.; Yaman, H.; Onguru, O.; Uysal, B.; Korkmaz, A.; Turel, S.; Kenar, L. The protective effect of melatonin and S-methylisothiourea treatments in nitrogen mustard induced lung toxicity in rats. *Environ. Toxicol. Pharmacol.* **2013**, *36*, 1283–1290. [CrossRef]
48. Kunak, Z.I.; Macit, E.; Yaren, H.; Yaman, H.; Cakir, E.; Aydin, I.; Turker, T.; Kurt, Y.G.; Ozcan, A.; Uysal, B.; et al. Protective Effects of Melatonin and S-Methylisothiourea on Mechlorethamine Induced Nephrotoxicity. *J. Surg. Res.* **2012**, *175*, e17–e23. [CrossRef]
49. Shokrzadeh, M.; Chabra, A.; Naghshvar, F.; Ahmadi, A.; Jafarinejhad, M.; Hasani-Nourian, Y. Protective Effects of Melatonin against Cyclophosphamide-induced Oxidative Lung Toxicity in Mice. *Drug Res.* **2015**, *65*, 281–286. [CrossRef]
50. Pita, R.; Marco-Contelles, J.; Ramos, E.; Del Pino, J.; Romero, A. Toxicity induced by chemical warfare agents: Insights on the protective role of melatonin. *Chem. Biol. Interact.* **2013**, *206*, 134–142. [CrossRef]
51. Romero, A.; Ramos, E.; López-Muñoz, F.; Ríos, C.D.L.; Egea, J.; Gil-Martín, E.; Pita, R.; Torrado, J.; Serrano, D.; Juberias, A. Toxicology of Blister Agents: Is Melatonin a Potential Therapeutic Option? *Diseases* **2021**, *9*, 27. [CrossRef]
52. Reiter, R.J.; Tan, D.X.; Galano, A. Melatonin: Exceeding Expectations. *Physiology* **2014**, *29*, 325–333. [CrossRef] [PubMed]
53. Tan, D.-X.; Manchester, L.C.; Terron, M.P.; Flores, L.J.; Reiter, R.J. One molecule, many derivatives: A never-ending interaction of melatonin with reactive oxygen and nitrogen species? *J. Pineal Res.* **2007**, *42*, 28–42. [CrossRef] [PubMed]
54. Smith, K.J.; Smith, W.J.; Hamilton, T.; Skelton, H.G.; Graham, J.S.; Okerberg, C.; Moeller, R.; Hackley, B.E., Jr. Histopathologic and Immunohistochemical Features in Human Skin after Exposure to Nitrogen and Sulfur Mustard. *Am. J. Dermatopathol.* **1998**, *20*, 22–28. [CrossRef] [PubMed]
55. Shakarjian, M.P.; Heck, D.E.; Gray, J.P.; Sinko, P.J.; Gordon, M.K.; Casillas, R.P.; Heindel, N.D.; Gerecke, D.R.; Laskin, D.L.; Laskin, J.D. Mechanisms Mediating the Vesicant Actions of Sulfur Mustard after Cutaneous Exposure. *Toxicol. Sci.* **2010**, *114*, 5–19. [CrossRef] [PubMed]
56. Malaviya, R.; Sunil, V.R.; Cervelli, J.; Anderson, D.R.; Holmes, W.W.; Conti, M.L.; Gordon, R.E.; Laskin, J.D.; Laskin, D.L. Inflammatory effects of inhaled sulfur mustard in rat lung. *Toxicol. Appl. Pharmacol.* **2010**, *248*, 89–99. [CrossRef]

57. Basu, S.; Binder, R.J.; Suto, R.; Anderson, K.M.; Srivastava, P.K. Necrotic but not apoptotic cell death releases heat shock proteins, which deliver a partial maturation signal to dendritic cells and activate the NF-κB pathway. *Int. Immunol.* **2000**, *12*, 1539–1546. [CrossRef]
58. Minsavage, G.D.; Dillman, J.F. Bifunctional Alkylating Agent-Induced p53 and Nonclassical Nuclear Factor κB Responses and Cell Death Are Altered by Caffeic Acid Phenethyl Ester: A Potential Role for Antioxidant/Electrophilic Response-Element Signaling. *J. Pharmacol. Exp.* **2007**, *321*, 202–212. [CrossRef]
59. Jowsey, P.A.; Williams, F.M.; Blain, P.G. DNA damage responses in cells exposed to sulphur mustard. *Toxicol. Lett.* **2012**, *209*, 1–10. [CrossRef]
60. Kono, H.; Rock, K.L. How dying cells alert the immune system to danger. *Nat. Rev. Immunol.* **2008**, *8*, 279–289. [CrossRef]
61. Silva, M.T. When two is better than one: Macrophages and neutrophils work in concert in innate immunity as complementary and cooperative partners of a myeloid phagocyte system. *J. Leukoc. Biol.* **2010**, *87*, 93–106. [CrossRef]
62. Eming, S.A.; Krieg, T.; Davidson, J.M. Inflammation in Wound Repair: Molecular and Cellular Mechanisms. *J. Investig. Dermatol.* **2007**, *127*, 514–525. [CrossRef] [PubMed]
63. Sunil, V.R.; Shen, J.; Patel-Vayas, K.; Gow, A.J.; Laskin, J.D.; Laskin, D.L. Role of reactive nitrogen species generated via inducible nitric oxide synthase in vesicant-induced lung injury, inflammation and altered lung functioning. *Toxicol. Appl. Pharmacol.* **2012**, *261*, 22–30. [CrossRef] [PubMed]
64. Aghanouri, R.; Ghanei, M.; Aslani, J.; Keivani-Amine, H.; Rastegar, F.; Karkhane, A. Fibrogenic cytokine levels in bronchoalveolar lavage aspirates 15 years after exposure to sulfur mustard. Am. J. Physiol. *Lung Cell. Mol. Physiol.* **2004**, *287*, L1160–L1164. [CrossRef] [PubMed]
65. Jo, E.-K.; Kim, J.K.; Shin, D.-M.; Sasakawa, C. Molecular mechanisms regulating NLRP3 inflammasome activation. *Cell. Mol. Immunol.* **2016**, *13*, 148–159. [CrossRef]
66. Arioz, B.I.; Taştan, B.; Tarakcioglu, E.; Tufekci, K.U.; Olcum, M.; Ersoy, N.; Bagriyanik, A.; Genc, K.; Genc, S. Melatonin Attenuates LPS-Induced Acute Depressive-Like Behaviors and Microglial NLRP3 Inflammasome Activation Through the SIRT1/Nrf2 Pathway. *Front. Immunol.* **2019**, *10*, 1511. [CrossRef]
67. Farré-Alins, V.; Narros-Fernández, P.; Palomino-Antolín, A.; Decouty-Pérez, C.; Lopez-Rodriguez, A.B.; Parada, E.; Muñoz-Montero, A.; Gómez-Rangel, V.; López-Muñoz, F.; Ramos, E.; et al. Melatonin Reduces NLRP3 Inflammasome Activation by Increasing α7 nAChR-Mediated Autophagic Flux. *Antioxidants* **2020**, *9*, 1299. [CrossRef]
68. Cao, Z.; Fang, Y.; Lu, Y.; Tan, D.; Du, C.; Li, Y.; Ma, Q.; Yu, J.; Chen, M.; Zhou, C.; et al. Melatonin alleviates cadmium-induced liver injury by inhibiting the TXNIP-NLRP3 inflammasome. *J. Pineal Res.* **2017**, *62*, e12389. [CrossRef]
69. García, J.A.; Volt, H.; Venegas, C.; Doerrier, C.; Escames, G.; López, L.C.; Acuña-Castroviejo, D. Disruption of the NF-κB/NLRP3 connection by melatonin requires retinoid-related orphan receptor-α and blocks the septic response in mice. *FASEB J.* **2015**, *29*, 3863–3875. [CrossRef]
70. Cao, S.; Shrestha, S.; Li, J.; Yu, X.; Chen, J.; Yan, F.; Ying, G.; Gu, C.; Wang, L.; Chen, G. Melatonin-mediated mitophagy protects against early brain injury after subarachnoid hemorrhage through inhibition of NLRP3 inflammasome activation. *Sci. Rep.* **2017**, *7*, 2417. [CrossRef]
71. Che, H.; Wang, Y.; Li, H.; Li, Y.; Sahil, A.; Lv, J.; Liu, Y.; Yang, Z.; Dong, R.; Xue, H.; et al. Melatonin alleviates cardiac fibrosis via inhibiting lncRNA MALAT1/miR-141-mediated NLRP3 inflammasome and TGF-β1/Smads signaling in diabetic cardiomyopathy. *FASEB J.* **2020**, *34*, 5282–5298. [CrossRef]
72. Kehe, K.; Raithel, K.; Kreppel, H.; Jochum, M.; Worek, F.; Thiermann, H. Inhibition of poly(ADP-ribose) polymerase (PARP) influences the mode of sulfur mustard (SM)-induced cell death in HaCaT cells. *Arch. Toxicol.* **2008**, *82*, 461–470. [CrossRef]
73. Ghasemi, H.; Ghazanfari, T.; Ghassemi-Broumand, M.; Javadi, M.A.; Babaei, M.; Soroush, M.R.; Yaraee, R.; Faghihzadeh, S.; Poorfarzam, S.; Owlia, P.; et al. Long-term ocular consequences of sulfur mustard in seriously eye-injured war veterans. *Cutan. Ocul. Toxicol.* **2009**, *28*, 71–77. [CrossRef]
74. Amini, H.; Solaymani-Dodaran, M.; Mousavi, B.; Alam Beladi, S.N.; Soroush, M.R.; Abolghasemi, J.; Vahedian-Azimi, A.; Salesi, M.; Guest, P.C.; Sahebkar, A.; et al. Long-term Health Outcomes among Survivors Exposed to Sulfur Mustard in Iran. *JAMA Netw. Open* **2020**, *3*, e2028894. [CrossRef] [PubMed]
75. Tan, D.-X.; Manchester, L.C.; Esteban-Zubero, E.; Zhou, Z.; Reiter, R.J. Melatonin as a Potent and Inducible Endogenous Antioxidant: Synthesis and Metabolism. *Molecules* **2015**, *20*, 18886–18906. [CrossRef] [PubMed]
76. Pérez-González, A.; Castañeda-Arriaga, R.; Álvarez-Idaboy, J.R.; Reiter, R.J.; Galano, A. Melatonin and its metabolites as chemical agents capable of directly repairing oxidized DNA. *J. Pineal Res.* **2019**, *66*, e12539. [CrossRef] [PubMed]
77. Menendez-Pelaez, A.; Reiter, R.J. Distribution of melatonin in mammalian tissues: The relative importance of nuclear versus cytosolic localization. *J. Pineal Res.* **1993**, *15*, 59–69. [CrossRef]
78. Venegas, C.; García, J.A.; Escames, G.; Ortiz, F.; López, A.; Doerrier, C.; García-Corzo, L.; López, L.C.; Reiter, R.J.; Acuña-Castroviejo, D. Extrapineal melatonin: Analysis of its subcellular distribution and daily fluctuations. *J. Pineal Res.* **2012**, *52*, 217–227. [CrossRef]
79. Pita, R.; Marco-Contelles, J.; Ramos, E.; del Pino, J.; Romero, A. Melatonin as potential candidate to prevent the toxicity induced by chemical warfare agents. *Arch. Toxicol.* **2014**, *88*, 3–4. [CrossRef]

80. Reiter, R.J. The Indoleamine Melatonin as a Free Radical Scavenger, Electron Donor, and Antioxidant: In vitro and in vivo studies. *Adv. Exp. Med. Biol.* **1996**, *398*, 307–313. [CrossRef]
81. Korkmaz, A.; Tan, D.-X.; Reiter, R.J. Acute and delayed sulfur mustard toxicity; novel mechanisms and future studies. *Interdiscip. Toxicol.* **2008**, *1*, 22–26. [CrossRef]
82. Weinberger, B.; Laskin, J.D.; Sunil, V.R.; Sinko, P.J.; Heck, D.E.; Laskin, D.L. Sulfur mustard-induced pulmonary injury: Therapeutic approaches to mitigating toxicity. *Pulm. Pharmacol. Ther.* **2011**, *24*, 92–99. [CrossRef]
83. Gilad, E.; Cuzzocrea, S.; Zingarelli, B.; Salzman, A.L.; Szabó, C. Melatonin is a scavenger of peroxynitrite. *Life Sci.* **1997**, *60*, PL169–PL174. [CrossRef]
84. Teixeira, A.; Morfim, M.P.; de Cordova, C.A.S.; Charão, C.C.T.; de Lima, V.R.; Creczynski-Pasa, T.B. Melatonin protects against pro-oxidant enzymes and reduces lipid peroxidation in distinct membranes induced by the hydroxyl and ascorbyl radicals and by peroxynitrite. *J. Pineal Res.* **2003**, *35*, 262–268. [CrossRef]
85. Zhang, H.-M.; Zhang, Y. Melatonin: A well-documented antioxidant with conditional pro-oxidant actions. *J. Pineal Res.* **2014**, *57*, 131–146. [CrossRef]
86. Yang, S.-G.; Joe, S.-Y.; Bae, J.-W.; Heo, G.-D.; Park, H.-J.; Koo, D.-B. Melatonin Protects Against Mdivi-1-Induced Abnormal Spindle Assembly and Mitochondrial Superoxide Production during Porcine Oocyte Maturation. *Front. Cell Dev. Biol.* **2021**, *9*, 693969. [CrossRef] [PubMed]
87. Majidinia, M.; Sadeghpour, A.; Mehrzadi, S.; Reiter, R.J.; Khatami, N.; Yousefi, B. Melatonin: A pleiotropic molecule that modulates DNA damage response and repair pathways. *J. Pineal Res.* **2017**, *63*, e12416. [CrossRef] [PubMed]
88. Galano, A.; Tan, D.-X.; Reiter, R.J. Melatonin: A Versatile Protector against Oxidative DNA Damage. *Molecules* **2018**, *23*, 530. [CrossRef]
89. Romero, A.; Gil Martín, E.; Rios, C.D.L.; Egea, J.; Ramos, E.; López-Muñoz, F.; Pita, R.; Juberías, A. Impact of melatonin effects on toxicology of vesicant chemical warfare agents: When science meets reality. *Melatonin Res.* **2020**, *3*, 101–119. [CrossRef]
90. Khateri, S.; Ghanei, M.; Keshavarz, S.; Soroush, M.; Haines, D. Incidence of Lung, Eye, and Skin Lesions as Late Complications in 34,000 Iranians with Wartime Exposure to Mustard Agent. *J. Occup. Environ. Med.* **2003**, *45*, 1136–1143. [CrossRef] [PubMed]
91. Vahedi, E.; Varzaneh, A.R.F.; Ghanei, M.; Afsharpaiman, S.; Poursaleh, Z. The effect of nightly nasal CPAP treatment on nocturnal hypoxemia and sleep disorders in mustard gas-injured patients. *Sleep Breath.* **2014**, *18*, 741–748. [CrossRef]
92. Mousavi, S.S.; Vahedi, E.; Shohrati, M.; Panahi, Y.; Parvin, S. Nocturnal serum melatonin levels in sulfur mustard exposed patients with sleep disorders. *J. R. Army Med. Corps* **2017**, *163*, 411–415. [CrossRef] [PubMed]
93. Mousavi, S.S.; Shohrati, M.; Vahedi, E.; Alitappeh, M.A.; Panahi, Y. Effect of melatonin administration on sleep quality in sulfur mustard exposed patients with sleep disorders. *Iran. J. Pharm. Res.* **2018**, *17*, 136–144. [CrossRef] [PubMed]
94. Boens, B.; Azouz, M.; Ouk, T.-S.; Zerrouki, R. Synthesis and Biological Evaluation of Nitrogen Mustard Derivatives of Purine Bases. *Nucleosides Nucleotides Nucleic Acids* **2013**, *32*, 69–80. [CrossRef] [PubMed]
95. Inturi, S.; Tewari-Singh, N.; Agarwal, C.; White, C.W.; Agarwal, R. Activation of DNA damage repair pathways in response to nitrogen mustard-induced DNA damage and toxicity in skin keratinocytes. *Mutat. Res.* **2014**, *763–764*, 53–63. [CrossRef]
96. Zhang, X.; Zhang, Z.; Chen, S.; Zhao, D.; Zhang, F.; Hu, Z.; Xiao, F.; Ma, X. Nitrogen mustard hydrochloride-induced acute respiratory failure and myelosuppression: A case report. *Exp. Ther. Med.* **2015**, *10*, 1293–1296. [CrossRef]
97. Rosenthal, D.S.; Simbulan-Rosenthal, C.M.; Iyer, S.; Smulson, M.E.; Spoonde, A.; Smith, W.; Ray, R. Sulfur Mustard Induces Markers of Terminal Differentiation and Apoptosis in Keratinocytes Via a Ca^{2+}-Calmodulin and Caspase-Dependent Pathway. *J. Investig. Dermatol.* **1998**, *111*, 64–71. [CrossRef]
98. Iliakis, G.; Wang, Y.; Guan, J.; Wang, H. DNA damage checkpoint control in cells exposed to ionizing radiation. *Oncogene* **2003**, *22*, 5834–5847. [CrossRef]
99. Sancar, A.; Lindsey-Boltz, L.A.; Ünsal-Kaçmaz, K.; Linn, S. Molecular Mechanisms of Mammalian DNA Repair and the DNA Damage Checkpoints. *Annu. Rev. Biochem.* **2004**, *73*, 39–85. [CrossRef]
100. Jain, A.K.; Tewari-Singh, N.; Inturi, S.; Kumar, D.; Orlicky, D.J.; Agarwal, C.; White, C.W.; Agarwal, R. Flavanone silibinin treatment attenuates nitrogen mustard-induced toxic effects in mouse skin. *Toxicol. Appl. Pharmacol.* **2015**, *285*, 71–78. [CrossRef]
101. Stefan, J.; Kim, T.-K.; Schedel, F.; Janjetovic, Z.; Crossman, D.; Steinbrink, K.; Slominski, R.; Zmijewski, J.; Tulic, M.; Reiter, R.; et al. Differential and Overlapping Effects of Melatonin and Its Metabolites on Keratinocyte Function: Bioinformatics and Metabolic Analyses. *Antioxidants* **2021**, *10*, 618. [CrossRef]
102. Goswami, D.G.; Tewari-Singh, N.; Dhar, D.; Kumar, D.; Agarwal, C.; Ammar, D.A.; Kant, R.; Enzenauer, R.W.; Petrash, J.M.; Agarwal, R. Nitrogen Mustard-Induced Corneal Injury Involves DNA Damage and Pathways Related to Inflammation, Epithelial-Stromal Separation, and Neovascularization. *Cornea* **2016**, *35*, 257–266. [CrossRef] [PubMed]
103. Santoro, R.; Marani, M.; Blandino, G.; Muti, P.; Strano, S. Melatonin triggers p53Ser phosphorylation and prevents DNA damage accumulation. *Oncogene* **2012**, *31*, 2931–2942. [CrossRef]
104. Santoro, R.; Mori, F.; Marani, M.; Grasso, G.; Cambria, M.A.; Blandino, G.; Muti, P.; Strano, S. Blockage of melatonin receptors impairs p53-mediated prevention of DNA damage accumulation. *Carcinogenesis* **2013**, *34*, 1051–1061. [CrossRef] [PubMed]
105. Sourdeval, M.; Lemaire, C.; Deniaud, A.; Taysse, L.; Daulon, S.; Breton, P.; Brenner, C.; Boisvieux-Ulrich, E.; Marano, F. Inhibition of caspase-dependent mitochondrial permeability transition protects airway epithelial cells against mustard-induced apoptosis. *Apoptosis* **2006**, *11*, 1545–1559. [CrossRef] [PubMed]

106. Reiter, R.J.; Rosales-Corral, S.A.; Tan, D.-X.; Acuna-Castroviejo, D.; Qin, L.; Yang, S.-F.; Xu, K. Melatonin, a Full Service Anti-Cancer Agent: Inhibition of Initiation, Progression and Metastasis. *Int. J. Mol. Sci.* **2017**, *18*, 843. [CrossRef]
107. Farhood, B.; Goradel, N.H.; Mortezaee, K.; Khanlarkhani, N.; Najafi, M.; Sahebkar, A. Melatonin and cancer: From the promotion of genomic stability to use in cancer treatment. *J. Cell. Physiol.* **2019**, *234*, 5613–5627. [CrossRef]
108. Gery, S.; Komatsu, N.; Baldjyan, L.; Yu, A.; Koo, D.; Koeffler, H.P. The Circadian Gene Per1 Plays an Important Role in Cell Growth and DNA Damage Control in Human Cancer Cells. *Mol. Cell* **2006**, *22*, 375–382. [CrossRef]
109. Sancar, A.; Lindsey-Boltz, L.; Kang, T.-H.; Reardon, J.T.; Lee, J.H.; Ozturk, N. Circadian clock control of the cellular response to DNA damage. *FEBS Lett.* **2010**, *584*, 2618–2625. [CrossRef]
110. Chuffa, L.G.D.A.; Seiva, F.; Cucielo, M.S.; Silveira, H.; Reiter, R.J.; Lupi, L.A. Clock genes and the role of melatonin in cancer cells: An overview. *Melatonin Res.* **2019**, *2*, 133–157. [CrossRef]
111. Zhang, J.; Jiang, H.; Du, K.; Xie, T.; Wang, B.; Chen, C.; Reiter, R.J.; Cen, B.; Yuan, Y. Pan-cancer analyses reveal genomics and clinical characteristics of the melatonergic regulators in cancer. *J. Pineal Res.* **2021**, *71*, e12758. [CrossRef]
112. Flynn-Evans, E.E.; Stevens, R.G.; Tabandeh, H.; Schernhammer, E.S.; Lockley, S.W. Total visual blindness is protective against breast cancer. *Cancer Causes Control.* **2009**, *20*, 1753–1756. [CrossRef]
113. Liu, R.; Fu, A.; Hoffman, A.E.; Zheng, T.; Zhu, Y. Melatonin enhances DNA repair capacity possibly by affecting genes involved in DNA damage responsive pathways. *BMC Cell Biol.* **2013**, *14*, 1. [CrossRef] [PubMed]
114. Lialiaris, T.; Lyratzopoulos, E.; Papachristou, F.; Simopoulou, M.; Mourelatos, C.; Nikolettos, N. Supplementation of melatonin protects human lymphocytes in vitro from the genotoxic activity of melphalan. *Mutagenesis* **2008**, *23*, 347–354. [CrossRef] [PubMed]
115. Tan, D.-X.; Pöeggeler, B.; Reiter, R.J.; Chen, L.-D.; Chen, S.; Lucien, M.C.; Barlow-Walden, L.R. The pineal hormone melatonin inhibits DNA-adduct formation induced by the chemical carcinogen safrole in vivo. *Cancer Lett.* **1993**, *70*, 65–71. [CrossRef] [PubMed]
116. Tan, D.-X.; Reiter, R.J.; Chen, L.-D.; Poeggeler, B.; Manchester, L.C.; Barlow-Walden, L.R. Both physiological and pharmacological levels of melatonin reduce DNA adduct formation induced by the carcinogen safrole. *Carcinogenesis* **1994**, *15*, 215–218. [CrossRef]
117. Bagheri, H.; Jafarpour, S.M.; Shekarchi, B.; Farhood, B. The radioprotective effects of melatonin and nanoselenium on DNA double-strand breaks in peripheral lymphocytes caused by I-131. *Indian J. Nucl. Med.* **2021**, *36*, 134–139. [CrossRef]
118. Abdullaev, S.A.; Glukhov, S.I.; Gaziev, A.I. Radioprotective and Radiomitigative Effects of Melatonin in Tissues with Different Proliferative Activity. *Antioxidants* **2021**, *10*, 1885. [CrossRef]
119. Kumar, A.; Choudhary, S.; Kumar, S.; Adhikari, J.S.; Kapoor, S.; Chaudhury, N.K. Role of melatonin mediated G-CSF induction in hematopoietic system of gamma-irradiated mice. *Life Sci.* **2022**, *289*, 120190. [CrossRef]
120. Wang, Q.; Wang, Y.; Du, L.; Xu, C.; Liu, Y.; Liu, Q.; Fan, S. Quantitative proteomic analysis of the effects of melatonin treatment for mice suffered from small intestinal damage induced by γ-ray radiation. *Int. J. Radiat. Biol.* **2021**, *97*, 1206–1216. [CrossRef]
121. Succu, S.; Berlinguer, F.; Pasciu, V.; Satta, V.; Leoni, G.G.; Naitana, S. Melatonin protects ram spermatozoa from cryopreservation injuries in a dose-dependent manner. *J. Pineal Res.* **2011**, *50*, 310–318. [CrossRef]
122. Pool, K.; Rickard, J.; De Graaf, S. Melatonin improves the motility and DNA integrity of frozen-thawed ram spermatozoa likely via suppression of mitochondrial superoxide production. *Domest. Anim. Endocrinol.* **2021**, *74*, 106516. [CrossRef] [PubMed]
123. Fadl, A.M.; Ghallab, A.R.M.; Abou-Ahmed, M.M.; Moawad, A.R. Melatonin can improve viability and functional integrity of cooled and frozen/thawed rabbit spermatozoa. *Reprod. Domest. Anim.* **2021**, *56*, 103–111. [CrossRef] [PubMed]
124. Malmir, M.; Noreini, S.N.; Ghafarizadeh, A.; Faraji, T.; Asali, Z. Ameliorative effect of melatonin on apoptosis, DNA fragmentation, membrane integrity and lipid peroxidation of spermatozoa in the idiopathic asthenoteratospermic men: In vitro. *Andrologia* **2021**, *53*, e13944. [CrossRef] [PubMed]
125. Leem, J.; Bai, G.; Kim, J.; Oh, J.S. Melatonin protects mouse oocytes from DNA damage by enhancing nonhomologous end-joining repair. *J. Pineal Res.* **2019**, *67*, e12603. [CrossRef]
126. Wang, L.; Zhang, J.; Zhao, C.; Jia, Z.; Feng, X. Melatonin Reverses 10-Hydroxycamptothecin-Induced Apoptosis and Autophagy in Mouse Oocyte. *Reprod. Sci.* **2021**, *28*, 1839–1849. [CrossRef]
127. Liang, S.; Jin, Y.-X.; Yuan, B.; Zhang, J.-B.; Kim, N.-H. Melatonin enhances the developmental competence of porcine somatic cell nuclear transfer embryos by preventing DNA damage induced by oxidative stress. *Sci. Rep.* **2017**, *7*, 11114. [CrossRef]
128. Damiani, A.P.; Strapazzon, G.; Sardinha, T.T.D.O.; Rohr, P.; Gajski, G.; De Pinho, R.A.; De Andrade, V.M. Melatonin supplementation over different time periods until ageing modulates genotoxic parameters in mice. *Mutagenesis* **2020**, *35*, 465–478. [CrossRef]
129. Tewari-Singh, N.; Gu, M.; Agarwal, C.; White, C.W.; Agarwal, R. Biological and Molecular Mechanisms of Sulfur Mustard Analogue-Induced Toxicity in JB6 and HaCaT Cells: Possible Role of Ataxia Telangiectasia-Mutated/Ataxia Telangiectasia-Rad3-Related Cell Cycle Checkpoint Pathway. *Chem. Res. Toxicol.* **2010**, *23*, 1034–1044. [CrossRef]
130. Wahler, G.; Heck, D.E.; Heindel, N.D.; Laskin, D.L.; Laskin, J.D.; Joseph, L.B. Antioxidant/stress response in mouse epidermis following exposure to nitrogen mustard. *Exp. Mol. Pathol.* **2020**, *114*, 104410. [CrossRef]
131. Sunil, V.R.; Patel-Vayas, K.; Shen, J.; Gow, A.J.; Laskin, J.D.; Laskin, D.L. Role of TNFR1 in lung injury and altered lung function induced by the model sulfur mustard vesicant, 2-chloroethyl ethyl sulfide. *Toxicol. Appl. Pharmacol.* **2011**, *250*, 245–255. [CrossRef]

132. Keyser, B.M.; Andres, D.K.; Holmes, W.W.; Paradiso, D.; Appell, A.; Letukas, V.A.; Benton, B.; Clark, O.E.; Gao, X.; Ray, P.; et al. Mustard Gas Inhalation Injury: Therapeutic Strategy. *Int. J. Toxicol.* **2014**, *33*, 271–281. [CrossRef] [PubMed]
133. Tewari-Singh, N.; Agarwal, C.; Huang, J.; Day, B.J.; White, C.W.; Agarwal, R. Efficacy of Glutathione in Ameliorating Sulfur Mustard Analog-Induced Toxicity in Cultured Skin Epidermal Cells and in SKH-1 Mouse Skin In Vivo. *J. Pharmacol. Exp. Ther.* **2011**, *336*, 450–459. [CrossRef] [PubMed]
134. Tewari-Singh, N.; Agarwal, R. Mustard vesicating agent-induced toxicity in the skin tissue and silibinin as a potential countermeasure. *Ann. N. Y. Acad. Sci.* **2016**, *1374*, 184–192. [CrossRef] [PubMed]
135. Radi, R. Oxygen radicals, nitric oxide, and peroxynitrite: Redox pathways in molecular medicine. *Proc. Natl. Acad. Sci. USA* **2018**, *115*, 5839–5848. [CrossRef]
136. Korkmaz, A.; Topal, T.; Oter, S. Pathophysiological aspects of cyclophosphamide and ifosfamide induced hemorrhagic cystitis; implication of reactive oxygen and nitrogen species as well as PARP activation. *Cell Biol. Toxicol.* **2007**, *23*, 303–312. [CrossRef]
137. Yaren, H.; Mollaoglu, H.; Kurt, B.; Korkmaz, A.; Oter, S.; Topal, T.; Karayilanoglu, T. Lung toxicity of nitrogen mustard may be mediated by nitric oxide and peroxynitrite in rats. *Res. Vet. Sci.* **2007**, *83*, 116–122. [CrossRef]
138. Korkmaz, A.; Kurt, B.; Yildirim, I.; Basal, S.; Topal, T.; Sadir, S.; Öter, S. Effects of Poly(ADP-Ribose) Polymerase Inhibition in Bladder Damage Caused by Cyclophosphamide in Rats. *Exp. Biol. Med.* **2008**, *233*, 338–343. [CrossRef]
139. Szabó, C. Multiple pathways of peroxynitrite cytotoxicity. *Toxicol. Lett.* **2003**, *140–141*, 105–112. [CrossRef]
140. Virág, L.; Szabó, E.; Gergely, P.; Szabó, C. Peroxynitrite-induced cytotoxicity: Mechanism and opportunities for intervention. *Toxicol. Lett.* **2003**, *140–141*, 113–124. [CrossRef]
141. Pacher, P.; Beckman, J.S.; Liaudet, L. Nitric Oxide and Peroxynitrite in Health and Disease. *Physiol. Rev.* **2007**, *87*, 315–424. [CrossRef]
142. Virág, L.; Szabó, C. The Therapeutic Potential of Poly(ADP-Ribose) Polymerase Inhibitors. *Pharmacol. Rev.* **2002**, *54*, 375–429. [CrossRef] [PubMed]
143. Chien, Y.-H.; Bau, D.-T.; Jan, K.-Y. Nitric oxide inhibits DNA-adduct excision in nucleotide excision repair. *Free. Radic. Biol. Med.* **2004**, *36*, 1011–1017. [CrossRef] [PubMed]
144. Yermilov, V.; Yoshie, Y.; Rubio, J.; Ohshima, H. Effects of carbon dioxide/bicarbonate on induction of DNA single-strand breaks and formation of 8-nitroguanine, 8-oxoguanine and base-propenal mediated by peroxynitrite. *FEBS Lett.* **1996**, *399*, 67–70. [CrossRef] [PubMed]
145. Yoshie, Y.; Ohshima, H. Nitric Oxide Synergistically Enhances DNA Strand Breakage Induced by Polyhydroxyaromatic Compounds, but Inhibits That Induced by the Fenton Reaction. *Arch. Biochem. Biophys.* **1997**, *342*, 13–21. [CrossRef] [PubMed]
146. Brimfield, A.; Soni, S.; Trimmer, K.; Zottola, M.; Sweeney, R.; Graham, J. Metabolic activation of sulfur mustard leads to oxygen free radical formation. *Free. Radic. Biol. Med.* **2011**, *52*, 811–817. [CrossRef] [PubMed]
147. Chatgilialoglu, C.; D'Angelantonio, M.; Guerra, M.; Kaloudis, P.; Mulazzani, Q.G. A Reevaluation of the Ambident Reactivity of the Guanine Moiety Towards Hydroxyl Radicals. *Angew. Chem. Int. Ed.* **2009**, *48*, 2214–2217. [CrossRef]
148. Ferreira, S.; Peliciari-Garcia, R.A.; Takahashi-Hyodo, S.; Rodrigues, A.; Amaral, F.; Berra, C.; Bordin, S.; Curi, R.; Cipolla-Neto, J. Effects of melatonin on DNA damage induced by cyclophosphamide in rats. *Braz. J. Med. Biol. Res.* **2013**, *46*, 278–286. [CrossRef]
149. Wang, Z.; Zhou, F.; Dou, Y.; Tian, X.; Liu, C.; Li, H.; Shen, H.; Chen, G. Melatonin Alleviates Intracerebral Hemorrhage-Induced Secondary Brain Injury in Rats via Suppressing Apoptosis, Inflammation, Oxidative Stress, DNA Damage, and Mitochondria Injury. *Transl. Stroke Res.* **2017**, *9*, 74–91. [CrossRef]
150. Janjetovic, Z.; Jarrett, S.G.; Lee, E.F.; Duprey, C.; Reiter, R.J.; Slominski, A.T. Melatonin and its metabolites protect human melanocytes against UVB-induced damage: Involvement of NRF2-mediated pathways. *Sci. Rep.* **2017**, *7*, 1274. [CrossRef]
151. Dong, K.; Goyarts, E.; Rella, A.; Pelle, E.; Wong, Y.H.; Pernodet, N. Age Associated Decrease of MT-1 Melatonin Receptor in Human Dermal Skin Fibroblasts Impairs Protection Against UV-Induced DNA Damage. *Int. J. Mol. Sci.* **2020**, *21*, 326. [CrossRef]
152. Huang, J.; Shan, W.; Li, N.; Zhou, B.; Guo, E.; Xia, M.; Lu, H.; Wu, Y.; Chen, J.; Wang, B.; et al. Melatonin provides protection against cisplatin-induced ovarian damage and loss of fertility in mice. *Reprod. Biomed. Online* **2021**, *42*, 505–519. [CrossRef] [PubMed]
153. Cabrera, J.; Burkhardt, S.; Tan, D.-X.; Manchester, L.C.; Karbownik, M.; Reiter, R.J. Autoxidation and Toxicant-Induced Oxidation of Lipid and DNA in Monkey Liver: Reduction of Molecular Damage by Melatonin. *Basic Clin. Pharmacol. Toxicol.* **2001**, *89*, 225–230. [CrossRef] [PubMed]
154. Sliwinski, T.; Rozej, W.; Morawiec-Bajda, A.; Morawiec, Z.; Reiter, R.; Blasiak, J. Protective action of melatonin against oxidative DNA damage—Chemical inactivation versus base-excision repair. *Mutat. Res.* **2007**, *634*, 220–227. [CrossRef] [PubMed]
155. Shih, Y.-H.; Chiu, K.-C.; Wang, T.-H.; Lan, W.-C.; Tsai, B.-H.; Wu, L.-J.; Hsia, S.-M.; Shieh, T.-M. Effects of melatonin to arecoline-induced reactive oxygen species production and DNA damage in oral squamous cell carcinoma. *J. Formos. Med. Assoc.* **2021**, *120*, 668–678. [CrossRef]
156. Martínez-Alfaro, M.; Hernández-Cortés, D.; Wrobel, K.; Cruz-Jiménez, G.; Leyva, J.C.R.; Piña-Zentella, R.M.; Trejo, A.C. Effect of melatonin administration on DNA damage and repair responses in lymphocytes of rats subchronically exposed to lead. *Mutat. Res.* **2012**, *742*, 37–42. [CrossRef]
157. Bernardini, L.; Barbosa, E.; Charão, M.F.; Goethel, G.; Muller, D.; Bau, C.; Steffens, N.A.; Stein, C.S.; Moresco, R.N.; Garcia, S.C.; et al. Oxidative damage, inflammation, genotoxic effect, and global DNA methylation caused by inhalation of formaldehyde and the purpose of melatonin. *Toxicol. Res.* **2020**, *9*, 778–789. [CrossRef]

158. Wu, H.-J.; Liu, C.; Duan, W.-X.; Xu, S.-C.; He, M.-D.; Chen, C.-H.; Wang, Y.; Zhou, Z.; Yu, Z.-P.; Zhang, L.; et al. Melatonin ameliorates bisphenol A-induced DNA damage in the germ cells of adult male rats. *Mutat. Res.* **2013**, *752*, 57–67. [CrossRef]
159. Fadda, L.M.; Ali, H.M.; Mohamed, A.M.; Hagar, H. Prophylactic administration of carnosine and melatonin abates the incidence of apoptosis, inflammation, and DNA damage induced by titanium dioxide nanoparticles in rat livers. *Environ. Sci. Pollut. Res.* **2020**, *27*, 19142–19150. [CrossRef]

antioxidants

MDPI

Article

Pro-Apoptotic and Anti-Migration Properties of a Thiazoline-Containing Platinum(II) Complex in MDA-MB-231 Breast Cancer Cells: The Role of Melatonin as a Synergistic Agent

Samuel Estirado [1], Elena Fernández-Delgado [1], Emilio Viñuelas-Zahínos [2], Francisco Luna-Giles [2], Ana B. Rodríguez [1], José A. Pariente [1,*] and Javier Espino [1,*]

1 Neuroimmunophysiology and Chrononutrition Research Group, Department of Physiology, Faculty of Science, University of Extremadura, 06006 Badajoz, Spain
2 Coordination Chemistry Research Group, Department of Organic and Inorganic Chemistry, Faculty of Science, University of Extremadura, 06006 Badajoz, Spain
* Correspondence: pariente@unex.es (J.A.P.); jespino@unex.es (J.E.); Tel.: +34-924289796 (J.E.)

Abstract: Triple-negative breast cancer (TNBC) is an aggressive cancer insensitive to hormonal and human epidermal growth factor receptor 2 (HER2)-targeted therapies and has a poor prognosis. Therefore, there is a need for the development of convenient anticancer strategies for the management of TNBC. In this paper, we evaluate the antitumoral potential of a platinum(II) complex coordinated with the ligand 2-(3,5-diphenylpyrazol-1-yl)-2-thiazoline (DPhPzTn), hereafter PtDPhPzTn, against the TNBC cell line MDA-MB-231, and compared its effect with both cisplatin and its less lipophilic counterpart PtPzTn, the latter containing the ligand 2-(pyrazol-1-yl)-2-thiazoline (PzTn). Then, the putative potentiating actions of melatonin, a naturally occurring antioxidant with renowned antitumor properties, on the tumor-killing ability of PtDPhPzTn were also checked in TNBC cells. Our results show that PtDPhPzTn presented enhanced cytotoxicity compared to both the classical drug cisplatin and PtPzTn. In addition, PtDPhPzTn was able to induce apoptosis, being more selective for MDA-MB-231 cells when compared to non-tumor breast epithelial MCF10A cells. Likewise, PtDPhPzTn produced moderate S phase arrest and greatly impaired the migration ability of MDA-MB-231 cells. Most importantly, the co-stimulation of TNBC cells with PtDPhPzTn and melatonin substantially enhanced apoptosis and markedly improved the anti-migratory action compared to PtDPhPzTn alone. Altogether, our findings provide evidence that PtDPhPzTn and melatonin could be potentially applied to breast cancer treatment as powerful synergistic agents.

Keywords: melatonin; metallodrugs; platinum(II) complex; thiazoline; apoptosis; breast cancer

Citation: Estirado, S.; Fernández-Delgado, E.; Viñuelas-Zahínos, E.; Luna-Giles, F.; Rodríguez, A.B.; Pariente, J.A.; Espino, J. Pro-Apoptotic and Anti-Migration Properties of a Thiazoline-Containing Platinum(II) Complex in MDA-MB-231 Breast Cancer Cells: The Role of Melatonin as a Synergistic Agent. *Antioxidants* **2022**, *11*, 1971. https://doi.org/10.3390/antiox11101971

Academic Editor: Stanley Omaye

Received: 1 August 2022
Accepted: 24 September 2022
Published: 1 October 2022

1. Introduction

Triple-negative breast cancer (TNBC) is the least common type of breast cancers, accounting for 10–15% of all cases, but is often more aggressive and difficult to treat due to the lack of efficient molecular targets, such as estrogen receptor (ER), progesterone receptor (PR), and human epidermal growth factor receptor 2 (HER2) [1]. Chemotherapy is one of the most widely used therapeutic modalities in cancer treatment, which includes a wide variety of drugs. Chemotherapeutic agents induce apoptotic cell death not only in tumor cells but also in normal non-tumor cells, thus producing both therapeutic effects (i.e., killing tumor cells) and undesirable side effects (e.g., nephrotoxicity and cardiotoxicity).

Most metal-based antitumor therapies, such as *cis*-diamminedichloroplatinum(II) (cisplatin), have been successfully employed for more than 40 years. However, the use of cisplatin may be limited by the intrinsic or acquired resistance of various types of cancer and its toxic side effects. With the idea of overcoming limitations of cisplatin therapy, platinum metallodrugs with a similar structure to cisplatin have been synthesized in recent years. To design new platinum(II) anticancer drugs, several structural characteristics can be modified by changing the nature of the ligands bound to platinum(II), which influences its chemical and biological properties. For instance, platinum(II) complexes have been synthesized with bioactive bidentate ligands derived from heterocyclic N-donors, such as the heterocycles of pyrazole [2] and thiazoline [3], which have shown promising antitumor properties and may lay the foundation for designing new ligands that coordinate the platinum(II) ion. In this sense, our research group has recently synthesized complexes of platinum(II) with ligands of this nature that reportedly produced cytotoxic effects in human leukemic HL-60 [4], histiocytic lymphoma U-937 [5], ovarian adenocarcinoma SK-OV-3, and cervical epithelial HeLa cells [6,7], these effects being dependent on apoptosis induction, caspase cascade activation, increased intracellular reactive oxygen species (ROS) generation, and/or DNA oxidative damage [4,5,7].

Melatonin (*N*-acetyl-5-methoxytryptamine) is a neurohormone mainly produced by the pineal gland that plays a crucial role in regulating cancer growth, especially in hormone-dependent breast tumors [8]. Melatonin applies its anticancer abilities through multiple mechanisms, including promoting apoptosis and/or autophagy, modulation of angiogenesis [9], and selective induction of oxidative stress due to its pro-oxidant effects in cancer cells [10,11]. Of note, there are various aspects of melatonin research on cancer treatment that could have immediate clinical applications. For instance, many studies have shown that melatonin administered together with conventional drugs (e.g., cisplatin, 5-fluorouracil, and doxorubicin) enhances the sensitivity of cancers (e.g., colorectal, cervix, and pancreatic) to the induction of cell death [12–17]. Melatonin is also able to interfere with tumor metastases in a variety of cancers by employing multiple mechanisms, such as epithelial–mesenchymal transition, cytoskeleton reorganization, and extracellular matrix remodeling [18]. Furthermore, indoleamine causes treatment-resistant malignancies to become susceptible to both endocrine therapy and chemotherapy [19,20]. In addition, melatonin's ability to reduce the side effects of cancer therapies also merits further attention as indoleamine has been shown to mitigate cell damage in numerous experimental paradigms, e.g., oral mucositis due to ionizing radiation [21] or the cardio-hepatic and renal toxicity of different drugs [22]. Therefore, mounting evidence suggests the convenience of using melatonin as adjuvant therapy for cancer treatment, which would far exceed improvements in the wellbeing of the patients.

Building upon previous data from our research group indicating that the enhanced lipophilicity of metallodrugs resulted in an improved ability to kill tumor cells [7], we evaluate in this paper the antitumoral potential of a recently synthesized platinum(II) complex containing the ligand 2-(3,5-diphenylpyrazol-1-yl)-2-thiazoline (DPhPzTn), hereafter PtDPhPzTn, against the TNBC cell line MDA-MB-231, and compare its effect with the classical chemotherapeutic agent cisplatin and its less lipophilic counterpart PtPzTn, the latter containing the ligand 2-(pyrazol-1-yl)-2-thiazoline (PzTn). In addition, we hypothesize that the anti-proliferative and pro-oxidant properties of melatonin could potentiate the tumor-killing ability of PtDPhPzTn in MDA-MB-231 cells.

2. Materials and Methods

2.1. General Procedures

The human epithelial breast adenocarcinoma MDA-MB-231 cell line, human histiocytic lymphoma U-937 cell line (No. 85011440), human promyelocytic leukemia HL-60 15–12 cell line (No. 88120805), and human epithelial cervix adenocarcinoma HeLa cell line (No. 93021013) were acquired from the European Collection of Authenticated Cell Cultures (ECACC) (Dorset, UK). The human mammary epithelial MCF10A cell line (No. CRL-10317) was obtained from the American Type Culture Collection (ATCC) (Barcelona, Spain). Roswell Park Memorial Institute (RPMI)-1640, Dulbecco's modified Eagle's medium (DMEM), DMEM F-12, penicillin/streptomycin, fetal bovine serum (FBS), and L-glutamine were purchased from ThermoFisher Scientific (Barcelona, Spain). *cis-*

Diamminedichloroplatinum(II) (cisplatin), melatonin, hydrocortisone, insulin, epidermal growth factor (EGF), and cholera toxin were bought from Sigma Aldrich (Madrid, Spain). All other reagents were of analytical/commercial grade and were used without further purification. The ligands PzTn and DPhPzTn [23,24] and the precursor [$PtCl_2(DMSO)_2$] [25] were synthesized as described in the literature.

For the elemental analysis, a Leco CHNS-932 microanalyzer was used. IR spectra were recorded on a Perkin-Elmer 100 FTIR spectrophotometer from KBr pellets in the 4000–400 cm^{-1} range. ^1H NMR spectra were obtained with a Bruker Avance III 500 instrument at 500 MHz in DMF-d7. ^1H NMR signals were referenced to residual proton resonances in deuterated solvents.

2.2. General Preparation of Platinum(II) Complexes

To perform the synthesis (Scheme 1), (300.0 mg, 0.71 mmol) the [$PtCl_2(DMSO)_2$] precursor was dissolved in 30 mL of ethanol in an Elenmeyer flask. On the other hand, the ligand was dissolved in 20 mL of ethanol. Once dissolved, the ligand corresponding was added to the precursor and it was refluxed and stirred for 24 h in the dark. The powder formed was vacuum filtered off and then recrystallized in DMF (for PtPzTn) or acetonitrile (for PtDPhPzTn). After that, crystals were separated by filtration, washed with distilled water and cold diethyl ether, and air-dried.

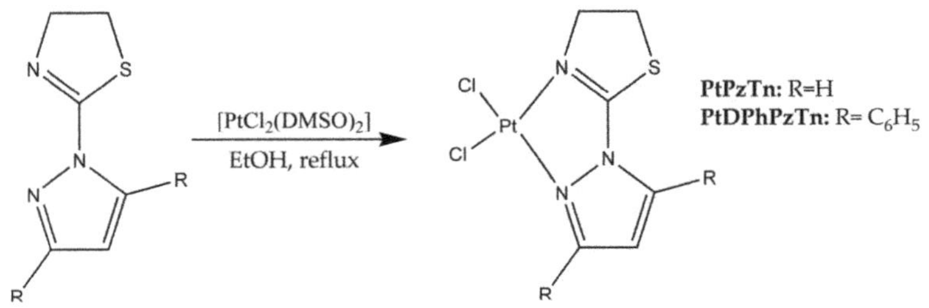

Scheme 1. Synthesis of PtPzTn and PtDPhPzTn.

2.2.1. Synthesis of [$PtCl_2$(PzTn)] (PtPzTn)

This compound was obtained by the general procedure using PzTn (109 mg, 0.71 mmol). Yield 240.2 mg (80.6%). Anal. Calc. (%) for $C_6H_7Cl_2PtN_3S$: C, 17.19; H, 1.68; N, 10.02; S, 7.65%. Found: C, 17.41; H, 1.91; N, 9.80; S, 7.59%. IR (KBr): 3137, 3094, 2934, 1596, 1518, 1439, 1413, 1359, 1312, 1131, 1071, 1004, 781, 601 cm^{-1}. ^1H NMR (500 MHz, DMF-d7) δ 9.08 (d, 1H, J = 3.0 Hz), 8.29 (d, 1H, J = 2.5 Hz), 7.11 (t, 1H, J = 2.8 Hz), 4.47 (t, 2H, J = 8.5 Hz), 4.20 (t, 2H, J = 8.3 Hz), ppm.

2.2.2. Synthesis of [$PtCl_2$(DPhPzTn)] (PtDPhPzTn)

This compound was obtained by the general procedure using DPhPzTn (227 mg, 0.71 mmol). Yield 243.5 mg (59.9%). Anal. Calc. (%) for $C_{18}H_{15}Cl_2N_3PtS$: C, 37.84; H, 2.65; N, 7.35; S, 5.61%. Found: C, 37.78; H, 2.57; N, 7.45; S, 5.45%. IR (KBr): 3095, 3053, 1603, 1587, 1555, 1485, 1445, 1381, 1319, 1018, 998, 833, 768, 699 cm^{-1}. ^1H NMR (500 MHz, DMF-d7) δ 7.86 (m, 4H), 7.69 (m, 3H), 7.49 (m, 3H), 7.19 (d, 1H, J = 1.0 Hz), 4.42 (t, 2H, J = 8.5 Hz), 3.87 (t, 2H, J = 8.5 Hz) ppm.

2.3. Cell Culture and Treatments

The cell lines displayed in Table 1 were cultured into 25 cm^2 culture flasks with the indicated media and supplements. Cells were kept under a humidified atmosphere containing 5% CO_2 at 37 °C, and their viability was routinely tested by the Trypan-blue exclusion method. Cells were challenged with the platinum(II) complexes PtDPhPzTn and PtPzTn, the reference drug cisplatin, melatonin, or vehicle for 24 h at the indicated concentrations. Dimethylformamide (DMF) or dimethylsulfoxide (DMSO) were utilized as vehicle and its final concentration did not exceed 0.2% (*v*/*v*).

Table 1. List of cell lines, cell culture media, and supplements.

Cell Lines	Disease/Tissue	Growth Medium	Supplements
MDA-MB-231	Human breast adenocarcinoma	DMEM	10% heat inactivated FBS 2 mM L-glutamine 100 U/mL penicillin/streptomycin
HeLa	Human cervix adenocarcinoma		
U-937	Human histiocytic lymphoma	RPMI-1640	
HL-60	Human promyelocytic leukemia		
MCF10A	Human mammary epithelial cells	DMEM F-12	5% horse serum 20 ng/mL EGF 10 µg/mL insulin 100 ng/mL cholera toxin 0.5 µg/mL hydrocortisone 100 U/mL penicillin/streptomycin

DMEM: Dulbecco's modified Eagle's medium; EGF: epidermal growth factor; FBS: fetal bovine serum; RPMI: Roswell Park Memorial Institute.

2.4. Cytotoxicity Assay

To examine the cytotoxic effects of the drugs on the different cell lines, the MTS method was performed by means of the CellTiter 96 AQueous One Solution Cell Proliferation assay (Promega, Madrid, Spain), as previously reported [6,7]. The experiments were run in triplicate and cells were challenged with the different drugs for 24 h.

Additionally, the effects (% cell death) of individual drugs and drug combinations were used to calculate combination index (CI) by using Compusyn Software (ComboSyn, Inc., Paramus, NJ, USA), following the Chou-Talalay method [26]. CI values indicate how the selected drug combinations influence their therapeutic efficacy (CI > 1–antagonistic; CI = 1–additive; CI < 1–synergistic) at different combination dosing intervals [27].

2.5. Determination of Apoptotic Cells

The population of apoptotic cells was studied by analyzing phosphatidylserine externalization via annexin V-FITC/propidium iodide (PI) assay (ThermoFisher Scientific, Barcelona, Spain), as previously described [14]. The percentages of live, early apoptotic, late apoptotic, and secondary necrotic cells were computed.

2.6. Caspase-3 and -9 Analysis

Caspase-3 and -9 activation was measured with CaspGLOW fluorescein active caspase-3 staining kit and Caspase-9 staining kit (Red), respectively, as published elsewhere [4]. The analysis of caspase activation was restricted to live cells, which were detected by using the vital dye Hoechst 33258 (10 µg/mL). Data are presented as the percentage of caspase-positive cells.

2.7. Measurement of Reactive Oxygen Species (ROS)

Reactive oxygen species (ROS) detection was performed by incubating cells (30 min, 37 °C) with 0.4 µM 2′,7′-dichlorodihydrofluorescein diacetate (DCFH-DA), as previously reported [4]. Again, the analysis of ROS production was restricted to live cells, which were detected by using the vital dye Hoechst 33258 (10 µg/mL). Data are portrayed as the percentage of DCF-stained cells.

2.8. Cell Cycle Analysis

To investigate changes in cell cycle progression upon treatment, a cell cycle distribution analysis was carried out in ethanol-fixed, PI-stained cells, as published elsewhere [4]. The populations of cells in G0/G1 transition, S phase, and G2/M transition, as well as sub-G1 (hypodiploid) cells are reported.

2.9. Wound-Healing Assay

Cells were seeded into 24-well plates and incubated at 37 °C until they reached 75–80% confluency. To evaluate cell migration, cells were equally wounded with a sterile 100 μL pipette tip and were subsequently challenged with the different treatments. The experiments were run in duplicate. The relative cell migration rates over 12 and 24 h were quantified based on the changes in wound area at time 0 h and after 12 or 24 h using an epifluorescence microscope (Nikon Eclipse TS100). The images were analyzed with the plugin "Wound healing size tool" available in Fiji/ImageJ software version 1.49 (https://github.com/AlejandraArnedo/Wound-healing-size-tool/wiki, accessed on 1 July 2022) [28].

2.10. Transwell Migration Assay

To assess cell migration, transwell migration assay was carried out in transwell chambers with non-coated membrane (24-well insert, pore size: 8 mm, Millipore, Madrid, Spain). Cells (1×10^5) were plated in serum free medium in the upper chamber. The medium containing 10% FBS was added in the lower chamber to act as chemoattractant. After 24 h, the cells that migrated through the pores to the lower surface of the inserts were fixed in 70% ethanol solution for 10 min and afterwards stained with crystal violet for additional 10 min. Inserts were then imaged and, subsequently, the membranes of the inserts were cut, placed in a 24-well plate, and incubated with 5% sodium dodecyl sulfate (SDS) to extract crystal violet (5 min on orbital shaker). Finally, the absorbance was read at 560 nm for quantification.

2.11. Statistical Analysis

Data represent the mean ± standard deviation (S.D.). To compare between treatments, statistical significance was calculated by a one-way analysis of variance (ANOVA) followed by Dunnett's test, unless otherwise specified. The half maximal inhibitory concentration (IC_{50}) values of the dose–response curves were calculated by using nonlinear regression. $p < 0.05$ was considered to indicate a statistically significant difference. The statistics software used was GraphPad Prism 7.04 for Windows (GraphPad Software, San Diego, CA, USA).

3. Results and Discussion

The biological activity of a thiazoline-containing platinum(II) complex PtDPhPzTn was first assessed. As depicted in Figure 1C, PtDPhPzTn exhibited a dose-dependent cytotoxic effect against the TNBC cell line MDA-MB-231 with an IC_{50} value of 10.47 ± 1.02 μM (Figure 1D), which was more than 5-fold lower than that of the reference drug cisplatin (56.80 ± 2.41 μM, Figure 1A,D) and more than 3-fold lower than that of its less lipophilic counterpart PtPzTn (33.51 ± 3.83 μM, Figure 1B,D). The same trend was observed when the metallodrugs were tested in other cell lines, namely U-937, HL-60, and HeLa (Figure 1D), thereby demonstrating that PtDPhPzTn outperforms the reference drug cisplatin (and PtPzTn) in terms of cytotoxicity, though this fact was especially striking in the case of MDA-MB-231 cells, which were moderately resistant to cisplatin (IC_{50} > 50 μM). These findings agree with our previous research [7], thus indicating that the incorporation of aromatic groups in the ligand can lead to the improved cytotoxicity of metallodrugs.

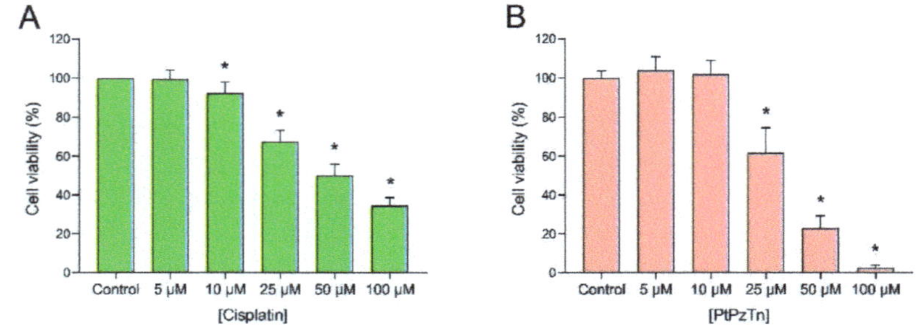

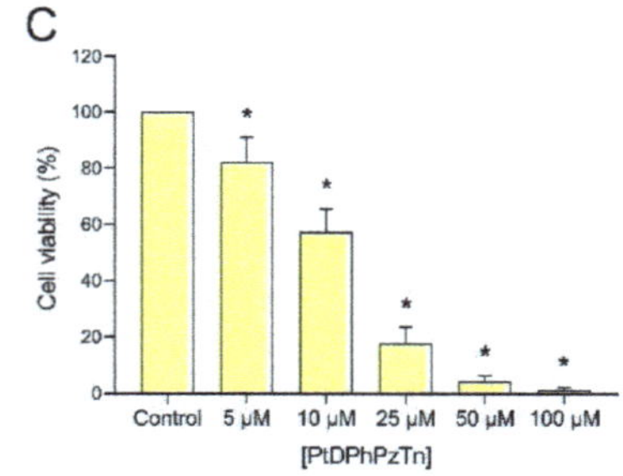

D

Cell lines	Metallodrugs CisPt	PtPzTn	PtDPhPzTn
MDA-MB-231	56.80 ± 2.41 c	33.51 ± 3.83 b	10.47 ± 1.02 a
U-937	8.88 ± 0.24 c	6.48 ± 0.81 b	3.19 ± 0.41 a
HL-60	12.16 ± 0.88 b	15.77 ± 0.73 c	5.36 ± 0.12 a
HeLa	16.21 ± 0.93 a	81.29 ± 2.95 b	12.17 ± 0.56 a

Figure 1. Dose–response curves of metallodrugs on cell viability. MDA-MB-231 cells were treated for 24 h with increasing concentrations (0–100 µM, as indicated) of cisplatin (**A**), the platinum(II) complexes PtPzTn (**B**) and PtDPhPzTn (**C**), or the vehicle (DMF, control). Data represent means ± S.D. of 6 independent experiments and are expressed as percentage of control values. * $p < 0.05$ compared to their corresponding control values (Dunnett's test). (**D**) IC_{50} values (mean ± S.D., µM) of cisplatin (CisPt), PtPzTn, and PtDPhPzTn against MDA-MB-231, U-937, HL-60, and HeLa cell lines. Within each row, values followed by a diverse lowercase letter (a–c) are significantly different ($p < 0.05$; Tukey's test).

Given that the platinum(II) complex PtDPhPzTn was the most effective drug in terms of cytotoxicity and the well-known anti-proliferative and pro-oxidant properties of the indoleamine melatonin in tumor cells, the apoptosis-promoting effect of the combinatory treatment with PtDPhPzTn and melatonin was determined in TNBC cells by analyzing phosphatidylserine externalization in the presence of PI. For this purpose, MDA-MB-231 cells were exposed at the fixed IC_{50} dose level of PtDPhPzTn (i.e., 10.4 µM) in the absence or presence of 1 mM melatonin. As shown in Figure 2, the platinum(II) complex caused a remarkable increase (33.25 ± 4.34% vs. 2.78 ± 1.04%; $p < 0.05$) in the population of secondary necrotic cells at the expense of a significant drop (32.13 ± 11.98% vs. 73.63 ± 8.38%; $p < 0.05$) in the proportion of live cells with respect to the control cells (Figure 2A,B), which is in line with our preliminary findings, indicating that the cytotoxic effect of PtDPhPzTn was associated with the induction of ROS-mediated apoptosis [7]. Previous studies have reported that pyrazole-platinum(II) complexes, especially those containing methyl substituents at the pyrazole ring, exhibited strong pro-apoptotic activity in MDA-MB-231 and MCF-7 breast cancer cell lines [29]. Likewise, dinuclear berenil-platinum(II) complexes were also described to trigger cellular oxidative modifications and, subsequently, apoptosis in MDA-MB-231 and MCF-7 breast cancer cell lines [30,31]. Melatonin per se did not produce any significant pro-apoptotic effect on MDA-MB-231 cells (Figure 2A,B). Interestingly, the combinatory treatment with PtDPhPzTn and melatonin further promoted apoptosis in TNBC cells, as demonstrated by a notable rise (49.11 ± 10.86% vs. 33.25 ± 4.34%; $p < 0.05$) in the percentage of secondary necrotic cells compared to the PtDPhPzTn-treated cells (Figure 2A,B). In this sense, melatonin has been previously shown to synergistically enhance antitumor function in tunicamycin-challenged MDA-MB-231 cells [32], which is in line with the results reported in this paper. The potentiating actions of melatonin on apoptosis induction by classical drugs, such as doxorubicin, lapatinib, and arsenic trioxide, has been also documented in breast cancer cell lines [33–36] and may be related to its pro-oxidant capacity in tumor cells [36]. To corroborate the combinatorial effect of PtDPhPzTn and melatonin in MDA-MB-231 cells, 5, 10, and 25 μM of the platinum(II) complex and 0.5, 1, and 2 mM of the indoleamine were combined and cell death was determined after 24 h to calculate the CI values (Figure S1). Our results show a synergistic effect of the combination of PtDPhPzTn and melatonin, especially when combining 1 mM melatonin and the three different concentrations of the platinum(II) complex (Table S1).

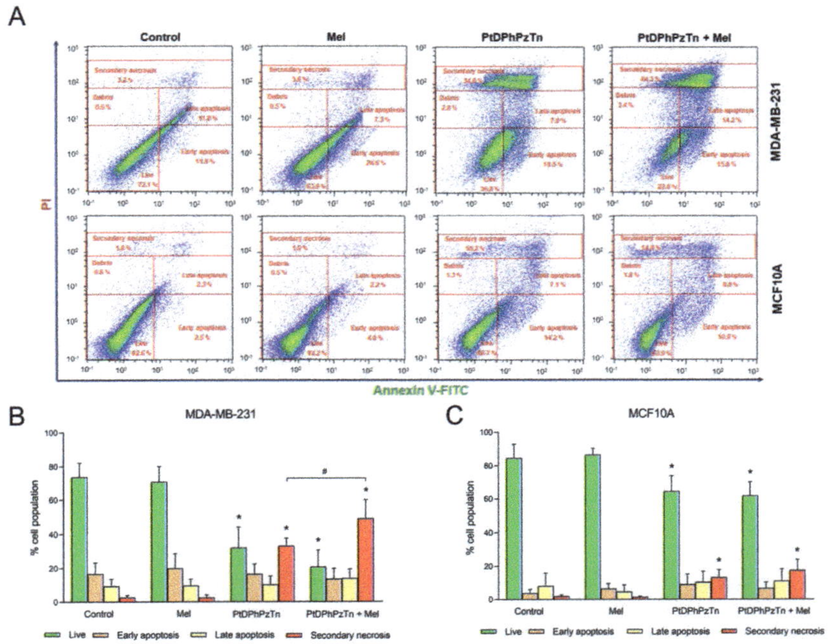

Figure 2. Combinatory treatment of PtDPhPzTn and melatonin differentially affects apoptosis of MDA-MB-231 and MCF10A cell lines. Cells were treated with 10.4 μM PtDPhPzTn in the presence or absence of 1 mM melatonin (Mel), or the vehicle (DMF/DMSO, control) for 24 h. (**A**) Representative cytograms depicting the redistribution of phosphatidylserine (annexin V-FITC staining) in the presence of propidium iodide (PI) after challenging cells with the indicated drugs. Histogram bars show percentages of live, early apoptotic, late apoptotic, and secondary necrotic MDA-MB-231 (**B**) and MCF10A (**C**) cells. Values represent means ± S.D. of 5 independent experiments. * $p < 0.05$ compared to control values (Dunnett's test). # $p < 0.05$ compared to PtDPhPzTn-treated cells (Tukey's test).

On the other hand, the complex PtDPhPzTn turned out to be less harmful to MCF-10A human breast epithelial cells than to tumor cells, as the very same dose of PtDPhPzTn (10.4 μM) only provoked a moderate reduction in the proportion of live cells (64.58 ± 9.31% vs. 84.76 ± 8.14%; $p < 0.05$; Figure 2A,C) with respect to control cells. Other platinum(II) complexes, including cyclometalated complexes and metal complexes with berenil and nitroimidazole, have been informed to show lower toxicity towards normal MCF-10A cells as well [37,38]. In addition, melatonin per se did not affect apoptosis of MCF10A cells (Figure 2A,C), which agrees with previous findings documenting no effect on non-tumoral breast epithelial MCF10A cells [39,40]. Concerning the combinatory treatment, and contrary to what happened on MDA-MB-231 breast cancer cells, it did not display any potentiating effect on the PtDPhPzTn-induced apoptosis of MCF10A cells (Figure 2A,C), thus indicating that this combined treatment exhibits lower apoptosis-promoting actions on normal mammary epithelial cells than on breast cancer cells.

To confirm that the main mode of cell death elicited by the platinum(II) complex was apoptosis, the activation of caspases, which are enzymes critically involved in the initiation and execution of such process, was investigated. When MDA-MB-231 cells were treated with 10.4 μM PtDPhPzTn for 24 h, a remarkable rise in the proportion of caspase-3-positive (60.17 ± 4.27% vs. 14.57 ± 9.08%; $p < 0.05$; Figure 3A,D) and caspase-9-positive (70.63 ± 6.37% vs. 15.07 ± 4.67%; $p < 0.05$; Figure 3B,E) cells was noted, thus indicating that the platinum(II) complex triggered intrinsic apoptosis in the TNBC cell line. In this sense, earlier studies have informed that both pyrazole-platinum(II) complexes and dinuclear berenil-platinum(II) complexes also caused caspase-3 and caspase-9-dependent apoptosis in breast cancer cells [29–31]. Of note, the combinatorial treatment with PtDPhPzTn and melatonin moderately potentiated the activation of caspases, though this effect was not statistically significant compared to PtDPhPzTn alone (78.28 ± 10.30% vs. 60.17 ± 4.27% and 86.30 ± 8.43% vs. 70.63 ± 6.37% for caspase-3 and -9, respectively; Figure 3A,B,D,E). Again, melatonin per se produced negligible effect on caspase activation of MDA-MB-231 cells (Figure 3A,B,D,E). On the other hand, the involvement of ROS in the apoptotic process evoked by the platinum(II) complex was also explored. Our findings show that PtDPhPzTn notably increased intracellular ROS production after 4 h (20.70 ± 3.66 vs. 7.35 ± 1.15% DCF-positive cells; $p < 0.05$; Figure 3C,F), which ultimately led to apoptotic cell death. These results agree with previous research describing that other platinum(II)-based also exerted their cytotoxic effect in breast cancer cells through oxidative stress-mediated apoptosis [30]. Additionally, the combined treatment with PtDPhPzTn and melatonin further augmented intracellular ROS production, this effect not being statistically significant compared to PtDPhPzTn-treated cells (26.29 ± 5.39% vs. 20.70 ± 3.66 DCF-positive cells; $p < 0.05$; Figure 3C,F). As for melatonin alone, it did not induce any change in the ROS production of MDA-MB-231 cells (Figure 3C,F).

As the platinum(II) complex displayed pro-apoptotic actions against MDA-MB-231 cells, its impact on cell cycle progression was anticipated and, therefore, cell cycle distribution analysis was carried out. Thus, when TNBC cells were challenged with 10.4 μM PtDPhPzTn, a moderate S phase arrest (55.32 ± 4.28% vs. 38.30 ± 5.44% in control cells; $p < 0.05$; Figure 4) was observed at the expense of the percentage of cells in G2/M transition. Additionally, the platinum(II) complex caused a rise in the subpopulation of cells with hypodiploid DNA content (sub-G1 cells; Figure 4), which is indicative of DNA fragmentation. In this line, earlier studies have proven that cyclometalated platinum(II) complexes also affect S phase of the cell cycle in breast cancer cells [37]. Nonetheless, other investigations have demonstrated that the inhibition of cell survival induced by pyrazole-platinum(II) complexes occurs by arresting the G1 phase of the cell cycle [29]. The same trend was observed for the combinatory treatment, though no potentiating effect was noticed. In fact, when TNBC cells were treated with 10.4 μM PtDPhPzTn in the presence of 1 mM melatonin, a significant accumulation of MDA-MB-231 cells at S phase (53.18 ± 6.39% vs. 38.30 ± 5.44% in control cells; $p < 0.05$; Figure 4) was ascertained. Likewise, the co-incubation with the platinum(II) complex and the indoleamine evoked an increment in the percentage of cells with hypodiploid DNA content (Figure 4). Nevertheless, melatonin per se did not induce any remarkable change in the cell cycle distribution of MDA-MB-231 cells (Figure 4), which is in line with previous research indicating no effect of indoleamine on the cell cycle of breast cancer cells [41].

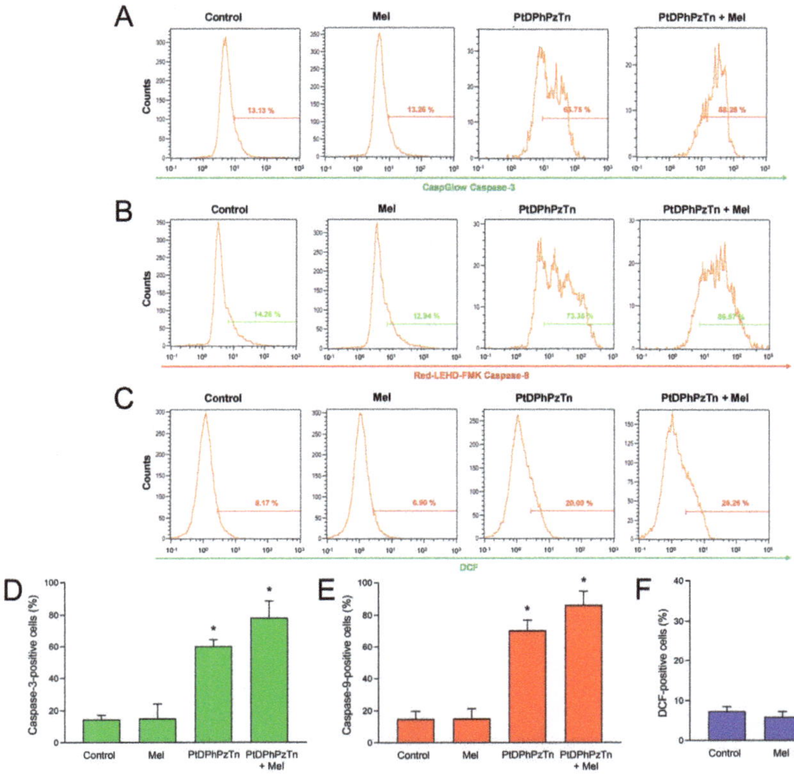

Figure 3. Caspase activation and reactive oxygen species (ROS) production triggered by combinatory treatment of PtDPhPzTn and melatonin in MDA-MB-231 cells. Cells were treated with 10.4 μM PtDPhPzTn in the presence or absence of 1 mM melatonin (Mel), or the vehicle (DMF/DMSO, control) for 24 h (except for ROS production, wherein cells were challenged for 4 h). Representative cytograms show caspase-3 (**A**) and -9 (**B**) activation as well as ROS production (**C**) after challenging cells with the indicated drugs. Histogram bars represent percentages of caspase-3-positive (**D**), caspase-9-positive (**E**), and DCF-stained (**F**) MDA-MB-231 cells, which were analyzed by flow cytometry (for details, see Sections 2.6 and 2.7). Values are presented as means ± SD of four independent experiments. * $p < 0.05$ compared to control values (Dunnett's test).

Finally, the ability of the platinum(II) complex PtDPhPzTn to modulate cell migration capacity of TNBC cells was analyzed by wound-healing and transwell migration assays. As shown in Figure 5A,B, the PtDPhPzTn complex greatly impaired the ability of MDA-MB-231 cells to migrate after both 12 h (29.76 ± 13.11% vs. 66.79 ± 8.40% in control cells; $p < 0.05$) and 24 h post-scratch (66.19 ± 24.20% vs. 98.44 ± 3.26% in control cells; $p < 0.05$), thus suggesting that this compound may be useful in blocking the metastatic migration of breast cancer cells. The same trend was observed in the transwell migration assay, wherein the platinum(II) complex almost completely blunted the migration capacity of MDA-MB-231 cells ($p < 0.05$; Figure 5C). In this sense, other authors have also synthesized metal complexes with potential to suppress the cell migration ability of TNBC cells, including *cis*-diphenyl pyridineamine platinum(II) complexes and platinum(II) complexes with tridentate ligands [42,43]. As for the combined treatment, the exposure of MDA-MB-231 cells at 10.4 μM PtDPhPzTn in the presence of 1 mM melatonin markedly enhanced the antimigration effect of the platinum(II) complex after both 12 h (15.57 ± 7.53% vs. 29.76 ± 13.11% in PtDPhPzTn-treated cells; $p < 0.05$; Figure 5A,B) and 24 h post-scratch (37.85 ± 12.93% vs.

66.19 ± 24.20% in PtDPhPzTn-treated cells; $p < 0.05$; Figure 5A,B). Regarding the transwell migration assay, the combinatory treatment diminished ($p < 0.05$) the proportion of migrating cells to the same extent as the platinum(II) complex alone (Figure 5C). Other studies has also proven that indoleamine synergistically improves the antimigration properties of apatinib, a tyrosine kinase inhibitor thought to inhibit angiogenesis, in breast cancer stem cells derived from MDA-MB-231 cells [44], whereas novel melatonin–tamoxifen drug conjugates were shown to effectively inhibit migration in different breast cancer cell lines, including tamoxifen-resistant cells [45]. Nonetheless, even though melatonin has been described to inhibit cell invasion potential in MDA-MB-231 and MCF-7 breast cancer cell lines [46], indoleamine per se only evoked a moderate, non-significant impairment in the ability of TNBC cells to migrate (Figure 5A,B) and showed no effect in the transwell migration assay.

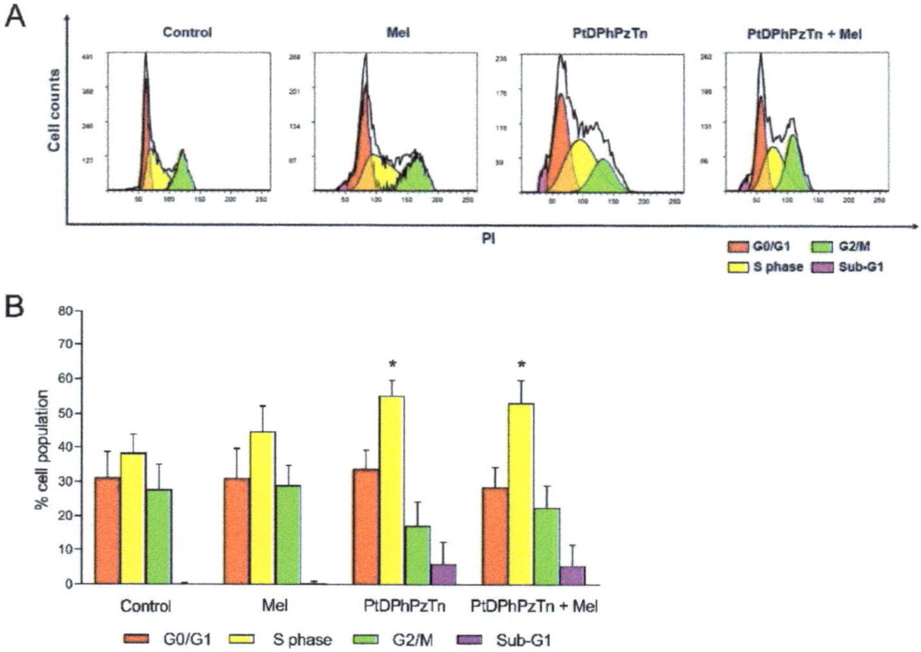

Figure 4. Combinatory treatment of PtDPhPzTn and melatonin disturbed cell cycle profile of MDA-MB-231 cell line. Cells were treated with 10.4 µM PtDPhPzTn in the presence or absence of 1 mM melatonin (Mel) or the vehicle (DMF/DMSO, control) for 24 h. (**A**) Representative cytograms displaying cell cycle distribution after challenging cells with the indicated drugs. (**B**) Histogram bars show percentages of cells in G0/G1 transition, S phase, G2/M transition and sub-G1 population (hypodiploid cells). Values represent means ± S.D. of 5 independent experiments. * $p < 0.05$ compared to control values (Dunnett's test).

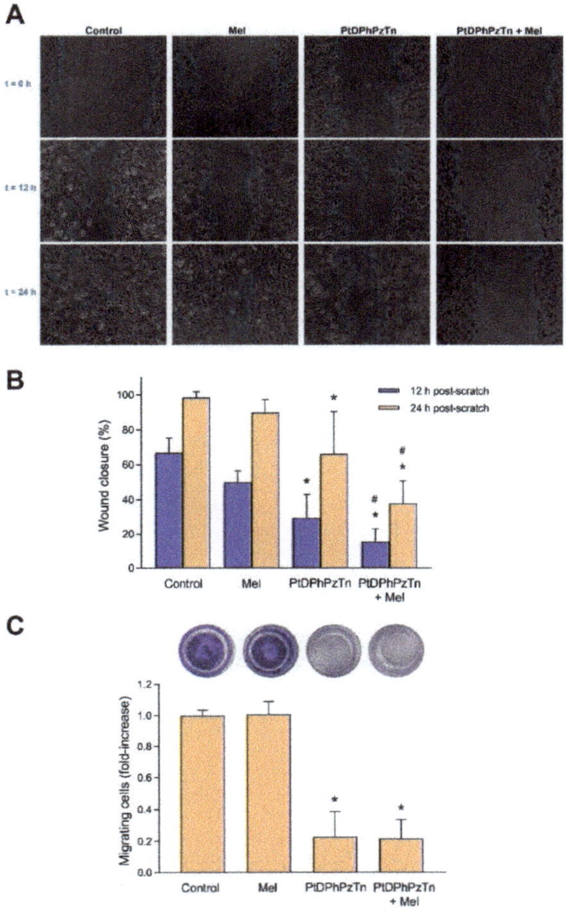

Figure 5. Combinatory treatment of PtDPhPzTn and melatonin altered migration properties of MDA-MB-231 cell line. Cells were treated with 10.4 μM PtDPhPzTn in the presence or absence of 1 mM melatonin (Mel) or the vehicle (DMF/DMSO, control) for 24 h. (**A**) Representative images depicting wound-healing assay in cells challenged with the indicated drugs. (**B**) Histogram bars show the percentage of wound closure after 12 and 24 h, where 100% represents a fully closed wound. Values represent means ± S.D. of 12 independent experiments. (**C**) Proportion of migrating cells after 24 h of exposure to the indicated drugs determined by transwell migration assay (for details, see Section 2.10). Representative images of transwell inserts for each condition are depicted in the upper part. Data represent means ± S.D. of 5 independent experiments and are expressed as fold-increase of control values. * $p < 0.05$ compared to its corresponding control value (Dunnett's test). # $p < 0.05$ compared to its corresponding PtDPhPzTn value (Tukey's test).

4. Conclusions

Platinum(II)-based complexes are undoubtedly an interesting alternative to canonical chemotherapy agents as they can help avoiding drug resistance and off target toxicity by promoting the apoptosis of tumor cells via multiple mechanisms, including oxidative stress, mitochondrial injury, and DNA damage. In this paper, we described that the thiazoline-containing platinum(II) complex PtDPhPzTn, which reportedly induces oxidative stress and apoptosis in tumor cells [7], outperformed, by far, the reference drug cisplatin and its less lipophilic counterpart PtPzTn in terms of cytotoxicity against MDA-MB-231 cells. More importantly, the combinatory treatment with PtDPhPzTn and the renowned antioxidant melatonin potentiated the pro-apoptotic, cell cycle arrest and anti-migration properties of the platinum(II) complex alone against MDA-MB-231 breast cancer cells. Therefore, our findings provide evidence that a combined therapy with PtDPhPzTn and melatonin could be potentially applied to breast cancer management as powerful synergistic agents.

Supplementary Materials: The following are available online at https://www.mdpi.com/article/10.3390/antiox11101971/s1, Figure S1. Synergistic effect of the combination of PtDPhPzTn and melatonin in MDA-MB-231 cells, Table S1. Calculation of the combination index (CI) data for non-constant combination of PtDPhPzTn and melatonin in cell death induction.

Author Contributions: Conceptualization, J.E.; methodology, S.E., E.F.-D. and J.E.; formal analysis, J.E., F.L.-G. and E.V.-Z.; investigation, S.E., E.F.-D., J.E., F.L.-G. and E.V.-Z.; data curation, S.E., E.F.-D., J.E., F.L.-G. and E.V.-Z.; writing—original draft preparation, S.E., J.E., F.L.-G., E.V.-Z., A.B.R. and J.A.P.; writing—review and editing, S.E., J.E., F.L.-G., E.V.-Z., A.B.R. and J.A.P.; visualization, S.E., E.F.-D., J.E., F.L.-G., E.V.-Z., A.B.R. and J.A.P.; supervision, J.E., F.L.-G., E.V.-Z., A.B.R. and J.A.P.; project administration, F.L.-G. and J.A.P.; funding acquisition, F.L.-G. and J.A.P. All authors have read and agreed to the published version of the manuscript.

Funding: This research was funded by Junta de Extremadura grants (ref. GR21042, GR21075, and IB18013). E. Fernández-Delgado holds a research pre-doctoral fellowship from Junta de Extremadura (jointly financed by European Social Fund, ref. PD18020). J. Espino holds a research post-doctoral fellowship from Junta de Extremadura (ref. TA18002).

Institutional Review Board Statement: Not applicable.

Informed Consent Statement: Not applicable.

Data Availability Statement: The data are contained within this article and the Supplementary File.

Acknowledgments: The authors appreciate the technical and human support provided by the facility of Bioscience Applied Techniques of SAIUEx (financed by UEx, Junta de Extremadura, MICINN, FEDER, and FSE).

Conflicts of Interest: The authors declare no conflict of interest.

References

1. Carey, L.; Winer, E.; Viale, G.; Cameron, D.; Gianni, L. Triple-negative breast cancer: Disease entity or title of convenience? *Nat. Rev. Clin. Oncol.* **2010**, 7, 683–692. [PubMed]
2. Shaw, A.Y.; Liau, H.H.; Lu, P.J.; Yang, C.N.; Lee, C.H.; Chen, J.Y.; Xu, Z.; Flynn, G. 3,5-Diaryl-1H-pyrazole as a molecular scaffold for the synthesis of apoptosis-inducing agents. *Bioorg. Med. Chem.* **2010**, 18, 3270–3278. [CrossRef]
3. Havrylyuk, D.; Zimenkovsky, B.; Vasylenko, O.; Zaprutko, L.; Gzella, A.; Lesyk, R. Synthesis of novel thiazolone-based compounds containing pyrazoline moiety and evaluation of their anticancer activity. *Eur. J. Med. Chem.* **2009**, 44, 1396–1404. [CrossRef] [PubMed]
4. Espino, J.; Fernández-Delgado, E.; Estirado, S.; de la Cruz-Martinez, F.; Villa-Carballar, S.; Viñuelas-Zahínos, E.; Luna-Giles, F.; Pariente, J.A. Synthesis and structure of a new thiazoline-based palladium(II) complex that promotes cytotoxicity and apoptosis of human promyelocytic leukemia HL-60 cells. *Sci. Rep.* **2020**, 10, 16745. [CrossRef]
5. Fernández-Delgado, E.; de la Cruz-Martínez, F.; Galán, C.; Franco, L.; Espino, J.; Viñuelas-Zahínos, E.; Luna-Giles, F.; Bejarano, I. Pt(II) and Pd(II) complexes with a thiazoline derivative ligand: Synthesis, structural characterization, antiproliferative activity and evaluation of pro-apoptotic ability in tumor cell lines HT-29 and U-937. *J. Inorg. Biochem.* **2020**, 202, 110870. [CrossRef] [PubMed]
6. Gutiérrez-Tarriño, S.; Espino, J.; Luna-Giles, F.; Rodríguez, A.B.; Pariente, J.A.; Viñuelas-Zahínos, E. Synthesis, characterization and antiproliferative evaluation of Pt(II) and Pd(II) complexes with a thiazine-pyridine derivative ligand. *Pharmaceuticals* **2021**, 14, 395. [CrossRef] [PubMed]
7. Fernández-Delgado, E.; Estirado, S.; Espino, J.; Viñuelas-Zahínos, E.; Luna-Giles, F.; Rodríguez Moratinos, A.B.; Pariente, J.A. Influence of ligand lipophilicity in Pt(II) complexes on their antiproliferative and apoptotic activities in tumour cell lines. *J. Inorg. Biochem.* **2022**, 227, 111688. [CrossRef] [PubMed]
8. Reiter, R.J.; Rosales-Corral, S.A.; Tan, D.X.; Acuna-Castroviejo, D.; Qin, L.; Yang, S.F.; Xu, K. Melatonin, a full service anti-cancer agent: Inhibition of initiation, progression and metastasis. *Int. J. Mol. Sci.* **2017**, 18, 843. [CrossRef]

9. Mehrzadi, S.; Pourhanifeh, M.H.; Mirzaei, A.; Moradian, F.; Hosseinzadeh, A. An updated review of mechanistic potentials of melatonin against cancer: Pivotal roles in angiogenesis, apoptosis, autophagy, endoplasmic reticulum stress and oxidative stress. *Cancer Cell Int.* **2021**, *21*, 1–28. [CrossRef]
10. Petronek, M.S.; Stolwijk, J.M.; Murray, S.D.; Steinbach, E.J.; Zakharia, Y.; Buettner, G.R.; Spitz, D.R.; Allen, B.G. Utilization of redox modulating small molecules that selectively act as pro-oxidants in cancer cells to open a therapeutic window for improving cancer therapy. *Redox Biol.* **2021**, *42*, 101864.
11. Zhang, H.M.; Zhang, Y. Melatonin: A well-documented antioxidant with conditional pro-oxidant actions. *J. Pineal Res.* **2014**, *57*, 131–146. [PubMed]
12. Uguz, A.C.; Cig, B.; Espino, J.; Bejarano, I.; Naziroglu, M.; Rodríguez, A.B.; Pariente, J.A. Melatonin potentiates chemotherapy-induced cytotoxicity and apoptosis in rat pancreatic tumor cells. *J. Pineal Res.* **2012**, *53*, 91–98. [CrossRef] [PubMed]
13. Pariente, R.; Bejarano, I.; Rodríguez, A.B.; Pariente, J.A.; Espino, J. Melatonin increases the effect of 5-fluorouracil-based chemotherapy in human colorectal adenocarcinoma cells in vitro. *Mol. Cell. Biochem.* **2018**, *440*, 43–51. [CrossRef] [PubMed]
14. Pariente, R.; Pariente, J.A.; Rodríguez, A.B.; Espino, J. Melatonin sensitizes human cervical cancer HeLa cells to cisplatin-induced cytotoxicity and apoptosis: Effects on oxidative stress and DNA fragmentation. *J. Pineal Res.* **2016**, *60*, 55–64. [CrossRef]
15. Pariente, R.; Bejarano, I.; Espino, J.; Rodríguez, A.B.; Pariente, J.A. Participation of MT3 melatonin receptors in the synergistic effect of melatonin on cytotoxic and apoptotic actions evoked by chemotherapeutics. *Cancer Chemother. Pharmacol.* **2017**, *80*, 985–998. [CrossRef]
16. Jadid, M.F.S.; Aghaei, E.; Taheri, E.; Seyyedsani, N.; Chavoshi, R.; Abbasi, S.; Khorrami, A.; Goleij, P.; Hajazimian, S.; Taefehshokr, S.; et al. Melatonin increases the anticancer potential of doxorubicin in Caco-2 colorectal cancer cells. *Environ. Toxicol.* **2021**, *36*, 1061–1069. [CrossRef]
17. González-González, A.; González, A.; Rueda, N.; Alonso-González, C.; Menéndez, J.M.; Martínez-Campa, C.; Mitola, S.; Cos, S. Usefulness of melatonin as complementary to chemotherapeutic agents at different stages of the angiogenic process. *Sci. Rep.* **2020**, *10*, 4790. [CrossRef]
18. Su, S.C.; Hsieh, M.J.; Yang, W.E.; Chung, W.H.; Reiter, R.J.; Yang, S.F. Cancer metastasis: Mechanisms of inhibition by melatonin. *J. Pineal Res.* **2017**, *62*, 12370. [CrossRef]
19. Dauchy, R.T.; Xiang, S.; Mao, L.; Brimer, S.; Wren, M.A.; Yuan, L.; Anbalagan, M.; Hauch, A.; Frasch, T.; Rowan, B.G.; et al. Circadian and melatonin disruption by exposure to light at night drives intrinsic resistance to tamoxifen therapy in breast cancer. *Cancer Res.* **2014**, *74*, 4099–4110. [CrossRef]
20. Xiang, S.; Dauchy, R.T.; Hauch, A.; Mao, L.; Yuan, L.; Wren, M.A.; Belancio, V.P.; Mondal, D.; Frasch, T.; Blask, D.E.; et al. Doxorubicin resistance in breast cancer is driven by light at night-induced disruption of the circadian melatonin signal. *J. Pineal Res.* **2015**, *59*, 60–69. [CrossRef]
21. Ortiz, F.; Acuña-Castroviejo, D.; Doerrier, C.; Dayoub, J.C.; Lõpez, L.C.; Venegas, C.; García, J.A.; Lõpez, A.; Volt, H.; Luna-Sánchez, M.; et al. Melatonin blunts the mitochondrial/NLRP3 connection and protects against radiation-induced oral mucositis. *J. Pineal Res.* **2015**, *58*, 34–49. [CrossRef] [PubMed]
22. Kilic, U.; Kilic, E.; Tuzcu, Z.; Tuzcu, M.; Ozercan, I.H.; Yilmaz, O.; Sahin, F.; Sahin, K. Melatonin suppresses cisplatin-induced nephrotoxicity via activation of Nrf-2/HO-1 pathway. *Nutr. Metab.* **2013**, *10*, 7. [CrossRef] [PubMed]
23. Bernalte-García, A.; Lozano-Vila, A.M.; Luna-Giles, F.; Pedrero-Marín, R. Structural characterization of a thiazoline-pyrazole ligand and its complexes with cobalt(II) and copper(II). *Polyhedron* **2006**, *25*, 1399–1407. [CrossRef]
24. Torres-García, P.; Pedrero-Marín, R.; Luna-Giles, F.; Huertas-Sánchez, A.V.; Viñuelas-Zahínos, E. Influence of steric strain of S,N-heterocycles derivative ligands on the coordination geometry in cadmium(II) nitrato complexes. *Polyhedron* **2012**, *31*, 307–318. [CrossRef]
25. Price, J.H.; Williamson, A.N.; Schramm, R.F.; Wayland, B.B. Palladium(II) and platinum(II) alkyl sulfoxide complexes. Examples of sulfur-bonded, mixed sulfur- and oxygen-bonded, and totally oxygen-bonded complexes. *Inorg. Chem.* **1972**, *11*, 1280–1284. [CrossRef]
26. Chou, T.C. Drug combination studies and their synergy quantification using the chou-talalay method. *Cancer Res.* **2010**, *70*, 440–446. [CrossRef]
27. Kommineni, N.; Saka, R.; Bulbake, U.; Khan, W. Cabazitaxel and thymoquinone co-loaded lipospheres as a synergistic combination for breast cancer. *Chem. Phys. Lipids* **2019**, *224*, 104707. [CrossRef]
28. Suarez-Arnedo, A.; Figueroa, F.T.; Clavijo, C.; Arbeláez, P.; Cruz, J.C.; Muñoz-Camargo, C. An image J plugin for the high throughput image analysis of in vitro scratch wound healing assays. *PLoS ONE* **2020**, *15*, e0232565. [CrossRef]
29. Czarnomysy, R.; Surażyński, A.; Muszynska, A.; Gornowicz, A.; Bielawska, A.; Bielawski, K. A novel series of pyrazole-platinum(II) complexes as potential anti-cancer agents that induce cell cycle arrest and apoptosis in breast cancer cells. *J. Enzym. Inhib. Med. Chem.* **2018**, *33*, 1006–1023. [CrossRef]
30. Bielawski, K.; Czarnomysy, R.; Muszyńska, A.; Bielawska, A.; Popławska, B. Cytotoxicity and induction of apoptosis of human breast cancer cells by novel platinum(II) complexes. *Environ. Toxicol. Pharmacol.* **2013**, *35*, 254–264. [CrossRef]
31. Agnieszka, G.; Ewa, A.; Anna, B.; Krzysztof, B.; Monika, C.; Elzbieta, S. Dinuclear berenil-platinum (II) complexes as modulators of apoptosis in human MCF-7 and MDA-MB231 breast cancer cells. *Anticancer. Agents Med. Chem.* **2014**, *14*, 1179–1186. [CrossRef] [PubMed]

32. Woo, S.M.; Min, K.J.; Kwon, T.K. Melatonin-mediated Bim up-regulation and cyclooxygenase-2 (COX-2) down-regulation enhances tunicamycin-induced apoptosis in MDA-MB-231 cells. *J. Pineal Res.* **2015**, *58*, 310–320. [CrossRef] [PubMed]
33. Nooshinfar, E.; Bashash, D.; Safaroghli-Azar, A.; Bayati, S.; Rezaei-Tavirani, M.; Ghaffari, S.H.; Akbari, M.E. Melatonin promotes ATO-induced apoptosis in MCF-7 cells: Proposing novel therapeutic potential for breast cancer. *Biomed. Pharmacother.* **2016**, *83*, 456–465. [CrossRef]
34. Koşar, P.A.; Nazıroğlu, M.; Övey, İ.S.; Çiğ, B. Synergic Effects of Doxorubicin and Melatonin on Apoptosis and Mitochondrial Oxidative Stress in MCF-7 Breast Cancer Cells: Involvement of TRPV1 Channels. *J. Membr. Biol.* **2016**, *249*, 129–140. [CrossRef] [PubMed]
35. Tran, Q.H.; Hoang, D.H.; Song, M.; Choe, W.; Kang, I.; Kim, S.S.; Ha, J. Melatonin and doxorubicin synergistically enhance apoptosis via autophagy-dependent reduction of AMPKα1 transcription in human breast cancer cells. *Exp. Mol. Med.* **2021**, *53*, 1413–1422. [CrossRef] [PubMed]
36. Sang, X.; Li, L.; Rui, C.; Liu, Y.; Liu, Z.; Tao, Z.; Cheng, H.; Liu, P. Induction of EnR stress by melatonin enhances the cytotoxic effect of lapatinib in HER2-positive breast cancer. *Cancer Lett.* **2021**, *518*, 82–93. [CrossRef]
37. Hajipour, F.; Mahdavinia, M.; Fereidoonnezhad, M. Half-lantern cyclometalated platinum(II) complexes as anticancer agents: Molecular docking, apoptosis, cell cycle analysis, and cytotoxic activity evaluations. *Anticancer. Agents Med. Chem.* **2022**, *22*, 1149–1158. [CrossRef]
38. Czarnomysy, R.; Radomska, D.; Muszynska, A.; Hermanowicz, J.M.; Prokop, I.; Bielawska, A.; Bielawski, K. Evaluation of the anticancer activities of novel transition metal complexes with berenil and nitroimidazole. *Molecules* **2020**, *25*, 2860. [CrossRef]
39. Jardin, I.; Diez-Bello, R.; Falcon, D.; Alvarado, S.; Regodon, S.; Salido, G.M.; Smani, T.; Rosado, J.A. Melatonin downregulates TRPC6, impairing store-operated calcium entry in triple-negative breast cancer cells. *J. Biol. Chem.* **2021**, *296*, 100254. [CrossRef]
40. Gelaleti, G.B.; Borin, T.F.; Maschio-Signorini, L.B.; Moschetta, M.G.; Jardim-Perassi, B.V.; Calvinho, G.B.; Facchini, M.C.; Viloria-Petit, A.M.; de Campos Zuccari, D.A.P. Efficacy of melatonin, IL-25 and siIL-17B in tumorigenesis-associated properties of breast cancer cell lines. *Life Sci.* **2017**, *183*, 98–109. [CrossRef]
41. Panzer, A.; Lottering, M.L.; Bianchi, P.; Glencross, D.K.; Stark, J.H.; Seegers, J.C. Melatonin has no effect on the growth, morphology or cell cycle of human breast cancer (MCF-7), cervical cancer (HeLa), osteosarcoma (MG-63) or lymphoblastoid (TK6) cells. *Cancer Lett.* **1998**, *122*, 17–23. [CrossRef]
42. Gamboa Varela, J.; De Chatterjee, A.; Guevara, P.; Ramirez, V.; Metta-Magaña, A.J.; Villagrán, D.; Varela-Ramirez, A.; Das, S.; Nuñez, J.E. Synthesis, characterization, and evaluation of *cis*-diphenyl pyridineamine platinum(II) complexes as potential anti-breast cancer agents. *J. Biol. Inorg. Chem.* **2014**, *19*, 967–979. [CrossRef] [PubMed]
43. Bai, X.; Ali, A.; Lv, Z.; Wang, N.; Zhao, X.; Hao, H.; Zhang, Y.; Rahman, F.U. Platinum complexes inhibit HER-2 enriched and triple-negative breast cancer cells metabolism to suppress growth, stemness and migration by targeting PKM/LDHA and CCND1/BCL2/ATG3 signaling pathways. *Eur. J. Med. Chem.* **2021**, *224*, 113689. [CrossRef] [PubMed]
44. Maroufi, N.F.; Rashidi, M.; Vahedian, V.; Jahanbazi, R.; Mostafaei, S.; Akbarzadeh, M.; Kazemzadeh, H.; Nejabati, H.R.; Isazadeh, A.; Rashidi, M.R.; et al. Effect of Apatinib plus melatonin on vasculogenic mimicry formation by cancer stem cells from breast cancer cell line. *Breast Cancer* **2022**, *29*, 260–273. [CrossRef] [PubMed]
45. Hasan, M.; Marzouk, M.A.; Adhikari, S.; Wright, T.D.; Miller, B.P.; Matossian, M.D.; Elliott, S.; Wright, M.; Alzoubi, M.; Collins-Burow, B.M.; et al. Pharmacological, mechanistic, and pharmacokinetic assessment of novel_ melatonin-tamoxifen drug conjugates as breast cancer drugs. *Mol. Pharmacol.* **2019**, *96*, 272–296. [CrossRef] [PubMed]
46. El-Sokkary, G.H.; Ismail, I.A.; Saber, S.H. Melatonin inhibits breast cancer cell invasion through modulating DJ-1/KLF17/ID-1 signaling pathway. *J. Cell Biochem.* **2019**, *120*, 3945–3957. [CrossRef]

antioxidants

Article

Dose-Dependent Effect of Melatonin on BAT Thermogenesis in Zücker Diabetic Fatty Rat: Future Clinical Implications for Obesity

Samira Aouichat [1], Enrique Raya [2], Antonio Molina-Carballo [3,4], Antonio Munoz-Hoyos [3,4], Abdelkarim Saleh Aloweidi [5], Ehab Kotb Elmahallawy [6] and Ahmad Agil [1,*]

1 Department of Pharmacology, Biohealth Institute Granada (IBs Granada) and Neuroscience Institute, School of Medicine, University of Granada, 18016 Granada, Spain
2 Department of Rheumatology, University Hospital Clinic San Cecilio, 18016 Granada, Spain
3 Department of Pediatrics, School of Medicine, University of Granada, 18016 Granada, Spain
4 Unit of Pediatric Neurology and Neurodevelopment, University Hospital Clinic San Cecilio, Andalusian Health Service, 18016 Granada, Spain
5 Department of Anesthesia and Intensive Care, Faculty of Medicine, University of Jordan, Amman 11942, Jordan
6 Department of Zoonoses, Faculty of Veterinary Medicine, Sohag University, Sohag 82524, Egypt
* Correspondence: aagil@ugr.es; Tel.: +34-958-2435

Abstract: Experimental data have revealed that melatonin at high doses reduced obesity and improved metabolic outcomes in experimental models of obesity, mainly by enhancing brown adipose tissue (BAT) thermogenesis. A potential dose-response relationship has yet to be performed to translate these promising findings into potential clinical therapy. This study aimed to assess the effects of different doses of melatonin on interscapular BAT (iBAT) thermogenic capacity in Zücker diabetic fatty (ZDF) rats. At 6 wk of age, male ZDF rats were divided into four groups ($n = 4$ per group): control and those treated with different doses of melatonin (0.1, 1, and 10 mg/kg of body weight) in their drinking water for 6 wk. Body weight (BW) was significantly decreased at doses of 1 and 10 mg/kg of melatonin, but not at 0.1 mg/kg compared with the control, with a similar rate of BW decrease being reached at the dose of 1 mg/kg (by ~11%) and 10 mg/kg (by ~12%). This effect was associated with a dose-dependent increase in the thermal response to the baseline condition or acute cold challenge in the interscapular area measurable by infrared thermography, with the highest thermal response being recorded at the 10 mg/kg dose. Upon histology, melatonin treatment markedly restored the typical brownish appearance of the tissue and promoted a shift in size distribution toward smaller adipocytes in a dose-dependent fashion, with the most pronounced brownish phenotype being observed at 10 mg/kg of melatonin. As a hallmark of thermogenesis, the protein level of uncoupled protein 1 (UCP1) from immunofluorescence and Western blot analysis increased significantly and dose-dependently at all three doses of melatonin, reaching the highest level at the dose of 10 mg/kg. Likewise, all three doses of melatonin modulated iBAT mitochondrial dynamics by increasing protein expression of the optic atrophy protein type 1 (OPA1) fusion marker and decreasing that of the dynamin-related protein1 (DRP1) fission marker, again dose-dependently, with the highest and lowest expression levels, respectively, being reached at the 10 mg/kg dose. These findings highlight for the first time the relevance of the dose-dependency of melatonin toward BW control and BAT thermogenic activation, which may have potential therapeutic implications for the treatment of obesity. To clinically apply the potential therapeutic of melatonin for obesity, we consider that the effective animal doses that should be extrapolated to obese individuals may be within the dose range of 1 to 10 mg/kg.

Keywords: melatonin; dose-response; BAT; thermogenesis; UCP1; mitochondria; obesity; ZDF rat

Citation: Aouichat, S.; Raya, E.; Molina-Carballo, A.; Munoz-Hoyos, A.; Aloweidi, A.S.; Elmahallawy, E.K.; Agil, A. Dose-Dependent Effect of Melatonin on BAT Thermogenesis in Zücker Diabetic Fatty Rat: Future Clinical Implications for Obesity. *Antioxidants* **2022**, *11*, 1646. https://doi.org/10.3390/antiox11091646

Academic Editors: Marina Garcia-Macia, Adrián Santos-Ledo and Ana B Rodríguez Moratinos

Received: 2 July 2022
Accepted: 22 August 2022
Published: 25 August 2022

1. Introduction

The prevalence of obesity has dramatically increased over the past 40 years and is expected to increase further to ~30.3 billion by 2030 [1]. Obesity develops when energy intake chronically exceeds energy expenditure (EE), with the deposition of this excess energy primarily as triglycerides in white adipose tissue [2]. Most of the current available anti-obesity pharmacological approaches act by reducing caloric intake or impairing fat absorption [3]. However, because these approaches are associated with multiple side effects [3], safe and effective alternatives for treating and overcoming obesity are urgently needed. In recent years, targeting BAT thermogenesis to increase EE has emerged as a promising therapeutic strategy to combat obesity [4].

Unlike white adipose tissue, which is involved in storing excess energy as fat that can be mobilized in times of need, BAT can convert stored fat into heat energy through a process termed adaptive non-shivering thermogenesis, which plays an important role in temperature regulation in rodents and newborn babies [5]. Morphologically, BAT is composed of multilocular lipid droplets and high contents of mitochondria that contain a specific protein in the inner mitochondrial membrane called uncoupling protein 1 (UCP1) [6]. The UCP1 plays a critical role in BAT-mediated thermogenesis by causing a proton gradient leak across the mitochondrial membrane and thereby uncoupling oxidative phosphorylation to generate heat instead of ATP synthesis [7]. The UCP1 in brown adipocytes is induced by a wide range of metabolite sources, among them fatty acids and glucose, which are major contributors and play a pivotal role in BAT thermogenesis [5,8]. Activated BAT can use glucose and fatty acids to maintain a high oxidative metabolic capacity, and when it is highly active, it exerts beneficial metabolic effects [5,8]. The oxidative metabolism and uncoupled respiration in brown adipocytes are mainly accomplished by mitochondria, which hold a central position in thermogenesis [5]. These organelles undergo dynamic processes of fusion and fission that can serve as a mechanism for metabolic adaptation to mitochondrial bioenergetic efficiency and energy substrate oxidation [5]. Therefore, normal mitochondrial function is essential for maintaining BAT function and metabolic homeostasis.

Recently, BAT research has dramatically increased after the rediscovery of metabolically active BAT in adult humans using ^{18}fluoro-2-deoxyglucose positron emission tomography (^{18}FDG-PET). Importantly, the prevalence and activity of adult human BAT, as assessed by 18FDG-PET, have been found to correlate negatively with body mass index (BMI) in humans [9–14], and individuals with detectable metabolically active BAT have a lower risk of developing type 2 diabetes and heart diseases [15]. It is therefore expected that targeting BAT thermogenesis could be a promising strategy and a hopeful target for counteracting obesity and related metabolic disorders. In support of this perspective, an earlier prospective cohort study has shown an increased prevalence of BAT activity after weight loss in morbidly obese patients undergoing gastric banding surgery, suggesting the effectiveness of targeting BAT activation in humans as an anti-obesity therapy [16]. Several candidates are now being developed as possible therapeutic targets in the process of BAT activation, but their potential adverse or toxic effects may limit their long-term use for this purpose [3,17]. In this context, there is an urgent and highly demanded need for developing effective and safe new drugs to reduce obesity by targeting BAT. In recent decades, melatonin has attracted increasing attention as a promising therapeutic for obesity and related diseases mainly due to its high efficacy and remarkable lack of toxicity and side effects at any dose, even for long-term use in humans [18,19].

Melatonin (N-acetyl-5-methoxytryptamine) is a naturally occurring substance mainly produced and secreted from the pineal gland and is well known for its role in controlling the circadian rhythm [20]. Exogenous melatonin is widely used to remedy sleep disorders and jet lag, either as a dietary supplement or as a drug, in many European countries and the USA [21,22]. The extraordinary intrinsic properties of melatonin, including high cell permeability, the ability to reach subcellular compartments, and its role as a robust anti-inflammatory, antioxidant and free radical scavenging, and anti-apoptotic agent, as well as its acceptable safety profile, make it a promising approach for the treatment of various diseases, including obesity and related diseases [23]. There is growing evidence that exogenous melatonin exerts beneficial effects on obesity and related metabolic disorders in animals; nevertheless, clinical evidence of its effectiveness in obese individuals is still insufficient and conflicting [4], which makes it difficult to draw a firm conclusion and thus further clinical trials are warranted to clarify this issue. The BAT thermogenic mechanism has been proposed to account for the anti-obesogenic effect of melatonin [24–26]. In this context, our previous studies carried out in ZDF rats have shown that melatonin treatment exhibited promising effects in increasing BAT mass and thermogenic activity, increasing UCP1 expression, improving mitochondrial respiration, and reducing brown adipocyte susceptibility to mitochondrial oxidative/nitrosative stress and apoptosis and, in addition, prevented BW gain and glucose–lipid metabolic disorders [26–29]. Although the observed beneficial effects of melatonin on BAT have not been translated yet to clinical experimentation, a small-scale proof of concept study in patients with melatonin deficiency due to radiotherapy or surgical removal of the pineal gland has shown that melatonin replacement therapy increased BAT volume and activity and improved cholesterol and triglyceride blood levels [30].

On the above basis, and given the ability of BAT to dissipate energy as heat and to contribute to EE, we are optimistic that melatonin can be used as a BAT activator for future therapeutic applications in obesity. Thus, information on the potentially effective dose of melatonin is urgently needed to underscore the clinical meaning of this potential therapy for obesity treatment in the future. To this end, the present study aimed to perform an effective dose-setting study to evaluate the effects of three different doses of melatonin (0.1, 1, and 10 mg/kg) on iBAT thermogenic capacity in ZDF rats. The determination of the dose-response relationship will have important implications on dosing regimens in clinical application, not only to determine the effective dose that produces the desired therapeutic effect but also because of its implications for economical and safety aspects, since lower doses may have the same desired therapeutic endpoint.

2. Materials and Methods

2.1. Ethical Statement

Ethical approval of this study was obtained from the Ethical Committee at the University of Granada (Granada, Spain) under the reference number 4-09-2016-CEEA, according to the European Union guidelines for animal care and protection.

2.2. Reagents

All reagents used were of the highest purity available. Melatonin was purchased from Sigma-Aldrich (Madrid, Spain).

2.3. Animals and Experimental Protocol

Male Zücker diabetic fatty rats (ZDF; fa/fa) were obtained at 5 wk of age from Charles River Laboratory (Charles River Laboratories, SA, Barcelona, Spain). Animals were maintained on Purina 5008 rat chow (protein 23%, fat 6.5%, carbohydrates 58.5%, fibre 4%, and ash 8%; Charles River Laboratories, SA, Barcelona, Spain) and housed two per clear plastic cage under a 12 h light/dark cycle (lights on at 07:00 a.m.) and controlled environmental conditions of temperature (28–30 °C) and relative humidity (30–40%). The animals were acclimated to room conditions for one week before the experiments, and water intake was recorded.

Upon initiation of the experiment, the animals were randomly divided into four groups (n = 4 per group) as follows: the first three groups were treated with melatonin in their drinking water at the doses of 0.1, 1, and 10 mg/kg b.w/day (Mel-0.1, Mel-1, and Mel-10, respectively) for 6 consecutive weeks, and the remaining group served as a vehicle-treated control (Control). Melatonin was dissolved in a minimum volume of absolute ethanol and then diluted in the drinking water to yield doses of 0.1, 1, and 10 mg/kg b.w, with a final ethanol concentration of 0.066% (*w*/*v*). The control group received the vehicle only in the drinking water at a comparable concentration and treatment duration. Fresh melatonin and vehicle solutions were prepared three times a week, and melatonin doses were adjusted for BWs throughout the experimental period. To carefully determine the treatment dose throughout the experimental period, water intake and BWs were recorded three times a week. All water bottles were covered with aluminium foil to prevent melatonin photodegradation. The BW of all animals was recorded at the beginning and the end of the experiment. At the end of the experiment, the animals were sacrificed under sodium thiobarbital (thiopental) anaesthesia, and the interscapular brown fat pads of each rat were immediately dissected and then frozen at −80 °C for further analysis. A portion of the iBAT pads was fixed in 4% paraformaldehyde for histological analysis.

2.4. Infrared Thermal Imaging Measurments

After overnight fasting, ZDF rats from the four groups were either kept at 28–30 °C (room temperature) or exposed to 4 °C for 5 min for an acute cold challenge between 09:00 and 11:00 h. The acute cold challenge was performed by placing the rat on a hot/cold plate analgesia meter precooled to 4 °C (Panlab SLU, Barcelona, Spain).

Thermal images of the BAT in the interscapular region were recorded with a thermal imaging camera (FLIR B425, FLIR Systems AB, Danderyd, Sweden) with a range limit of −20 °C to 120 °C. Dorsal thermal images at a perpendicular distance of 20 cm were regularly captured before (at room temperature) and immediately after the acute cold challenge. The images were analysed with a specific software package (FLIR Systems software).

2.5. Purification of Mitochondria

Mitochondrial protein extract was isolated from iBAT by serial centrifugation according to the instructions previously described by our research group [31]. Briefly, about 300 mg of iBAT was washed with cold saline and homogenized (1:10 *w/v*) in an isolation medium (10 mM Tris, 250 mM sucrose, 0.5 mM Na_2EDTA, and 1 g/L free fatty acid bovine serum albumin (BSA); pH 7.4; 4 °C) with a Teflon pestle. The homogenate was centrifuged at a low speed (1000× *g*) for 10 min at 4 °C, and the resultant supernatant was decanted to fresh tubes and subjected to high-speed centrifugation (15,000× *g*) for 20 min at 4 °C. The resultant pellet containing the mitochondria was resuspended in 1 mL of isolation medium (BSA free) and centrifuged again at high speed (15,000× *g*) for 20 min at 4 °C to pellet the pure mitochondria. The pelleted pure mitochondria were resuspended in 1 mL of respiration buffer (20 mM HEPES, 0.5 mM EGTA, 3 mM $MgCl_2$, 20 mM taurine, 10 mM KH_2PO_4, 200 mM sucrose, and 1 g/L fatty-acid-free BSA) and incubated on ice for 10–15 min to allow the rearrangement of the membranes. Aliquots of pure mitochondrial extracts were stored at −80 °C for Western blotting. Protein concentration in the mitochondrial extracts was measured by the Bradford method [32] using BSA as a standard.

2.6. Western Blot Analysis

Western blot analysis was performed according to the instructions previously described by our research group [31]. Equal amounts of mitochondrial protein extracts were resolved on SDS-PAGE (sodium dodecyl sulphate polyacrylamide gel electrophoresis). The gels for immunoblot analyses were transferred to a nitrocellulose membrane (Bio-Rad Trans-Blot SD, Bio-Rad Laboratories). The membranes were then blocked with 5% non-fat dry milk in tris-buffered saline (TBS) containing 0.05% Tween-20 (TBS-T) for 1 h at 37 °C and incubated overnight at 4 °C with primary antibodies against UCP1 (cat#U-6382), OPA1 (cat#SAB-5700860), and DRP1 (cat#SAB-5700783) at 1:500–1:2000 dilutions. All the antibodies were obtained from Sigma-Aldrich (Sigma-Aldrich, Madrid, Spain). Equal protein loading was demonstrated by incubating the membranes with mouse β-actin antibody (cat#SC-81178; Santa Cruz Biotechnology, Santa Cruz, CA, USA) at 1:1000 dilution. After the incubation, the membranes were washed three times for 20 min in TBS-T and incubated for 1 h at room temperature with respective horseradish-peroxidase-conjugated secondary antibodies (Sigma-Aldrich, Madrid, Spain) at 1:1000 dilution. The membrane was washed three times for 20 min in the TBS-T, and then a chemiluminescence assay system (ECL kit, GE Healthcare Life Sciences, Buckinghamshire, UK) was used to develop the immunoreactivity bands. Finally, the protein band densities were quantitatively analysed using Image J 1.33 software (National Institutes of Health, Bethesda, MD, USA). The results were normalized to β-actin as a loading control. All experiments were performed in triplicate.

2.7. Histological Analysis

Following anaesthesia, excised interscapular brown fat was fixed overnight in 10% (*w*/*v*) buffered formaldehyde at 4 °C for 24 h, rinsed with 0.1 M phosphate buffer (PBS (phosphate-buffered saline) pH 7.4, dehydrated in a graded series of ethanol (from 80% to absolute alcohol), cleared in xylene, and embedded in paraffin. The paraffin blocks from each group were cut with a microtome into 4 μm-thick sections, stained with haematoxylin and eosin (H&E), and inspected under a light microscope (Olympus, Germany) equipped with a digital camera system (Carl Zeiss camera, model Axiocam ERc 5s. Göttingen, Germany). Images of H&E-stained tissue sections were digitized, and adipocyte size and frequency distribution of adipocyte size was determined using Image J 1.33 software (National Institutes of Health, Bethesda, MD, USA). The average adipocyte size was expressed as the average cross-sectional area per cell (μm^2/cell) of the tissue sample, which was calculated based on the values of about 200 random adipocytes per group.

2.8. Immunofluorescence Staining for UCP1

For UCP1 immunofluorescence, unstained deparaffinised sections were treated with citrate buffer pH 6.0 for 20 min at 60 °C for antigen retrieval and then blocked with 5% BSA, followed by overnight incubation at 4 °C with the same UCP1 antibody used for Western blot analysis (cat# U-6382; Sigma-Aldrich, Madrid, Spain) at 1:1000 dilution. After incubation, the tissue sections were washed three times with a 1 × PBS solution containing Tween-20 and then incubated with secondary anti-rabbit cyanine3 antibody (Cy3) (cat# A-10520; Invitrogen, Molecular Probes, Carlsbad, CA, USA). Fluorescence photomicrographs were captured by an Olympus IX2 fluorescence microscope at a magnification of ×200. All photomicrographic images from each group were captured under the same camera and microscope settings. The immunofluorescence intensities were quantitatively analysed using Image J 1.33 software (National Institutes of Health, Bethesda, MD, USA).

2.9. Statistical Analysis

Statistical Package of Social Science (IBM SPSS Software, version 15, Michigan, IL, USA) was used for statistical analysis. All results are expressed as mean ± standard deviation (S.D.) values. Comparisons between experimental groups were analysed using one-way ANOVA followed by the Tukey post hoc test. Differences between group means were considered statistically significant if $p < 0.05$.

3. Results

3.1. Dose-Dependent Effects of Melatonin on Body Weight and iBAT Thermogenic Activity

We have previously shown that melatonin at a dose of 10 mg/kg for 6 wk protected against the excessive BW gain and increased the thermogenic function of iBAT in the ZDF rats [26,27]. Here, we investigated whether 0.1, 1, and 10 mg/kg of melatonin for the same treatment period and in the same animal strain showed dose-dependent effects on BW and iBAT thermogenic activity.

After 6 wk of treatment, melatonin significantly decreased the BW in the Mel-1 (by ~11%) and Mel-10 (by ~12%) groups compared with the control group ($p < 0.05$ and $p < 0.05$, respectively; Table 1), with no significant difference among them ($p > 0.05$; Table 1). In the Mel-0.1 group, the final BW showed a trend towards a nonsignificant decrease (by ~5%) compared with the control group ($p > 0.05$; Table 1). The final BW was found to be significantly lower in the Mel-10 and Mel-1 groups than in the Mel-0.1 group ($p < 0.05$ and $p < 0.05$, respectively; Table 1).

Table 1. Dose-dependent effects of melatonin on body weight in Zücker diabetic fatty rats.

Groups	Control	Mel-0.1	Mel-1	Mel-10
Final body weight (g)	495.7 ± 6.4	473.4 ± 4.6	443.0 ± 3.1 *×	436.1 ± 8.5 *#
Body weight gain (%)	74.7 ± 1.7	66.3 ± 2.4	55.2 ± 1.1 *×	54.7 ± 1.8 *#

Mel-0.1, Mel-1, and Mel-10: melatonin at a dose of 0.1, 1, and 10 mg/kg b.w, respectively. Values are means ± S.D ($n = 4$). * $p < 0.05$ Mel-1 and Mel-10 vs. Control; # $p < 0.05$ Mel-10 vs. Mel-0.1; × $p < 0.05$ Mel-1 vs. Mel-0.1 (One-way ANOVA with Tukey post hoc test).

To assess the degree of iBAT thermogenic activity, we first measured the interscapular skin temperature by infrared thermography at both baseline and acute cold conditions. The acute cold test was performed to gauge the adaptive thermogenesis in iBAT, which has been found to be blunted in obese and overweight individuals in the cold-activated state [33,34]. The interscapular fat depot was selected because it is the largest and most accessible BAT depot in rodents that may correspond to the supraclavicular BAT depot described in adult humans [35].

Based on the experiment, the baseline interscapular skin temperature was significantly higher in the Mel-1 (35.4 ± 0.1 °C) and Mel-10 (35.8 ± 0.2 °C) groups than in the control group (34.3 ± 0.2 °C; $p < 0.05$ and $p < 0.05$, respectively; Figure 1a), without a significant difference among them ($p > 0.05$; Figure 1a). There was no significant difference in the baseline interscapular skin temperature between the Mel-0.1 (34.5 ± 0.1 °C) and control groups ($p > 0.05$; Figure 1a). Significantly higher interscapular skin temperature degrees were found in the Mel-10 and Mel-1 groups than in the Mel-0.1 group ($p < 0.05$ and $p < 0.05$, respectively; Figure 1a).

After the cold challenge, the interscapular skin temperature rose significantly in the Mel-1 (0.26 ± 0.04 °C) and Mel-10 groups (0.38 ± 0.05 °C) in comparison with the control group (0.04 ± 0.03 °C; $p < 0.05$ and $p < 0.01$, respectively; Figure 1b), with no significant difference among them ($p > 0.05$; Figure 1b). A non-significant trend towards a higher increase in the interscapular skin temperature was observed in the Mel-0.1 group (0.10 ± 0.01 °C) when compared with the control group ($p > 0.05$; Figure 1b). The Mel-10 and Mel-1 groups displayed a significantly higher increase in the interscapular skin temperature than the Mel-0.1 group ($p < 0.05$ and $p < 0.05$, respectively; Figure 1b). The dose-dependent effect of melatonin treatment on the iBAT thermogenesis was also clearly apparent in the infrared thermal images shown in Figure 1c.

3.2. Dose-Dependent Effects of Melatonin on iBAT Morphology

To explore whether the dose-dependent effect of melatonin on iBAT thermogenesis was reflected in particular morphological characteristics, iBAT histological analysis was performed. As shown in the H&E-stained preparations (Figure 2a), the iBAT of the ZDF control rats consisted predominately of hypertrophic unilocular adipocytes with a large lipid droplet, giving a white-like appearance rather than the typical brownish phenotype. Consistent with the above-described infrared thermography results, all three doses of melatonin resulted in a gradual suppression of unilocular lipid droplet deposition, with the smallest sign of whitening being observed in the Mel-10 group. Notably, an apparent accumulation of smaller adipocytes with multilocular lipid droplets was observed in the iBAT of the Mel-1 and Mel-10 groups. In the Mel-0.1 group, although the brownish phenotype of the iBAT was not evident, it was clearly distinguishable from the control iBAT, with apparently smaller unilocular adipocytes.

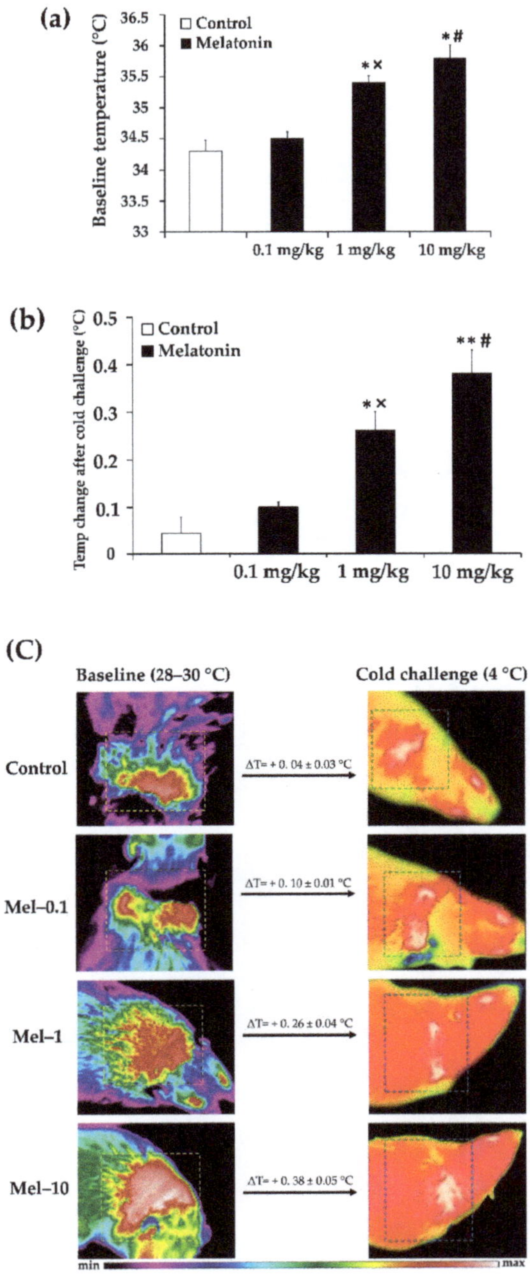

Figure 1. Dose-dependent effects of melatonin on iBAT thermogenesis in Zücker diabetic fatty rats as measured by infrared thermography. (**a**,**b**) Quantification of the interscapular skin temperature at baseline (**a**) and after acute cold challenge (**b**). (**c**) Representative infrared thermal images of the interscapular skin temperature at baseline and after acute cold challenge. Mel-0.1, Mel-1, and Mel-10: melatonin at a dose of 0.1, 1, and 10 mg/kg b.w, respectively. Values are means ± S.D ($n = 4$). * $p < 0.05$, ** $p < 0.01$ Mel-1 and Mel-10 vs. Control; # $p < 0.05$ Mel-10 vs. Mel-0.1; × $p < 0.05$ Mel-1 vs. Mel-0.1 (One-way ANOVA with Tukey post hoc test).

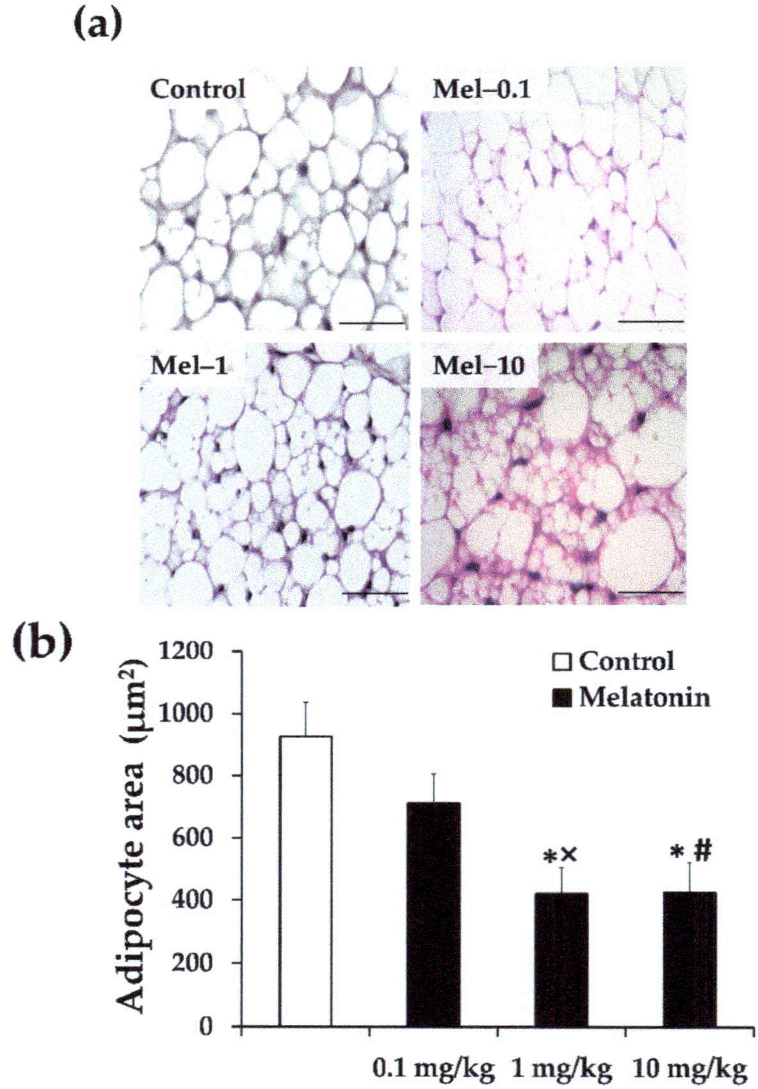

Figure 2. *Cont.*

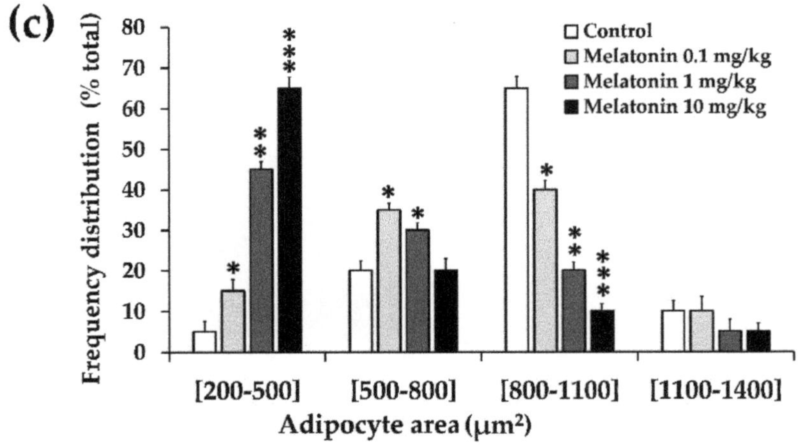

Figure 2. Dose-dependent effects of melatonin on iBAT morphology in Zücker diabetic fatty rats. (**a**) Representative haematoxylin and eosin staining of iBAT sections (original magnification ×400). (**b**,**c**) Mean adipocyte area (**b**) and frequency distribution of adipocyte area (**c**) were measured using a quantitative morphometric method with Image J software. Mel-0.1, Mel-1, and Mel-10: melatonin at a dose of 0.1, 1, and 10 mg/kg b.w, respectively. Values are means ± S.D (n = 200 adipocytes/group). * $p < 0.05$, ** $p < 0.01$, *** $p < 0.001$ Mel-0.1, Mel-1, and Mel-10 vs. Control; # $p < 0.05$ Mel-10 vs. Mel-0.1; × $p < 0.05$ Mel-1 vs. Mel-0.1 (One-way ANOVA Tukey with post hoc test). Scale bar: 50 μm.

The morphometric quantitative analysis shows that adipocyte size was significantly smaller in the Mel-1 (420.4 ± 83.9 μm^2) and Mel-10 (425.1 ± 95.0 μm^2) groups than the control group (926.7 ± 109.1 μm^2; $p < 0.05$ and $p < 0.05$, respectively; Figure 2b), with no significant difference among them ($p > 0.05$; Figure 2b). A non-significant tendency toward smaller adipocyte size was appreciated in the Mel-0.1 group (711.2 ± 93.8 μm^2) compared with the control group ($p > 0.5$; Figure 2b). The Mel-1 and Mel-10 groups exhibited significantly smaller adipocytes than the Mel-0.1 group ($p < 0.05$ and $p < 0.05$, respectively; Figure 2b). In addition, the frequency distribution analysis shows a shift in size distribution toward smaller adipocytes in all three melatonin-treated groups (Figure 2c). Notably, a significant and progressive decreased proportion of larger (800–1100 μm^2) and increased proportion of smaller (200–500 μm^2) adipocytes were noted in iBAT from the Mel-0.1 (larger: 40%; smaller: 15%), Mel-1 (larger: 20%; smaller: 45%), and Mel-10 (larger: 10%; smaller: 65%) groups, compared with the control group ($p < 0.05$, $p < 0.01$ and $p < 0.001$, respectively; Figure 2c).

3.3. Dose-Dependent Effects of Melatonin on Thermogenic and Mitochondrial Dynamic Markers

To characterize the dose-dependent effects of melatonin treatment on the iBAT thermogenic activity, we next assessed the thermogenic UCP1 protein expression in the iBAT using immunofluorescence and Western blotting. UCP1 is known as a key molecule for BAT thermogenesis [36].

By immunofluorescence assay, UCP1 protein expression showed weakened fluorescent immunoreactivity in the iBAT of the control group, whereas melatonin treatment progressively increased the UCP1 fluorescent immunoreactivity with increasing doses of melatonin and achieved the strongest fluorescent signal at the 10 mg/kg dose (Figure 3a). The quantitative analysis of the UCP1 fluorescent immunoreactivity intensity indicated significantly higher UCP1 expression in all melatonin-treated groups, Mel-0.1 (3.4-fold), Mel-1(8.0-fold), and Mel-10 (13.2-fold), relative to the control group ($p < 0.05$, $p < 0.01$ and $p < 0.001$, respectively; Figure 3b). The highest UCP1 immunoreactivity intensity was observed in the Mel-10 group, as compared with the Mel-1 (1.6-fold; $p < 0.05$; Figure 3b) and Mel-0.1 (3.9-fold; $p < 0.01$; Figure 3b) groups. The Mel-1 group showed a significantly higher UCP1 immunoreactivity than the Mel-0.1 group (2.3-fold; $p < 0.05$; Figure 3b).

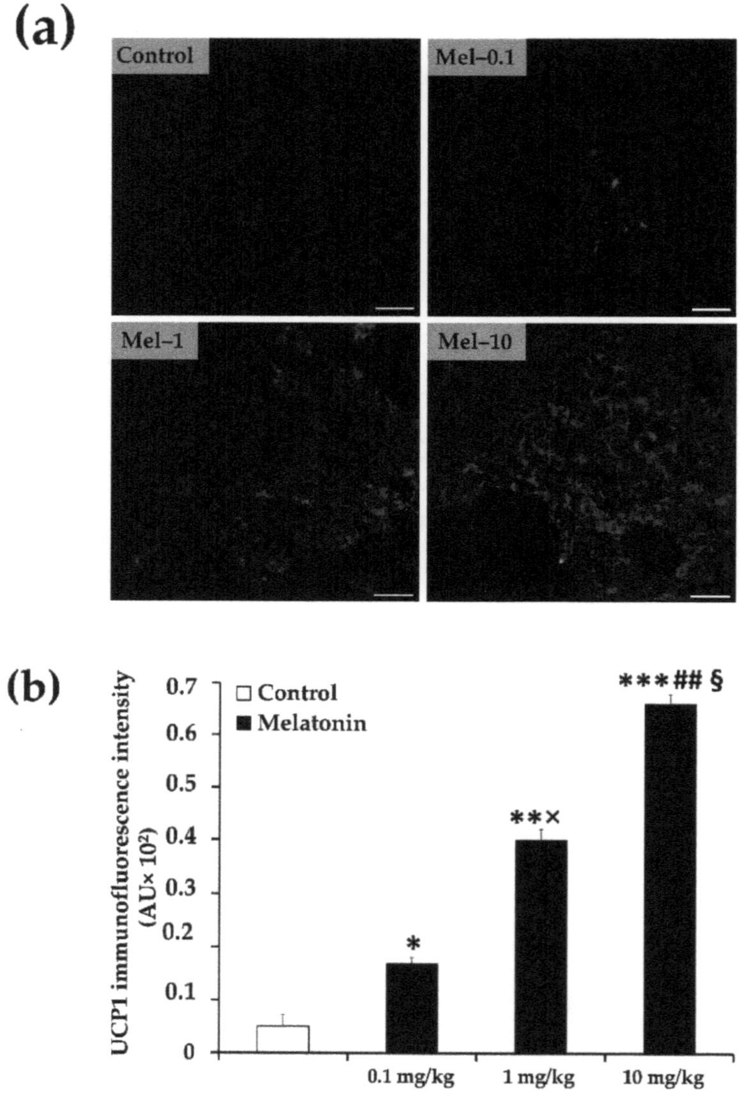

Figure 3. *Cont.*

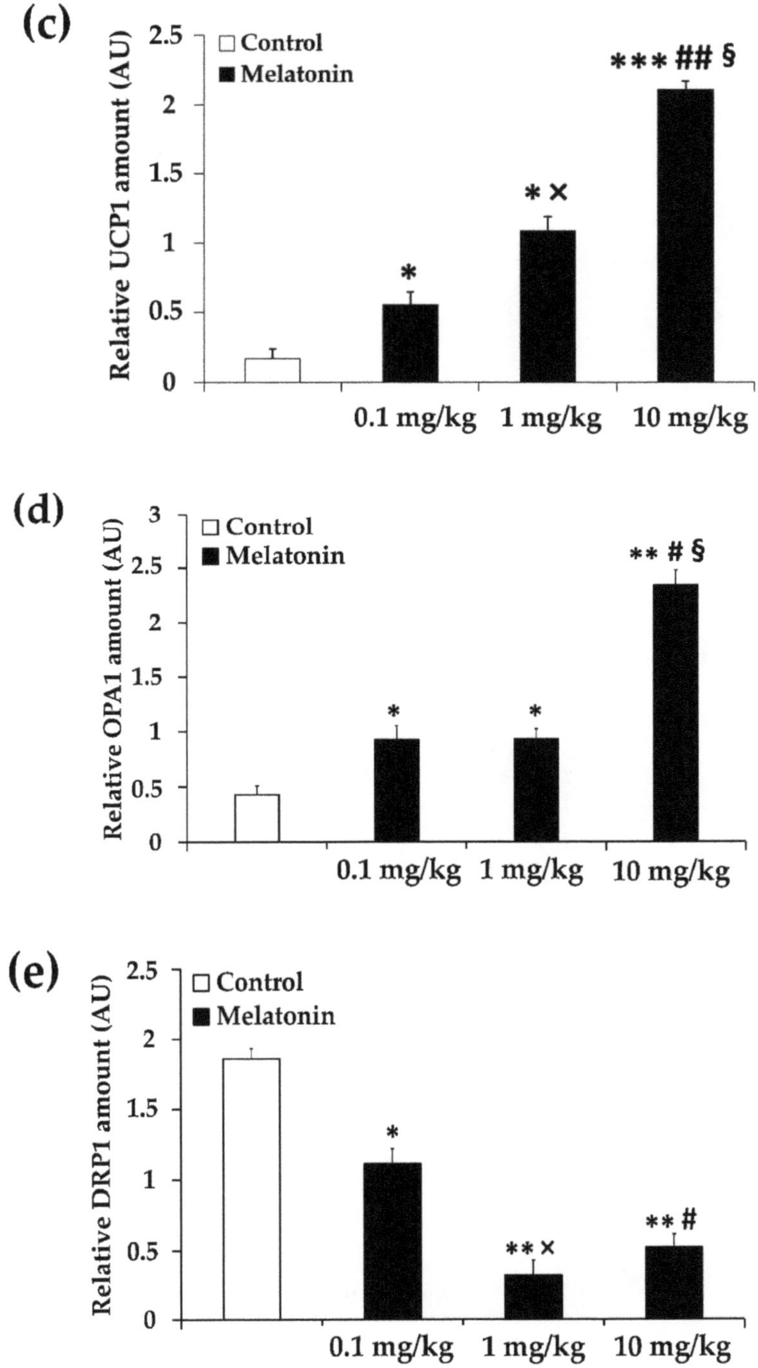

Figure 3. *Cont.*

(f)

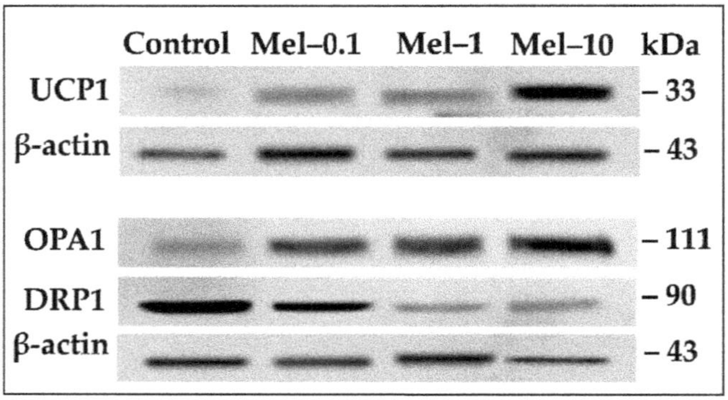

Figure 3. Dose-dependent effects of melatonin on hallmark thermogenic and mitochondrial dynamic markers in iBAT of Zücker diabetic fatty rats. (**a**) Representative microscopic immunofluorescent images of uncoupled protein 1 (UCP1) visualized with cyanine3 secondary antibody (original magnification ×200). (**b**) Immunofluorescence intensity quantification of UCP1. (**c–e**) Western blot densitometry quantification of UCP1, optic atrophy protein type 1 (OPA1), and dynamin-related protein1 (DRP1). (**f**) Representative Western blot images of UCP1, OPA1, and DRP1. Mel-0.1, Mel-1, and Mel-10: melatonin at a dose of 0.1, 1, and 10 mg/kg b.w, respectively. Values are means ± S.D (n = 3) of UCP1 fluorescence intensity and ratios of specific protein levels to β-actin (loading protein). * $p < 0.05$, ** $p < 0.01$, *** $p < 0.001$ Mel-0.1, Mel-1 and Mel-10 vs. Control; # $p < 0.05$, ## $p < 0.05$ Mel-10 vs. Mel-0.1; § $p < 0.05$ Mel-10 vs. Mel-1; × $p < 0.05$ Mel-1 vs. Mel-0.1 (One-way ANOVA with Tukey post hoc test). Scale bar: 50 μm.

Western blotting of UCP1 in the mitochondrial fractions of iBAT was also performed to further support the above findings. The protein level of UCP1 determined by Western blot was considered the most relevant parameter to estimate the BAT thermogenic capacity in response to chronic stimuli since it correlates quantitatively and temporally with the total thermogenic capacity [37]. Consistent with the results of the immunofluorescence, Western blot analysis indicated that the protein expression of UCP1 is also significantly upregulated in all melatonin-treated groups, Mel-0.1 (3.3-fold), Mel-1(6.4-fold), and Mel-10 (12.4-fold), as compared with that in the control group ($p < 0.05$, $p < 0.05$ and $p < 0.001$, respectively; Figure 3c). The highest protein level of UCP1 was observed in the Mel-10 group, as compared with the Mel-1 (1.9-fold; $p < 0.05$; Figure 3c) and Mel-0.1 (3.8-fold; $p < 0.01$; Figure 3c) groups. The Mel-1 group showed a significantly higher UCP1 protein amount than the Mel-0.1 group (1.9-fold; $p < 0.05$; Figure 3c).

Mitochondrial fusion and fission dynamics play a critical role in BAT thermogenesis, and the abnormal function of mitochondria dynamics leads to obesity and related diseases [38–40]. Accordingly, we explored the possible dose-dependent effects of melatonin treatment on the iBAT mitochondrial dynamic by assessing OPA1 and DRP1 levels in isolated iBAT mitochondria by Western blotting. OPA1 mediates mitochondrial fusion, whereas DRP1 is considered a key regulator of mitochondrial fission [41].

As shown in Figure 3d, OPA1 protein level was found to be significantly upregulated in all melatonin-treated groups, Mel-0.1 (2.2-fold), Mel-1 (2.2-fold), and Mel-10 (5.5-fold), compared with that in the control group ($p < 0.05$, $p < 0.05$ and $p < 0.01$, respectively). The highest protein level of OPA1 was observed in the Mel-10 group, as compared with the Mel-1 (2.5-fold; $p < 0.05$) and Mel-0.1 (2.5-fold; $p < 0.05$) groups. There was no statistically significant difference between the Mel-0.1 and Mel-1 groups ($p > 0.05$) in the protein expression of OPA1.

In addition, the protein level of DRP1 was found to be significantly repressed in all melatonin-treated groups, Mel-0.1 (1.7-fold), Mel-1 (5.8-fold), and Mel-10 (3.6-fold), compared with that in the control group ($p < 0.05$, $p < 0.01$ and $p < 0.01$, respectively;

Figure 3e), with no significant difference between the Mel-1 and Mel-10 groups ($p > 0.5$; Figure 3e). The Mel-1 and Mel-10 groups displayed a significantly lower DRP1 protein content than the Mel-0.1 group (3.5-fold and 2.1-fold; $p < 0.05$ and $p < 0.05$, respectively; Figure 3e).

4. Discussion

The present findings revealed for the first time that chronic daily oral administration of melatonin at doses of 1 and 10 mg/kg b.w, but not at a 0.1 mg/kg b.w dose, prevents BW gain and enhances the basal and adaptive iBAT thermogenesis in obese diabetic ZDF rats. At the molecular level, melatonin treatment showed successful dose-response effects on mitochondrial thermogenic and dynamic expression markers, even at a 0.1 mg/kg dose, suggesting a previously unrecognized potency and efficacy of melatonin in targeting mitochondria, which might have a valuable therapeutic role in the treatment of obesity.

The BW gain-lowering effect of melatonin has generated much interest in recent years and has been extensively studied in animal models and clinical trials. The data from animal studies using various models of obesity have convincingly shown that orally administered melatonin in doses ranging between 1 and 100 mg/kg b.w per day resulted in reduced BW gain [4,42–44]. Our present results confirm the existing evidence for the beneficial effect of melatonin on BW, showing that oral melatonin at doses of 1 and 10 mg/kg for 6 wk could subsequently reduce BW gain. Importantly, as our current data shows that 1 and 10 mg/kg of melatonin produces practically similar effects on the BW, it is reasonable to suggest that the dose of 1 mg/kg may be the maximum effective dose, which needs to be verified with an additional more precise effective dose setting study. If confirmed, an extrapolation of 1 mg/kg of melatonin would be sufficient to achieve maximum BW loss in obese individuals, which would be, on the one hand, of crucial importance to prevent unnecessary high dosing and, on the other hand, would be highly advantageous in terms of patient costs. In any case, since both doses were shown to be effective in reducing the BW gain, we estimate that the effective melatonin dose to achieve desirable effects on BW loss in ZDF rats is within the dose range of 1 to 10 mg/kg b.w. If we convert these animal doses to the human doses according to standard dose translation, based on dividing the surface area by factor 6.17, the calculated equivalent human dose is between 0.16 and 1.6 mg/kg per day ($1/6.17 = 0.16$ and $10/6.17 = 1.6$) [45]. Thus, we anticipate that the therapeutically effective dose to achieve significant weight loss in obese individuals is within the dose range of 0.16 to 1.6 mg/kg of total BW per day. For an adult individual of 75 kg, further clinical trials using melatonin with doses in the 12 to 120 mg/day range ($0.16 \times 75 = 12$ and $1.6 \times 75 = 120$) may be warranted and, in addition, appear to be reasonable since this dose range would not cause serious adverse events, based on the data of some clinical studies using melatonin in other health conditions [18,19,46–52]. Nevertheless, because there are fewer reports on the safety of high doses of melatonin ($\geq$100 mg/day), its potential efficacity in obese patients should be closely screened for relevant safety concerns. This anticipated range of doses is far from that usually given in clinical trials for obesity. Indeed, the melatonin doses used in obese subjects have ranged from 1 to 10 mg per day orally, and the data from various clinical trials on the effects of melatonin on BW loss outcomes are conflicting, ranging from a modest to no effect, depending on the dose, duration of administration, and whether used as adjunct or single therapy [53–57]. It is worth mentioning that the positive findings from the clinical trials reported here should not be used as a confirmation of melatonin's efficacity in weight loss since melatonin has been given as an adjunct to a low-calorie diet therapy that may confound the relationship between the melatonin and weight loss outcomes [53,54]. We are optimistic that the present dose-response data may help resolve the previous conflicting clinical findings of a possibly inadequate dose of melatonin. On the other hand, the important feature of the current study is that we used a relatively low dose of 0.1 mg/kg melatonin, far from the melatonin doses that have been usually offered to experimental animals. Compared with the control, such a melatonin dose failed to show any significant difference in BW over a 6 wk period. Given that this dose is equivalent to ~1.2 mg of melatonin for a 75 kg adult individual, this finding is consistent with a randomized clinical trial that has shown that daily oral administration of melatonin at a dose of 1 or 3 mg for one year in post-menopausal women weighing 66.6 –76.0 kg had no significant effect on BW and BMI compared with placebo [57].

Although the mechanism behind melatonin-induced BW reduction is unclear, increased BAT thermogenic activity to increase EE could account for the phenotype, as we proposed previously [26]. Since the rediscovery of functional BAT in adult humans, extensive studies have examined the role of active BAT in metabolism in human adult populations. Most clinical studies investigating BAT thermogenic capacity have been performed under cold challenge, which is well known as a powerful natural tool to detect BAT prevalence and activation [58]. In this scenario, using a standardized cooling protocol, several prospective studies have shown that obesity was associated with impaired cold-induced BAT activation, which could be indicative of a loss of BAT thermogenic capacity in obese subjects [9,16,33,34,59,60]. Coupled with this observation, defective BAT activities upon cold exposure have been reported in various experimental models of obesity [61]. Consistent with this observation, our previous study using infrared thermography to measure the temperature of the skin overlying the iBAT, has shown an impaired iBAT thermogenic response to acute cold challenge in ZDF rats and that melatonin at a dose of 10 mg/kg has been found to reverse this pattern to normal [26]. Our main goal in performing the cold challenge test was to gauge the thermogenic ability of iBAT and investigate whether it is restored after melatonin treatment since, at room temperature (baseline), the animals are chronically adapted to that temperature and have no physiological need to activate it. The present results confirm our previous data and, in addition, show an attractive dose-response relationship between melatonin treatment and cold-induced iBAT activation. Notably, melatonin at doses of 1 and 10 mg/kg resulted in a significant rise in the interscapular skin temperature, with the maximum effect being noticed at 10 mg/kg, whereas melatonin at a dose of 0.1 mg/kg did not seem to be as effective as the other doses. This finding would tend to indicate that melatonin restores obesity-induced iBAT thermogenic dysfunction, and this seems to be in a dose-dependent manner. In addition, even at baseline conditions, where the animals are adapted to that temperature and do not need higher heat production, melatonin at doses of 1 and 10 mg/kg, but not at 0.1 mg/kg, was found to increase the interscapular skin temperature, which may reflect the enhancement of the baseline BAT activity. The lack of a significant effect with a dose of 0.1 mg/kg of melatonin on interscapular skin temperature is related either to the short treatment duration (6 wk) or the relatively low melatonin dose, which may not be sufficient to produce a significant thermogenic response. Hence, further investigation with a prolonged duration of treatment is necessary before drawing any firm conclusion on this issue. These findings are in agreement with those of other studies showing that melatonin at doses ranging from 1 to 10 mg/kg plays a crucial role in the regulation of BAT thermogenesis in melatonin deficient experimental models and seasonal breeder species [62–66]. Notwithstanding that our iBAT temperature results are not original, they expand and advance previous results by supporting the potential use of melatonin to promote BAT thermogenesis in the obesity context in a dose-dependent manner.

The hallmark of dysfunctional BAT in obesity is a whitening phenotype that has been previously characterized in experimental models of obesity by enlarged lipid droplet accumulation, mitochondrial dysfunction, and functional loss [67–70]. Consistent with this notion, the iBAT from the control ZDF group assumed a striking whitening appearance, characterized by excessive accumulation of hypertrophied lipid droplets and pale eosinophilic staining, which strongly indicates fuel switching from thermogenesis to lipid storage. In line with this idea, the BAT whitening phenotype has also been described previously in the ZDF strain model, with impaired glucose uptake and an increase in the fatty acid synthesis enzyme [67]. We have previously shown that melatonin at a dose of 10 mg/kg successfully recovered the typical microscopic appearance of the iBAT, including depleted lipid content, multilocular adipocytes, and dense eosinophilic staining [26]. In the present study, 1 mg/kg of melatonin was found to be as effective as 10 mg/kg for recovering the typical iBAT appearance; however, the brownish microscopic appearance tends to be more pronounced at 10 mg/kg. In the case of 0.1 mg/kg of melatonin, the brownish phenotype of the tissue was not evident, but the adipocytes were remarkably smaller sized than the control. These results seem compatible with the infrared thermography results and thus support the idea that they might likely reflect the reversibility of iBAT thermogenic capacity in a dose-response manner. The reduced lipid deposition within the iBAT following the melatonin treatment most likely reflects increased fatty acid β-oxidation, given that mitochondrial fatty acid β-oxidation has been demonstrated to be critical for maintaining the brown adipocyte phenotype both during times of activation and quiescence (basal state) [71]. Fatty acid β-oxidation fuels the increase in uncoupled mitochondrial respiration and contributes to inducing the expression of thermogenic genes such as UCP1 [71]. UCP1 is a hallmark of brown adipocytes and plays a pivotal role in BAT thermogenic function since it confers to brown adipocytes their specific ability to dissipate the proton gradient as heat, and its downregulation or absence has been reported to be associated with impaired BAT thermogenic capacity and lower EE [72–74]. The efficacy of melatonin to induce UCP1 expression in iBAT has already been demonstrated by other groups in experimental ageing and melatonin deficient models, and the melatonin doses offered to animals have ranged from 1 to 10 mg/kg.bw [64–66,75]. This present study expanded and advanced the previous finding by showing that the iBAT UCP1 protein expression from immunofluorescence and Western blot analysis significantly increased at doses of 1 and 10 mg/kg of melatonin and, surprisingly, even at 0.1mg/kg of melatonin in an experimental model of diabetes and obesity. These findings argue again for a recovery of the iBAT thermogenic phenotype, which might be attributable to the recruitment and/or reactivation of BAT. The dose-response pattern of UCP1 protein expression in response to melatonin was consistent with that of the infrared thermal image data. Of note, UCP1 protein levels progressively increased with increasing doses of melatonin and reached the highest increase at a dose of 10 mg/kg, which could explain the highest iBAT temperature observed under acute cold exposure and basal state. Another point especially worthy of note, is that the significant increase in UCP1 level at a dose of 0.1 mg/kg of melatonin was not associated with changes in the iBAT temperature that were observed under both basal and cold conditions. This dissociation could be explained either by the fact that the amount of UCP1 encoded protein in the tissue at the dose of 0.1 mg/kg of melatonin was not sufficient enough to be translated to a significant iBAT thermogenic response or by the small number of animals used for the infrared thermograph test, as a larger number of the animals might be needed to demonstrate significant differences. In any case, this result might be advantageous since, on the one hand, it indicates that future clinical applications using this treatment for promoting the BAT thermogenic phenotype in obese individuals are less prone to any potential side effects and more cost-effective in economic terms. On the other hand, it would tend to indicate that the melatonin molecule is endowed with a higher potency than previously recognized.

One possible explanation for the BAT thermogenic repression in obese individuals could be related to the inflammatory and oxidative stress microenvironment in BAT that is known to occur in the adiposity state [61]. As previously reported in various obesity models, these deleterious conditions have been found to impair mitochondrial respiratory and antioxidant functions, which ultimately cause loss of BAT thermogenic signature and brown adipocyte apoptosis [61,70,76,77]. On this basis, and since mitochondria are widely recognized as a therapeutic target for melatonin, it is reasonable to speculate that melatonin might protect against obesity-induced BAT dysfunction by improving the performance of mitochondrial respiratory and antioxidant functions. In support of this speculation, our previous study has shown that chronic oral administration of melatonin improved the functionality of mitochondria isolated from iBAT of the ZDF rats by increasing mitochondrial respiration and reducing mitochondrial oxidative/nitrosative stress and susceptibility to apoptosis [29]. Mitochondria are dynamic organelles that continuously divide and fuse, and the mitochondrial fission and fusion process plays a key role in providing a basis for mitochondrial functions, including respiration and oxidative stress balance [7]. Mitochondrial fusion is believed to be beneficial because it increases mitochondrial efficiency, whereas mitochondrial fission seems to be associated with mitochondrial malfunctions and oxidative stress [7]. Given the potentially critical role of mitochondrial dynamics in regulating BAT thermogenesis [38–40,78,79], we elected to investigate the possible dose-dependent effect of melatonin on mitochondrial dynamics by Western blotting of OPA1 fusion and DRP1 fission markers. We found that both OPA1 and DRP1 were modulated by melatonin treatment in a dose-dependent fashion. Of note, the protein level of OPA1 increased significantly at all three doses of melatonin and reached the highest level at 10 mg/kg, whereas that of DRP1 significantly decreased at all three doses as well, with the lowest level equally being observed at doses of 1 and 10 mg/kg. This suggests that the balance between fusion and fission processes might be shifted toward mitochondrial fusion to maintain high-quality functional mitochondria, which can be essential to support efficient fatty acid β-oxidation and uncoupled mitochondrial respiration required for the maintenance of the brown adipocyte phenotype. Consistent with this suggestion, an earlier study in OPA1 BAT knockout mice has demonstrated that OPA1 deficiency impaired BAT activation and led to mitochondrial bioenergetic deficiency and thermogenic gene program downregulation, providing direct evidence that OPA1 plays a pivotal role in BAT thermogenic activation [40]. In vivo evidence in mice lacking the ATP-independent metalloprotease OMA1, which plays an essential role in the proteolytic inactivation of OPA1, reinforces the importance of OPA1-related fusion for BAT thermogenesis [39]. In addition, knockout models of BAT fusion-related markers in mice have been shown to lead to dysfunctional mitochondria and impaired BAT thermogenic activity [41,79]. Based on these findings, our results might implicate the fusion process as an important target for melatonin to enhance mitochondrial functions and preserve the BAT thermogenic signature and hence favour the maintenance of the BAT thermogenic phenotype under the deleterious conditions of obesity. The ability of melatonin to regulate mitochondrial dynamics is in line with other reports in different tissues under various pathological conditions [31,80–83]. On the other hand, these results imply that the increased fusion and decreased fission through melatonin treatment results in the inhibition of mitochondrial fragmentation. Excessive mitochondrial fragmentation due to imbalanced mitochondrial dynamics has often been reciprocally linked to mitochondrial oxidative stress, which is critical for inducing cell death [5,7]. We have previously analysed the mitochondrial fractions of ZDF rats' iBAT for superoxide dismutase (SOD) activity and nitrite level as markers of oxidative stress and mitochondrial permeability transition pore (mPTP) activity as a marker of apoptosis [29]. The results have shown reduced SOD activity associated with increased nitrite content and mPTP activity in untreated ZDF rats compared with their lean littermates [29]. Interestingly, treatment with melatonin has been found to reduce the mitochondrial nitrosative and oxidative status and susceptibility to apoptosis by decreasing nitrite levels, increasing SOD antioxidant enzyme activity and inhibiting mPTP activity [29]. Given the close interaction between redox status and mitochondrial dynamics [7], we were prompted to ask whether melatonin protects against iBAT thermogenic dysfunction via direct interaction with components of mitochondria fusion-fission pathways, as the beneficial role of melatonin in promoting fusion and inhibiting fission processes may not be the cause, but rather the consequence of its protective effects in reducing mitochondrial oxidative stress. Consistent with this idea, SOD2 has been reported to enhance the mitochondrial fusion process independent of its antioxidant activity [84]. Therefore, given that melatonin is acknowledged to have direct antioxidative protection, including directly scavenging free radicals and enhancing the activity of the antioxidant enzyme [23], it is likely that melatonin promoted OPA1-mediated mitochondrial fusion by activating the antioxidant pathway. Though the precise molecular mechanisms through which melatonin regulates OPA1 and DRP1 protein levels were beyond the scope of this study, our findings provide preliminary insight into the role of melatonin in the regulation of BAT mitochondrial fission-fusion, which may play a key role in the deregulation of BAT thermogenic function in obesity and consequently could be novel future therapeutic targets.

The potential effect of melatonin on mitochondrial dynamics could not rule out the possibility that melatonin might restore the BAT thermogenic ability by recruiting thermogenic brown adipocytes through differentiation of brown adipocyte progenitors, given that progenitor cells have been found to still exist in obese BAT [85]. In support of this possibility, sheep brown adipocyte precursor cells isolated from perirenal BAT exposed to in vitro melatonin treatment have promoted brown adipocyte formation and increased protein expression levels of brown adipogenic markers via AMK-activated protein kinase, thus showing melatonin involvement in brown adipocyte differentiation [86]. Melatonin administration may promote the recruitment of new brown adipocytes, possibly by acting centrally on MT1 receptors located on neurons of the suprachiasmatic nucleus to increase the sympathetic tone in BAT, or directly on melatonin receptors located on the brown adipocytes (MT1 and MT2), which results in a reduction of intracellular cAMP levels with a lowering of PKA activity and cAMP-response element-binding protein phosphorylation [4,24].

The present findings, if confirmed clinically, suggest melatonin would be advantageous over other drugs that are currently being tested in clinical trials or EMA- and FDA-approved drugs that have now been extended to treat obesity, by targeting BAT but without serious adverse events on the one hand due to its safety profile, and, on the other hand, due to the fact that melatonin may act by directly targeting mitochondria rather than by secondary off-target effects [17].

5. Conclusions

The results presented here reveal for the first time that chronic oral administration of melatonin to obese diabetic ZDF rats at doses of 0.1, 1, or 10 mg/kg b.w over 6 wk enhances the BW gain and iBAT thermogenic features in a dose-dependent manner, with the underlying mechanism for the iBAT thermogenesis possibly being through restoring brown adipocyte mitochondrial function, which could have great therapeutic value, especially for promoting BAT reactivation among obese individuals with blunted BAT activity. Therefore, to translate these promising findings into potential clinical anti-obesity therapy, we estimate that the therapeutically effective dose for promoting BAT thermogenesis and, consequently, weight loss is within the dose range of 12 to 120 mg orally per day for a 75 kg adult individual, based on human equivalent dose calculation. Future powered randomized clinical trials are warranted to confirm this. In addition, future in vitro and in vivo in-depth studies are required to fully understand the potential effect of melatonin on iBAT mitochondrial dynamics and decipher the related underlying molecular mechanism, as a full understanding of the mechanisms could facilitate the clinical application of melatonin for obesity.

Author Contributions: Conceptualization, A.A.; methodology, S.A.; software, S.A.; formal analysis, A.A. and E.K.E.; writing—original draft preparation, S.A.; writing—review and editing, A.A. and S.A.; critical revision and approval of the manuscript, A.M.-C., A.M.-H., A.S.A. and E.R.; final approval, A.A.; supervision, A.A.; project administration, A.A.; funding acquisition, A.A. All authors have read and agreed to the published version of the manuscript.

Funding: This research was funded by grant SAF2016-79794-R from Ministrio de Ciencia e Innovación (Spain) and European Regional Development Fund (ERDF) and B-CTS-102-UGR20 from University of Granada & FEDER Andalucía-UGR 2020.

Institutional Review Board Statement: The study was conducted according to the guidelines of the Declaration of Helsinki and approved by the Ethical Committee of the University of Granada (Granada, Spain) under the reference number 4-09-2016-CEEA, according to the European Union guidelines for animal care and protection.

Informed Consent Statement: Not applicable.

Data Availability Statement: Not applicable.

Acknowledgments: The authors thank Mohamed Tassi for processing the histological sections and quantitative analysis (Service of Microscopy, CIBM, University of Granada).

Conflicts of Interest: The authors declare no conflict of interest.

References

1. WHO. *Overweight and Obesity*; WHO: Geneva, Switzerland, 2018.
2. Blüher, M. Obesity: Global epidemiology and pathogenesis. *Nat. Rev. Endocrinol.* **2019**, *15*, 288–298. [CrossRef] [PubMed]
3. Jimenez-Munoz, C.M.; López, M.; Albericio, F.; Makowski, K. Targeting energy expenditure—Drugs for obesity treatment. *Pharmaceuticals* **2021**, *14*, 435. [CrossRef] [PubMed]
4. Guan, Q.; Wang, Z.; Cao, J.; Dong, Y.; Chen, Y. Mechanisms of Melatonin in Obesity: A Review. *Int. J. Mol. Sci.* **2022**, *23*, 218. [CrossRef]
5. Michurina, S.S.; Stafeev, I.S.; Menshikov, M.Y.; Parfyonova, Y.V. Mitochondrial dynamics keep balance of nutrient combustion in thermogenic adipocytes. *Mitochondrion* **2021**, *59*, 157–168. [CrossRef] [PubMed]
6. Wu, J.; Boström, P.; Sparks, L.M.; Ye, L.; Choi, J.H.; Giang, A.-H.; Khandekar, M.; Virtanen, K.A.; Nuutila, P.; Schaart, G.; et al. Beige Adipocytes Are a Distinct Type of Thermogenic Fat Cell in Mouse and Human. *Cell* **2012**, *150*, 366–376. [CrossRef] [PubMed]
7. Lee, J.H.; Park, A.; Oh, K.J.; Lee, S.C.; Kim, W.K.; Bae, K.H. The Role of Adipose Tissue Mitochondria: Regulation of Mitochondrial Function for the Treatment of Metabolic Diseases. *Int. J. Mol. Sci.* **2019**, *20*, 4924. [CrossRef]
8. Peres Valgas da Silva, C.; Hernández-Saavedra, D.; White, J.D.; Stanford, K.I. Cold and Exercise: Therapeutic Tools to Activate Brown Adipose Tissue and Combat Obesity. *Biology* **2019**, *8*, 9. [CrossRef]
9. van Marken Lichtenbelt, W.D.; Vanhommerig, J.W.; Smulders, N.M.; Drossaerts, J.M.A.F.L.; Kemerink, G.J.; Bouvy, N.D.; Schrauwen, P.; Teule, G.J.J. Cold-activated brown adipose tissue in healthy men. *N. Engl. J. Med.* **2009**, *360*, 1500–1508. [CrossRef]
10. Cypess, A.M.; Lehman, S.; Williams, G.; Tal, I.; Rodman, D.; Goldfine, A.B.; Kuo, F.C.; Palmer, E.L.; Tseng, Y.-H.; Doria, A.; et al. Identification and Importance of Brown Adipose Tissue in Adult Humans. *N. Engl. J. Med.* **2009**, *360*, 1509–1517. [CrossRef]
11. Pfannenberg, C.; Werner, M.K.; Ripkens, S.; Stef, I.; Deckert, A.; Schmadl, M.; Reimold, M.; Häring, H.U.; Claussen, C.D.; Stefan, N. Impact of age on the relationships of brown adipose tissue with sex and adiposity in humans. *Diabetes* **2010**, *59*, 1789–1793. [CrossRef]
12. Perkins, A.C.; Mshelia, D.S.; Symonds, M.E.; Sathekge, M. Prevalence and pattern of brown adipose tissue distribution of 18F-FDG in patients undergoing PET-CT in a subtropical climatic zone. *Nucl. Med. Commun.* **2013**, *34*, 168–174. [CrossRef] [PubMed]
13. Steinberg, J.D.; Vogel, W.; Vegt, E. Factors influencing brown fat activation in FDG PET/CT: A retrospective analysis of 15,000+ cases. *Br. J. Radiol.* **2017**, *90*, 20170093. [CrossRef] [PubMed]
14. Yoneshiro, T.; Aita, S.; Matsushita, M.; Kameya, T.; Nakada, K.; Kawai, Y.; Saito, M. Brown adipose tissue, whole-body energy expenditure, and thermogenesis in healthy adult men. *Obesity* **2011**, *19*, 13–16. [CrossRef] [PubMed]
15. Becher, T.; Palanisamy, S.; Kramer, D.J.; Eljalby, M.; Marx, S.J.; Wibmer, A.G.; Butler, S.D.; Jiang, C.S.; Vaughan, R.; Schöder, H.; et al. Brown adipose tissue is associated with cardiometabolic health. *Nat. Med.* **2021**, *27*, 58–65. [CrossRef] [PubMed]
16. Vijgen, G.H.E.J.; Bouvy, N.D.; Teule, G.J.J.; Brans, B.; Hoeks, J.; Schrauwen, P.; Van Marken Lichtenbelt, W.D. Increase in brown adipose tissue activity after weight loss in morbidly obese subjects. *J. Clin. Endocrinol. Metab.* **2012**, *97*, E1229–E1233. [CrossRef]
17. Mukherjee, J.; Baranwal, A.; Schade, K.N. Classification of Therapeutic and Experimental Drugs for Brown Adipose Tissue Activation: Potential Treatment Strategies for Diabetes and Obesity. *Curr. Diabetes Rev.* **2016**, *12*, 414–428. [CrossRef]
18. Menczel Schrire, Z.; Phillips, C.L.; Chapman, J.L.; Duffy, S.L.; Wong, G.; D'Rozario, A.L.; Comas, M.; Raisin, I.; Saini, B.; Gordon, C.J.; et al. Safety of higher doses of melatonin in adults: A systematic review and meta-analysis. *J. Pineal Res.* **2022**, *72*, e12782. [CrossRef]
19. Andersen, L.P.H.; Gögenur, I.; Rosenberg, J.; Reiter, R.J. The Safety of Melatonin in Humans. *Clin. Drug Investig.* **2016**, *36*, 169–175. [CrossRef]
20. Navarro-Alarcón, M.; Ruiz-Ojeda, F.J.; Blanca-Herrera, R.M.; A-Serrano, M.M.; Acuña-Castroviejo, D.; Fernández-Vázquez, G.; Agil, A. Melatonin and metabolic regulation: A review. *Food Funct.* **2014**, *5*, 2806–2832. [CrossRef]
21. Karamitri, A.; Jockers, R. Melatonin in type 2 diabetes mellitus and obesity. *Nat. Rev. Endocrinol.* **2019**, *15*, 105–125. [CrossRef]
22. Meng, X.; Li, Y.; Li, S.; Zhou, Y.; Gan, R.Y.; Xu, D.P.; Li, H. Bin Dietary sources and bioactivities of melatonin. *Nutrients* **2017**, *9*, 367. [CrossRef] [PubMed]
23. Ferlazzo, N.; Andolina, G.; Cannata, A.; Costanzo, M.G.; Rizzo, V.; Currò, M.; Ientile, R.; Caccamo, D. Is Melatonin the Cornucopia of the 21st Century? *Antioxidants* **2020**, *9*, 1088. [CrossRef] [PubMed]
24. Tan, D.X.; Manchester, L.C.; Fuentes-Broto, L.; Paredes, S.D.; Reiter, R.J. Significance and application of melatonin in the regulation of brown adipose tissue metabolism: Relation to human obesity. *Obes. Rev.* **2011**, *12*, 167–188. [CrossRef] [PubMed]
25. Richard, D.; Picard, F. Quebec Heart And Lung Institute Research Center and Laval University Interdisciplinary Group In Obesity Research, 2725 Chemin Sainte-Foy, Quebec G1V 4G5 Canada. *Hear. Lung* **2011**, *16*, 1233–1260.
26. Fernández Vázquez, G.; Reiter, R.J.; Agil, A. Melatonin increases brown adipose tissue mass and function in Zücker diabetic fatty rats: Implications for obesity control. *J. Pineal Res.* **2018**, *64*, e12472. [CrossRef]
27. Agil, A.; Navarro-Alarcõn, M.; Ruiz, R.; Abuhamadah, S.; El-Mir, M.Y.; Vázquez, G.F. Beneficial effects of melatonin on obesity and lipid profile in young Zucker diabetic fatty rats. *J. Pineal Res.* **2011**, *50*, 207–212. [CrossRef] [PubMed]

28. Agil, A.; Rosado, I.; Ruiz, R.; Figueroa, A.; Zen, N.; Fernández-Vázquez, G. Melatonin improves glucose homeostasis in young Zucker diabetic fatty rats. *J. Pineal Res.* **2012**, *52*, 203–210. [CrossRef] [PubMed]
29. Agil, A.; Navarro-Alarcon, M.; Ali, F.A.Z.; Albrakati, A.; Salagre, D.; Campoy, C.; Elmahallawy, E.K. Melatonin enhances the mitochondrial functionality of brown adipose tissue in obese—Diabetic rats. *Antioxidants* **2021**, *10*, 1482. [CrossRef]
30. Halpern, B.; Mancini, M.C.; Bueno, C.; Barcelos, I.P.; de Melo, M.E.; Lima, M.S.; Carneiro, C.G.; Sapienza, M.T.; Buchpiguel, C.A.; do Amaral, F.G.; et al. Melatonin Increases Brown Adipose Tissue Volume and Activity in Patients With Melatonin Deficiency: A Proof-of-Concept Study. *Diabetes* **2019**, *68*, 947–952. [CrossRef]
31. Agil, A.; Chayah, M.; Visiedo, L.; Navarro-alarcon, M.; Reiter, R.J. Melatonin improves mitochondrial dynamics and function in the kidney of Zücker diabetic fatty rats. *J. Clin. Med.* **2020**, *9*, 2916. [CrossRef]
32. MM, B. A rapid and sensitive method for the quantitation of microgram quantities of protein utilizing the principle of protein-dye binding. *Anal. Biochem.* **1976**, *72*, 248–254. [CrossRef]
33. Orava, J.; Nuutila, P.; Noponen, T.; Parkkola, R.; Viljanen, T.; Enerbäck, S.; Rissanen, A.; Pietiläinen, K.H.; Virtanen, K.A. Blunted metabolic responses to cold and insulin stimulation in brown adipose tissue of obese humans. *Obesity* **2013**, *21*, 2279–2287. [CrossRef]
34. Saari, T.J.; Raiko, J.; U-Din, M.; Niemi, T.; Taittonen, M.; Laine, J.; Savisto, N.; Haaparanta-Solin, M.; Nuutila, P.; Virtanen, K.A. Basal and cold-induced fatty acid uptake of human brown adipose tissue is impaired in obesity. *Sci. Rep.* **2020**, *10*, 1–11. [CrossRef] [PubMed]
35. Sacks, H.; Symonds, M.E. Anatomical locations of human brown adipose tissue: Functional relevance and implications in obesity and type 2 diabetes. *Diabetes* **2013**, *62*, 1783–1790. [CrossRef] [PubMed]
36. Liu, X. Common and distinct regulation of human and mouse brown and beige adipose tissues: A promising therapeutic target for obesity. *Protein Cell* **2017**, *8*, 446–454. [CrossRef]
37. Nedergaard, J.; Cannon, B. UCP1 mRNA does not produce heat. *Biochim. Biophys. Act Mol. Cell Biol. Lipids* **2013**, *1831*, 943–949. [CrossRef]
38. Pisani, D.F.; Barquissau, V.; Chambard, J.C.; Beuzelin, D.; Ghandour, R.A.; Giroud, M.; Mairal, A.; Pagnotta, S.; Cinti, S.; Langin, D.; et al. Mitochondrial fission is associated with UCP1 activity in human brite/beige adipocytes. *Mol. Metab.* **2018**, *7*, 35. [CrossRef]
39. Quirós, P.M.; Ramsay, A.J.; Sala, D.; Fernández-Vizarra, E.; Rodríguez, F.; Peinado, J.R.; Fernández-García, M.S.; Vega, J.A.; Enríquez, J.A.; Zorzano, A.; et al. Loss of mitochondrial protease OMA1 alters processing of the GTPase OPA1 and causes obesity and defective thermogenesis in mice. *EMBO J.* **2012**, *31*, 2117–2133. [CrossRef]
40. Pereira, R.O.; Marti, A.; Olvera, A.C.; Tadinada, S.M.; Bjorkman, S.H.; Weatherford, E.T.; Morgan, D.A.; Westphal, M.; Patel, P.H.; Kirby, A.K.; et al. OPA1 deletion in brown adipose tissue improves thermoregulation and systemic metabolism via FGF21. *Elife* **2021**, *10*, e66519. [CrossRef]
41. Boutant, M.; Kulkarni, S.S.; Joffraud, M.; Ratajczak, J.; Valera-Alberni, M.; Combe, R.; Zorzano, A.; Cantó, C. Mfn2 is critical for brown adipose tissue thermogenic function. *EMBO J.* **2017**, *36*, 1543–1558. [CrossRef]
42. Favero, G.; Stacchiotti, A.; Castrezzati, S.; Bonomini, F.; Albanese, M.; Rezzani, R.; Rodella, L.F. Melatonin reduces obesity and restores adipokine patterns and metabolism in obese (ob/ob) mice. *Nutr. Res.* **2015**, *35*, 891–900. [CrossRef] [PubMed]
43. Prunet-Marcassus, B.; Desbazeille, M.; Bros, A.; Louche, K.; Delagrange, P.; Renard, P.; Casteilla, L.; Pénicaud, L. Melatonin reduces body weight gain in Sprague Dawley rats with diet-induced obesity. *Endocrinology* **2003**, *144*, 5347–5352. [CrossRef] [PubMed]
44. Farias, T.D.; Cruz, M.M.; Sa, R.C.; Severi, I.; Perugini, J.; Senzacqua, M.; Cerutti, S.M.; Giordano, A.; Cinti, S.; Alonso-Vale, M.I. Melatonin Supplementation Decreases Hypertrophic Obesity and Inflammation Induced by High-Fat Diet in Mice. *Front. Endocrinol.* **2019**, *10*, 750. [CrossRef]
45. Shin, J.; Seol, I.; Son, C. Interpretation of Animal Dose and Human Equivalent Dose for Drug Development. *J. Korean Orient. Med.* **2010**, *31*, 1–7.
46. Weishaupt, J.H.; Bartels, C.; Pölking, E.; Dietrich, J.; Rohde, G.; Poeggeler, B.; Mertens, N.; Sperling, S.; Bohn, M.; Hüther, G.; et al. Reduced oxidative damage in ALS by high-dose enteral melatonin treatment. *J. Pineal Res.* **2006**, *41*, 313–323. [CrossRef]
47. Sánchez-López, A.L.; Ortiz, G.G.; Pacheco-Moises, F.P.; Mireles-Ramírez, M.A.; Bitzer-Quintero, O.K.; Delgado-Lara, D.L.C.; Ramírez-Jirano, L.J.; Velázquez-Brizuela, I.E. Efficacy of Melatonin on Serum Pro-inflammatory Cytokines and Oxidative Stress Markers in Relapsing Remitting Multiple Sclerosis. *Arch. Med. Res.* **2018**, *49*, 391–398. [CrossRef]
48. Nordlund, J.J.; Lerner, A.B. The effects of oral melatonin on skin color and on the release of pituitary hormones. *J. Clin. Endocrinol. Metab.* **1977**, *45*, 768–774. [CrossRef]
49. Lõpez-González, A.; Álvarez-Sánchez, N.; Lardone, P.J.; Cruz-Chamorro, I.; Martínez-Lõpez, A.; Guerrero, J.M.; Reiter, R.J.; Carrillo-Vico, A. Melatonin treatment improves primary progressive multiple sclerosis: A case report. *J. Pineal Res.* **2015**, *58*, 173–177. [CrossRef]
50. Mikalauskas, S.; Mikalauskene, L.; Zorn, M.; Markus, A. The use of high-dose melatonin in liver resection is safe: First clinical experience. *J. Pineal Res.* **2011**, *50*, 381–388. [CrossRef]
51. Kleszczyński, K.; Slominski, A.T.; Steinbrink, K.; Reiter, R.J. Clinical trials for use of melatonin to fight against COVID-19 are urgently needed. *Nutrients* **2020**, *12*, 2561. [CrossRef]

52. Galley, H.F.; Lowes, D.A.; Allen, L.; Cameron, G.; Aucott, L.S.; Webster, N.R. Melatonin as a potential therapy for sepsis: A phase I dose escalation study and an ex vivo whole blood model under conditions of sepsis. *J. Pineal Res.* **2014**, *56*, 427–438. [CrossRef] [PubMed]
53. Mohammadi, S.; Rastmanesh, R.; Jahangir, F.; Amiri, Z.; Djafarian, K.; Mohsenpour, M.A.; Hassanipour, S.; Ghaffarian-Bahraman, A. Melatonin Supplementation and Anthropometric Indices: A Randomized Double-Blind Controlled Clinical Trial. *Biomed Res. Int.* **2021**, *2021*, 3502325. [CrossRef] [PubMed]
54. Szewczyk-golec, K.; Rajewski, P.; Gackowski, M.; Mila-kierzenkowska, C.; Weso, R.; Sutkowy, P.; Paw, M.; Wo, A. Research Article Melatonin Supplementation Lowers Oxidative Stress and Regulates Adipokines in Obese Patients on a Calorie-Restricted Diet. *Oxidative Med. Cell. Longev.* **2017**, *2017*, 8494107. [CrossRef]
55. Mesri Alamdari, N.; Mahdavi, R.; Roshanravan, N.; Lotfi Yaghin, N.; Ostadrahimi, A.R.; Faramarzi, E. A double-blind, placebo-controlled trial related to the effects of melatonin on oxidative stress and inflammatory parameters of obese women. *Horm. Metab. Res.* **2014**, *47*, 504–508. [CrossRef]
56. Walecka-Kapica, E.; Klupińska, G.; Chojnacki, J.; Tomaszewska-Warda, K.; Błońska, A.; Chojnacki, C. The effect of melatonin supplementation on the quality of sleep and weight status in postmenopausal women. *Prz. Menopauzalny* **2014**, *13*, 334–338. [CrossRef]
57. Amstrup, A.K.; Sikjaer, T.; Pedersen, S.B.; Heickendorff, L.; Mosekilde, L.; Rejnmark, L. Reduced fat mass and increased lean mass in response to 1 year of melatonin treatment in postmenopausal women: A randomized placebo-controlled trial. *Clin. Endocrinol.* **2016**, *84*, 342–347. [CrossRef]
58. Van Der Lans, A.A.J.J.; Wierts, R.; Vosselman, M.J.; Schrauwen, P.; Brans, B.; van Marken Lichtenbelt, W.D. Cold-activated brown adipose tissue in human adults: Methodological issues. *Am. J. Physiol.-Regul. Integr. Comp. Physiol.* **2014**, *307*, 103–113. [CrossRef]
59. Wijers, S.L.J.; Saris, W.H.M.; Lichtenbelt, W.D.V.M. Cold-induced adaptive thermogenesis in lean and obese. *Obesity* **2010**, *18*, 1092–1099. [CrossRef]
60. Carpentier, A.C.; Blondin, D.P.; Virtanen, K.A.; Richard, D.; Haman, F.; Turcotte, É.E. Brown adipose tissue energy metabolism in humans. *Front. Endocrinol.* **2018**, *9*, 447. [CrossRef]
61. Alcalá, M.; Calderon-Dominguez, M.; Serra, D.; Herrero, L.; Viana, M. Mechanisms of impaired brown adipose tissue recruitment in obesity. *Front. Physiol.* **2019**, *10*, 94. [CrossRef]
62. Heldmaier, G.; Hoffmann, K. Melatonin stimulates growth of brown adipose tissue. *Nature* **1974**, *247*, 224–225. [CrossRef] [PubMed]
63. Holtorf, A.P.; Heldmaier, G.; Thiele, G.; Steinlechner, S. Diurnal Changes in Sensitivity to Melatonin in Intact and Pinealectomized Djungarian Hamsters: Effects on Thermogenesis, Cold Tolerance, and Gonads. *J. Pineal Res.* **1985**, *2*, 393–403. [CrossRef]
64. Halpern, B.; Mancini, M.C.; Mendes, C.; MacHado, C.M.L.; Prando, S.; Sapienza, M.T.; Buchpiguel, C.A.; Do Amaral, F.G.; Cipolla-Neto, J. Melatonin deficiency decreases brown adipose tissue acute thermogenic capacity of in rats measured by 18F-FDG PET. *Diabetol. Metab. Syndr.* **2020**, *12*, 82. [CrossRef] [PubMed]
65. Mendes, C.; Gomes, G.; Belpiede, L.T.; do Carmo Buonfiglio, D.; Motta-Teixeira, L.C.; Amaral, F.G.; Cipolla-Neto, J. The effects of melatonin daily supplementation to aged rats on the ability to withstand cold, thermoregulation and body weight. *Life Sci.* **2021**, *265*, 118769. [CrossRef] [PubMed]
66. Buonfiglio, D.; Parthimos, R.; Dantas, R.; Silva, R.C.; Gomes, G.; Andrade-Silva, J.; Ramos-Lobo, A.; Amaral, F.G.; Matos, R.; Sinésio, J.; et al. Melatonin Absence Leads to Long-Term Leptin Resistance and Overweight in Rats. *Front. Endocrinol.* **2018**, *9*, 122. [CrossRef]
67. Lapa, C.; Arias-Loza, P.; Hayakawa, N.; Wakabayashi, H.; Werner, R.A.; Chen, X.; Shinaji, T.; Herrmann, K.; Pelzer, T.; Higuchi, T. Whitening and Impaired Glucose Utilization of Brown Adipose Tissue in a Rat Model of Type 2 Diabetes Mellitus. *Sci. Rep.* **2017**, *7*, 5–10. [CrossRef]
68. De Jong, J.M.A.; Sun, W.; Pires, N.D.; Frontini, A.; Balaz, M.; Jespersen, N.Z.; Feizi, A.; Petrovic, K.; Fischer, A.W.; Bokhari, M.H.; et al. Human brown adipose tissue is phenocopied by classical brown adipose tissue in physiologically humanized mice. *Nat. Metab.* **2019**, *1*, 830–843. [CrossRef]
69. Kotzbeck, P.; Giordano, A.; Mondini, E.; Murano, I.; Severi, I.; Venema, W.; Cecchini, M.P.; Kershaw, E.E.; Barbatelli, G.; Haemmerle, G.; et al. Brown adipose tissue whitening leads to brown adipocyte death and adipose tissue inflammation. *J. Lipid Res.* **2018**, *59*, 784. [CrossRef]
70. Shimizu, I.; Aprahamian, T.; Kikuchi, R.; Shimizu, A. Vascular rarefaction mediates whitening of brown fat in obesity. *J. Clin. Investig.* **2014**, *124*, 2099–2112. [CrossRef]
71. Gonzalez-Hurtado, E.; Lee, J.; Choi, J.; Wolfgang, M.J. Fatty acid oxidation is required for active and quiescent brown adipose tissue maintenance and thermogenic programing. *Mol. Metab.* **2018**, *7*, 45–56. [CrossRef]
72. Fischer, A.W.; Behrens, J.; Sass, F.; Schlein, C.; Heine, M.; Pertzborn, P.; Scheja, L.; Heeren, J. Brown adipose tissue lipoprotein and glucose disposal is not determined by thermogenesis in uncoupling protein 1-deficient mice. *J. Lipid Res.* **2020**, *61*, 1377–1389. [CrossRef]
73. Kazak, L.; Chouchani, E.T.; Stavrovskaya, I.G.; Lu, G.Z.; Jedrychowski, M.P.; Egan, D.F.; Kumari, M.; Kong, X.; Erickson, B.K.; Szpyt, J.; et al. UCP1 deficiency causes brown fat respiratory chain depletion and sensitizes mitochondria to calcium overload-induced dysfunction. *Proc. Natl. Acad. Sci. USA* **2017**, *114*, 7981–7986. [CrossRef] [PubMed]

74. Luijten, I.H.N.; Feldmann, H.M.; Von Essen, G.; Cannon, B.; Nedergaard, J. In the absence of UCP1-mediated diet-induced thermogenesis, obesity is augmented even in the obesity-resistant 129S mouse strain. *Am. J. Physiol. Endocrinol. Metab.* **2019**, *316*, E729–E740. [CrossRef]
75. de Souza, C.A.P.; Gallo, C.C.; de Camargo, L.S.; de Carvalho, P.V.V.; Olesçuck, I.F.; Macedo, F.; da Cunha, F.M.; Cipolla-Neto, J.; do Amaral, F.G. Melatonin multiple effects on brown adipose tissue molecular machinery. *J. Pineal Res.* **2019**, *66*, e12549. [CrossRef] [PubMed]
76. Omran, F.; Christian, M. Inflammatory Signaling and Brown Fat Activity. *Front. Endocrinol.* **2020**, *11*, 156. [CrossRef]
77. Rebiger, L.; Lenzen, S.; Mehmeti, I. Susceptibility of brown adipocytes to pro-inflammatory cytokine toxicity and reactive oxygen species. *Biosci. Rep.* **2016**, *36*, e00306. [CrossRef]
78. Wikstrom, J.D.; Mahdaviani, K.; Liesa, M.; Sereda, S.B.; Si, Y.; Las, G.; Twig, G.; Petrovic, N.; Zingaretti, C.; Graham, A.; et al. Hormone-induced mitochondrial fission is utilized by brown adipocytes as an amplification pathway for energy expenditure. *EMBO J.* **2014**, *33*, 418–436. [CrossRef]
79. Mahdaviani, K.; Benador, I.Y.; Su, S.; Gharakhanian, R.A.; Stiles, L.; Trudeau, K.M.; Cardamone, M.; Enríquez-Zarralanga, V.; Ritou, E.; Aprahamian, T.; et al. Mfn2 deletion in brown adipose tissue protects from insulin resistance and impairs thermogenesis. *EMBO Rep.* **2017**, *18*, 1123–1138. [CrossRef]
80. Ding, M.; Feng, N.; Tang, D.; Feng, J.; Li, Z.; Jia, M.; Liu, Z.; Gu, X.; Wang, Y.; Fu, F.; et al. Melatonin prevents Drp1-mediated mitochondrial fission in diabetic hearts through SIRT1-PGC1α pathway. *J. Pineal Res.* **2018**, *65*, e12491. [CrossRef]
81. Ma, S.; Dong, Z. Melatonin Attenuates Cardiac Reperfusion Stress by Improving OPA1-Related Mitochondrial Fusion in a Yap-Hippo Pathway-Dependent Manner. *J. Cardiovasc. Pharmacol.* **2019**, *73*, 27–39. [CrossRef]
82. Zhong, M.; Zhang, Y. Melatonin prevents abnormal mitochondrial dynamics resulting from the neurotoxicity of cadmium by blocking calcium-dependent translocation of Drp1 to the mitochondria. *J. Pineal Res.* **2016**, *60*, 291–302. [CrossRef]
83. Chang, J.Y.; Yu, F.; Shi, L.; Ko, M.L.; Ko, G.Y. Melatonin Affects Mitochondrial Fission/Fusion Dynamics in the Diabetic Retina. *J. Diabetes Res.* **2019**, *2019*, 8463125. [CrossRef] [PubMed]
84. Bhaskar, S.; Sheshadri, P.; Joseph, J.P.; Potdar, C.; Prasanna, J.; Kumar, A. Mitochondrial Superoxide Dismutase Specifies Early Neural Commitment by Modulating Mitochondrial Dynamics. *Iscience* **2020**, *23*, 101564. [CrossRef] [PubMed]
85. Jespersen, N.Z.; Andersen, M.W.; Jensen, V.H.; Stærkær, T.W.; Severinsen, M.C.K.; Mandrup, S.; Pedersen, B.K.; Nielsen, S.; Scheele, C. Thermogenic genes are blunted whereas brown adipose tissue identity is preserved in human obesity. *bioRxiv* **2020**. [CrossRef]
86. Gao, X.-Y.; Deng, B.-H.; Li, X.-R.; Wang, Y.; Zhang, J.-X.; Hao, X.-Y.; Zhao, J.-X. Melatonin Regulates Differentiation of Sheep Brown Adipocyte Precursor Cells Via AMP-Activated Protein Kinase. *Front. Vet. Sci.* **2021**, *8*, 588. [CrossRef]

antioxidants

Review

Understanding the Mechanism of Action of Melatonin, Which Induces ROS Production in Cancer Cells

Javier Florido [1,2], César Rodriguez-Santana [1,2], Laura Martinez-Ruiz [1,2], Alba López-Rodríguez [1,2], Darío Acuña-Castroviejo [1,2,3], Iryna Rusanova [1,3,4] and Germaine Escames [1,2,3,*]

1 Biomedical Research Center, Health Sciences Technology Park, University of Granada, 18016 Granada, Spain
2 Department of Physiology, Faculty of Medicine, University of Granada, 18016 Granada, Spain
3 Centro de Investigación Biomédica en Red Fragilidad y Envejecimiento Saludable (CIBERFES), Ibs.Granada, San Cecilio University Hospital, 18016 Granada, Spain
4 Department of Biochemistry and Molecular Biology, Faculty of Science, University of Granada, 18019 Granada, Spain
* Correspondence: gescames@ugr.es; Tel.: +34-958-241-000 (ext. 20363)

Abstract: Reactive oxygen species (ROS) constitute a group of highly reactive molecules that have evolved as regulators of important signaling pathways. In this context, tumor cells have an altered redox balance compared to normal cells, which can be targeted as an antitumoral therapy by ROS levels and by decreasing the capacity of the antioxidant system, leading to programmed cell death. Melatonin is of particular importance in the development of innovative cancer treatments due to its oncostatic impact and lack of adverse effects. Despite being widely recognized as a pro-oxidant molecule in tumor cells, the mechanism of action of melatonin remains unclear, which has hindered its use in clinical treatments. The current review aims to describe and clarify the proposed mechanism of action of melatonin inducing ROS production in cancer cells in order to propose future anti-neoplastic clinical applications.

Keywords: melatonin; reactive oxygen species; apoptosis; mitochondria; cancer

Citation: Florido, J.; Rodriguez-Santana, C.; Martinez-Ruiz, L.; López-Rodríguez, A.; Acuña-Castroviejo, D.; Rusanova, I.; Escames, G. Understanding the Mechanism of Action of Melatonin, Which Induces ROS Production in Cancer Cells. *Antioxidants* **2022**, *11*, 1621. https://doi.org/10.3390/antiox11081621

Academic Editors: Marina Garcia-Macia and Adrián Santos-Ledo

Received: 29 July 2022
Accepted: 19 August 2022
Published: 20 August 2022

1. Introduction

Cancer continues to be a dreadful disease despite the continuous efforts made to develop novel treatment modalities. According to the GLOBOCAN report published in 2020, 19.3 million new cancer cases were expected in 2020, with an estimated 10 million cancer mortalities. Furthermore, with the number of cancer cases expected to reach 28.4 million around the world in 2040 [1], it is necessary to develop innovative treatments and to investigate new therapeutic targets.

Reactive oxygen species (ROS), which are generally considered byproducts of oxygen consumption and cellular metabolism, are formed by the partial reduction of molecular oxygen [2]. Thus, tumor cells have an altered redox balance compared to normal cells, which can be targeted by antitumoral therapies by increasing ROS levels and by decreasing antioxidant system capacity, leading to apoptosis. This is the case for certain chemotherapeutic compounds that impact the induction of ROS production, resulting in irreparable damage and cell death [3].

The hormone melatonin (aMT; N-acetyl-5-methoxytryptamine) is synthesized by the pineal gland as well as by several types of tissues. It is also well-known that melatonin is produced by mitochondria [4] at higher concentrations than in other cellular compartment [5,6]. Melatonin, together with its metabolites, are not only involved in multiple cellular processes [7] but are also potent free radical scavengers and broad-spectrum antioxidants with evolutionarily conserved properties [8,9]. Melatonin is capable of reducing free radical damage by acting directly as a free radical scavenger and indirectly by stimulating antioxidant enzyme activity [10]. In addition, melatonin plays an effective role in maintaining mitochondrial homeostasis, which protects against oxidative damage [11–15]. However, despite its widely recognized antioxidant role in protecting normal cells against cytotoxicity and apoptosis, melatonin, which induces apoptosis in several types of cancer cells, has also been shown to have pro-oxidant effects [16].

Various groups have, for many years, reported that high concentrations of melatonin can promote ROS generation, leading to cell death in a variety of cancers [17–22]; this suggests that melatonin can act as both an antioxidant and pro-oxidant in human cell lines, depending on the concentration and treatment duration. Furthermore, numerous studies have shown that melatonin enhances the cytotoxic effects of chemotherapeutic drugs on cancer cells, depending on the dose [23], thus suggesting that melatonin increases their chemotherapeutic effect [24].

In conclusion, melatonin is an ideal candidate for use in innovative cancer therapies. However, unlike its antioxidant effect, the way in which melatonin produces ROS in tumoral cells remains unclear. Thus, the identification of signaling pathways and key molecules associated with the pro-oxidant effect of melatonin is extremely important with regard to the effectiveness of clinical anti-neoplastic therapies.

2. Involvement of Receptors in Melatonin's Pro-Oxidant Activity

Melatonin appears to exert some of its effects in humans through the MT1 and MT2 membrane receptors [25] and also provides high-affinity binding for the nuclear receptors ROR/RZR [26]. In addition to regular receptors, melatonin binds other proteins such as calmodulin (CaM) and calreticuline (CALR) [27,28], which interact with melatonin at concentrations ranging from sub-nanomolar to millimolar in mammalian cells [29]. In addition, the level of melatonin to which these targets respond can have concentration ranges of over seven orders of magnitude [30]. In this context, it has been widely demonstrated that high levels of melatonin are necessary to induce ROS production in tumor cells [19,20,31,32], thus highlighting the independent impact of low-affinity melatonin targets such as MT1, MT2, and ROR receptors. This is confirmed by melatonin's pro-oxidant effect, which is unaffected by the MT1/MT2 antagonist luzindole and is not elicited by melatonin analogues with high affinity for MT1/MT2 receptors [29].

In contrast, it has been suggested that melatonin's pro-oxidant activity derives from binding to calmodulin, to which other enzymes, such as phospholipase A2 (PLA2), associated with oxidative stress, also bind [29].

The PLA2 enzyme cleaves membrane phospholipids, leading to the release of membrane-bound arachidonic acid (AA), which is processed by cyclooxygenases (COXs) and lipoxygenases (LOXs) to produce important inflammatory mediators, such as prostaglandins and leukotrienes, as well as to increase ROS production [33]. However, Ca^{2+}-independent phospholipase PLA2 (iPLA2) can bind to calmodulin that is inactivated. Radogna et al. [34] demonstrated that melatonin, at high concentrations, binds to calmodulin, thus inducing the release of iPLA2 and, consequently, increasing ROS production in human tumor monocytes (U937 cells; Figure 1). The increase in ROS levels by melatonin is inhibited by chlorpromazine, which prevents melatonin from binding to calmodulin, which is insensitive to calmidazolium, which does not prevent this interaction [34].

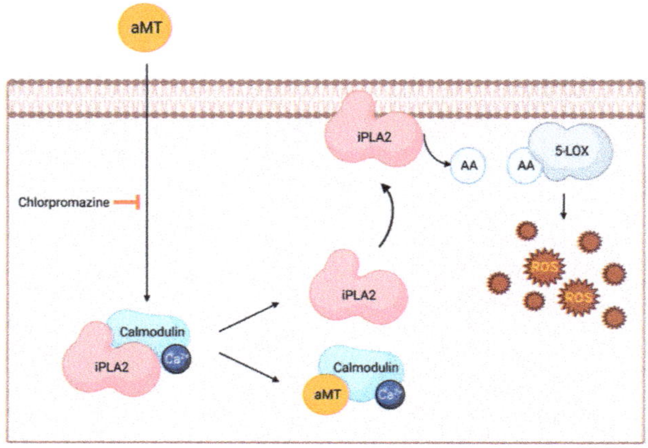

Figure 1. Melatonin induces ROS production in cancer cells through calmodulin binding. Melatonin binds to calmodulin, leading to the release of sequestered Ca^{2+}-independent PLA2, which is then free to move to membranes and to release high doses of AA; in turn, liberated AA feeds 5-LOX to produce free radicals. Melatonin (aMT); Ca^{2+}-independent PLA2 (iPLA2); arachidonic acid (AA); 5-lipoxygenase (5-LOX). Image created using BioRender.com(accessed on 16 July 2022).

Finally, iPLA2 and 5-LOX inhibitors also abolish melatonin's ability to stimulate ROS production, indicating that these two enzymes are involved in melatonin's pro-oxidant activity [34]. All these findings suggest that the binding of melatonin to calmodulin is required in order to induce ROS production.

3. Involvement of Molecular Pathways in Melatonin's Pro-Oxidative Activity

In addition to its effect through its proven affinity for receptors, melatonin can induce oxidative stress directly by increasing ROS levels and indirectly by regulating the expression of various proteins involved in metabolic pathways, as described below.

3.1. The Sirtuin Pathway

Sirtuins, which are class III histone deacetylase enzymes, are key molecular proteins involved in oxidative stress [35] and play an important role in both normal and cancer cells [36].

Sirtuin-3 (SIRT3), which is located primarily in the mitochondrial matrix, regulates intracellular metabolism, mainly by modulating mitochondrial oxidative stress [37]. It also regulates superoxide dismutase 2 (SOD2) by deacetylating and activating the SOD2 transcription factor FOXO3a as well as by directly deacetylating and activating SOD2 dismutase activity [38]. Furthermore, decreased SIRT3 expression increases ROS-mediated oxidative damage, suggesting that SIRT3 inhibitors could be therapeutically beneficial [39]. Li et al. [40] demonstrated that melatonin potentiates the cytotoxic effects of shikonin (SHK) in HeLa cancer cells by inducing oxidative stress through the inhibition of SIRT3/SOD2 expression and activity. The combination of melatonin and SHK promotes apoptosis, which increases ROS production in various types of cancer cells. All these effects are reversed by ROS scavengers, thus suggesting that melatonin induces apoptosis in cancer cells by increasing ROS production. This is probably explained by the excessive levels of ROS induced by melatonin, which releases mitochondrial cytochrome C, leading to apoptosis in cancer cells [41].

However, other authors, such as Chen et al. [42], have reported that melatonin increases SIRT3 activity in lung cancer cells. This results in the deacetylation of pyruvate dehydrogenase (PDH) to enhance complex I and IV activity, which increases ROS production and reverses the Warburg effect (Figure 2). Thus, melatonin induces cancer cellular apoptosis by elevating ROS generation due to an increase in OXPHOS through the stimulation of SIRT3 activity [42]. Further research is required to clarify melatonin's mechanism of action via SIRT3.

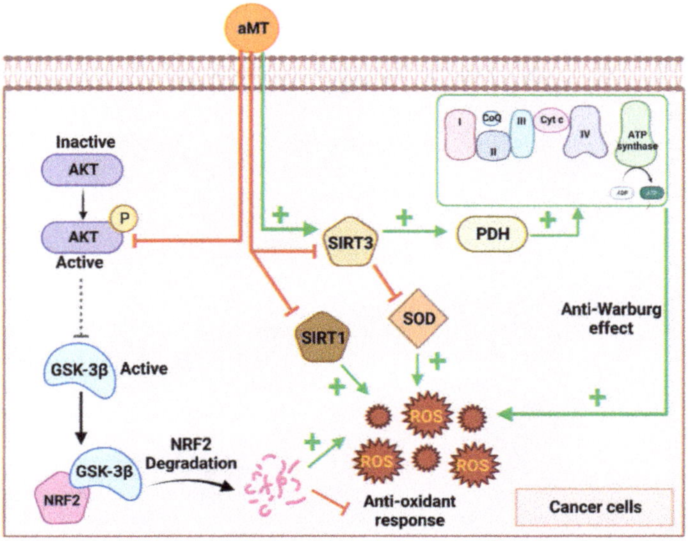

Figure 2. Different mechanisms by which melatonin induces ROS production in cancer cells. Melatonin inhibits the AKT pathway, leading to the activation of GSK-3β, which induces NRF2 degradation. On the other hand, melatonin regulates Sirtuin 3 (SIRT3) through its activation or inhibition, leading to an anti-Warburg effect or SOD inhibition, respectively. Finally, melatonin has been shown to inhibit SIRT1 in cancer cells. All these processes lead to an increase in ROS production and antitumor activity. Melatonin (aMT); glycogen synthase kinase-3β (GSK-3β); superoxide dismutase (SOD); pyruvate dehydrogenase (PDH). Image created using BioRender.com (accessed on 18 July 2022).

SIRT-1, a conserved nicotinamide adenine dinucleotide (NAD+)-dependent deacetylase, is associated with melatonin's antitumoral and pro-oxidative activity. Melatonin has been demonstrated to increase SIRT1 activity in normal cells, leading to a decrease in ROS production and the regulation of cell homeostasis [43,44]. However, SIRT1 overexpression in tumor cells correlates with the silencing of tumor suppressor genes and cancer resistance to chemotherapy. Melatonin in tumor cells, such as human osteosarcoma, has been shown to inhibit SIRT1, resulting in increased pro-oxidant and antitumor activity (Figure 2) [45]. The inhibition of SIRT1 through the use of the inhibitor sirtinol or SIRT1 siRNA also increases melatonin's antitumor activity. The upregulation of SIRT-1 by its activator, SRT1720, attenuates melatonin's antioxidant and antitumor activity, indicating that its induction of ROS production in tumor cells is activated by SIRT1 [45]. It has been suggested that melatonin is directly involved in controlling SIRT1, which inhibits its activity in cancer cells, in contrast to its stimulatory action in normal cells [46].

3.2. The Akt Pathway

Akt, which is a serine/threonine kinase that was previously known as protein kinase B (PKB), plays a crucial role in major cellular functions such as cell size, cell cycle progression, glucose metabolism regulation, genome stability, transcription, and protein synthesis. Akt promotes cell survival by mediating cellular growth factors and by blocking apoptosis through the inactivation of pro-apoptotic proteins as well as through the regulation of ROS balance [47]. A wide range of proteins are sensitive to phosphorylation by AKT, such as glycogen synthase kinase-3β (GSK-3β), a protein serine/threonine kinase involved in cell signaling that is phosphorylated on serine 9 and then inactivated [48].

Melatonin has been widely reported to modulate Akt in both tumoral and nontumoral cells [49–52]. In human melanoma cells (SK-MEL-1), Perdomo et al. [53] demonstrate that, as melatonin promotes GSK-3β dephosphorylation at serine 9 via Akt pathway inhibition, the activation of GSK-3β by melatonin results in an increase in ROS production (Figure 2). This was confirmed by the use of the GSK-3β inhibitor BIO, which partially abrogates the generation of ROS in response to melatonin. Perdomo et al. also reported that the pro-oxidative effects of GSK-3 are due to the degradation of NRF2, the master regulator of endogenous antioxidant responses [53]. This pattern was also reported in the glioblastoma cell line U87MG, where the inhibition of Akt triggers GSK-3β activity, which, in turn, switches off the antioxidant response of NRF2 [54].

3.3. *Involvement of OXPHOS Induction in the Conflict between Melatonin's Pro-Oxidative and Anti-WARBURG Activity*

The Warburg effect, which is commonly associated with solid tumors, contributes significantly to their hardness, invasiveness, and metastatic capability as well as rendering them resistant to radio- and chemotherapies [55–57]. In most mammalian cells, mitochondria are an important source of hydrogen peroxide (H_2O_2) and superoxide anion (O_2^-) [58]. The Warburg effect of glycolysis upregulation in cancer cell energy metabolism could reduce the production of H_2O_2 and O_2^- by decreasing OXPHOS activity [59]. Just as the inhibition of metabolic reprogramming induces ROS production in tumor cells, melatonin increases OXPHOS capacity and inhibits glycolysis in cancer cells, resulting in increased ROS production [20,42].

As shown in Figure 2, Chen et al. [42] reported that melatonin reverses the Warburg effect by stimulating the SIRT3/PDH axis in lung cancer cell lines. The pyruvate dehydrogenase complex (PDC) plays an important role in catalyzing the conversion of pyruvate to acetyl-CoA, which is associated with mitochondrial ATP production. Pyruvate dehydrogenase (PDH), which is the first and most important enzymic component of PDC, converts pyruvate to acetyl-CoA and then enters tricarboxylic acid (TCA) to produce ATP as well as electron donors such as NADH. In cancer cells, metabolic reprogramming results in the inhibition of PDH, leading to a decrease in acetyl-CoA. Pyruvate is then shunted away from the mitochondrial cancer cell metabolism through its reduction to lactate (Warburg effect) [60]. Chen et al. [42] showed that melatonin promotes SIRT3 expression and PDH deacetylation in order to enhance complex I and IV activity, leading to a reversal of the Warburg effect and ROS production, followed by cell death in lung cancer cells (A549, PC9, and LLC). This hypothesis was corroborated using 3-TYP, a selective SIRT3 inhibitor that abolishes melatonin's ability to stimulate PDH activity and ROS production, suggesting that the melatonin regulation of SIRT3 is required to reverse the Warburg effect. Thus, not only does SIRT3 inhibit SOD, as described above, but it also induces PDH, which together lead to ROS production and apoptosis in tumoral cells [42].

Our research group found that treatment with melatonin inhibits metabolic reprogramming by inhibiting glycolysis and increasing OXPHOS activity, leading to the production of ROS in head and neck squamous cell carcinoma (HNSCC) [20]. Thus, enhanced OXPHOS activity increases the levels of ROS, thereby inducing cancer cell death [61]. In primary human kidney mesangial cells, melatonin can also induce rapid ROS generation via the antimycin-A-sensitive site in mitochondrial complex III [9].

It may also be possible that melatonin not only reverses Warburg effects, leading to ROS production in cancer cells, by activating sirtuins but also regulates the malignancy-promoting transcription factor hypoxia-inducible factor-1α (HIF-1α). Therefore, it is also suggested that the underlying mechanism of the pro-oxidant effect of melatonin on cancer cells involves the inhibition of HIF-1α by melatonin. HIF-1α is a key transcription agent involved in mediating Warburg-type metabolisms in diseased cells [62,63]. HIF-1α is also part of an oxygen sensing system that is activated when the partial pressure (pO_2) of intracellular oxygen becomes depressed. HIF-1 plays a critical role in stimulating the mitochondrial pyruvate dehydrogenase kinase (PDK), which leads to the inactivation of the pyruvate dehydrogenase complex (PDC), thereby reducing the mitochondrial conversion of pyruvate to acetyl coenzyme A and stimulating the Warburg effect [64,65]. Research has demonstrated that melatonin, either directly or indirectly, inhibits HIF-1 in cancer cells [66–68], thus likely reversing the Warburg effect and inducing both ROS production and cell death. However, despite the proven close relationship between melatonin's anti-Warburg effect and increased ROS production, further research is required to understand the precise mechanisms involved.

4. Decreased Antioxidant Defenses

The pro-oxidant and oncostatic effects of melatonin can be explained by the increased levels of intracellular ROS and by the decrease in antioxidant capacity exhibited in the melatonin-treated cells, as described above. Melatonin decreases antioxidant enzymes such as catalase, glutathione peroxidase (GSH-Px), and SOD and also increases lipid peroxidation in different cancer types (Table 1). However, in nontumoral cells, melatonin reduces oxidative stress damage to scavenging free radicals and increases antioxidant enzyme activity [8,69,70].

Table 1. Studies of melatonin's effects on antioxidant defenses in cancer cells.

Type of Cancer Cell	Melatonin Dosage	Effects	Authors
Human colorectal cancer SW-480 cell line	300 μM	Decreased catalase and SOD activity	[71]
Human colorectal carcinoma HCT-116 cell line	10^{-6} M	Decreased catalase and GSH-Px activity and increased lipid peroxidation	[23]
Human histiocytic lymphoma U937 cell line	0.5, 1, and 2 mM	Decreased SOD2 activity	[40]
Human hepatocellular carcinoma HepG2 cell line	0 to 20 mmol/L	Decreased SOD2 activity	[72]
Human HNSCCCal-27 cell line	100, 500,and 1500 μM	Decreased SOD2 activity	[20]
Human HNSCC Cal-27 cell line	100, 500, and 1000 μM	Decreased GPx activity	[31]
Human HNSCC Cal-27 cell line	100, 500, 1000, and 1500 μM	IR + aMT at 100 μM: increased GPx activitya MT at 1000 μM alone or combined with CDDP: decreased GPx activity	[19]
Xenograft mouse colon cancer (CT26 cell line)	20 mg/kg	Melatonin improved SOD and GPx activity in nontargeted tissues and reduced these two enzymes in the tumor tissue.	[73]

It has also been suggested that melatonin regulates antioxidant enzymes through the cellular prion protein (PrPC)-dependent pathway. Normal PrPC is a ubiquitous glycoprotein involved in various physiological cellular processes, including proliferation, differentiation, stress protection, and signal transduction regulation. PrPC is also involved in tumor resistance in colorectal cancer cells [74]. It also protects cells against oxidative stress by increasing the activity of antioxidant enzymes such as SOD and catalase [75], while the silencing of PrPC decreases this antioxidant activity [76,77].

Some studies have shown that melatonin reduces PrPC levels, leading to a decrease in antioxidant defenses in cancer cells [78]. Lee et al. [79] reported that PrPC levels increased in human oxaliplatin-resistant cell lines (SNU-C5/Oxal-R), resulting in an increased antioxidant effect through an increase in SOD and catalase activity. On the other hand, cotreatment with oxaliplatin and melatonin reduced the level of PrPC and consequently led to the suppression of antioxidant enzyme activity and increased superoxide anion generation in these cancer cells. The increase in superoxide anion levels is related to the activation of the endoplasmic reticulum (ER) stress-mediated signaling pathway and the induction of apoptosis through the regulation of apoptosis-associated proteins [79].

Melatonin's induction of ROS and mitochondrial dysfunction via PrPC could be explained by the regulation of the expression of PTEN-induced putative kinase 1 (PINK1), a protein located in the outer mitochondrial membrane that maintains mitochondria homeostasis. Won Yun et al. [80] showed that melatonin suppresses PrPC and PINK1 expression, which leads to an increase in the production of mitochondrial superoxide in colorectal cancer cells. The impact of melatonin is greater when PrPC is silenced, indicating that the inhibition of PrPC expression enhances melatonin-mediated pro-oxidant activity [80].

5. Reverse Electron Transport (RET): Another Melatonin Mechanism That Could Induce ROS

In addition to the mechanism described above, melatonin may induce ROS production via reverse electron transport (RET). RET occurs in mitochondria when the pool of coenzyme Q becomes overly reduced by electrons from respiratory complex II or other enzymes as well as in the presence of high proton motive force (Δp). Under these conditions, CI activity reduces NAD+ to NADH with electrons from the ubiquinol pool, thus generating high levels of mtROS [81,82]. Complex I has also been found to produce ROS in a forward or reverse direction, depending on the substrates used to feed the respiratory chain, suggesting that a change in cell metabolism induces RET. We previously demonstrated that melatonin reverses metabolic reprogramming in HNSCC cells [20]. Given these findings and the melatonin-modified tumor metabolism, we determined whether melatonin increases mtROS via RET.

Thus, our research group reported [83] that melatonin increases mitochondrial CII activity, membrane potential, and CoQH2/CoQ, which are essential conditions for RET (Figure 3). Interestingly, mitochondrial complex inhibitors, including rotenone, which increases forward ROS production but decreases ROS production via RET, abolish the increase in ROS, indicating that melatonin increases ROS generation via RET. Under hypoxic conditions, HIF maintains ROS production as well as their integrity at physiologically low levels by decreasing respiratory activity [84]. However, it was demonstrated that melatonin has the ability to destabilize HIF-1α [85]. In previous research, we showed that melatonin inhibits tumor cells by reversing aerobic glycolysis [20], which is a key step in the destabilization of HIF-1α [86], either by suppressing synthesis or by promoting degradation [87]. We therefore hypothesize that the inhibition of HIF-1α is the principal regulator of melatonin's pro-oxidant activity in cancer cells. Nevertheless, further research is required to elucidate melatonin's mechanism of action, which induces ROS production via RET.

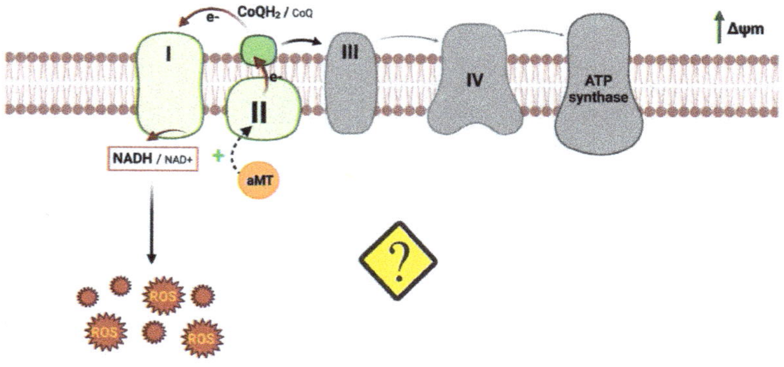

Figure 3. Possible mechanisms of action of melatonin to induce ROS production via RET. In our view, melatonin increases mitochondrial CII activity, membrane potential, and CoQH2/CoQ, leading to RET-ROS production. Image created using BioRender.com (accessed on 22 July 2022).

6. Debate around the Antioxidant Effects of Melatonin in Cancer Cells

The beneficial effects of melatonin's antioxidant activity are widely documented in the literature. However, although the majority of studies described the pro-oxidant effects of melatonin on cancer cells, as detailed above, melatonin can also reduce ROS production in some cancer types.

The antioxidant activity of melatonin is due to its direct effect as a radical scavenger, increasing the expression and activity of antioxidant enzymes, and regulating mitochondrial homeostasis [88]. It has been suggested that factors such as treatment duration and molecular concentrations as well as target cell type and conditions affect the outcome of melatonin on oxidative stress [89]. Melatonin's antioxidant activity has mainly been reported in three different states of tumorigenesis: cancer initiation, progression, and metastasis [90]. Prior to cellular malignancy, the antioxidant activity of melatonin maintains the genomic integrity of cells by preventing cellular tumorigenesis and by protecting DNA against oxidative damage. Melatonin prevents DNA mutations, either directly through free radical scavenging activity or indirectly through the inhibition of metal-induced DNA damage, by stimulating antioxidant enzymes, enhancing the DNA repair system, and suppressing pro-oxidative enzymes [91]. Once the oncostatic process has begun, some studies have reported that melatonin, at low doses, impairs the proliferation and apoptotic resistance of oral cancer cells by inactivating ROS-dependent Akt signaling, which is involved in the downregulation of cyclin D1, proliferating cell nuclear antigen (PCNA), and Bcl-2 as well as in Bax upregulation [92]. Interestingly, the antioxidative effects of melatonin are condition-dependent, as evidenced by the changing nature of its pro- and antioxidant activities in both in vitro cellular and acellular studies [88].

Other research has demonstrated that melatonin's antioxidant activity has antiangiogenic and antimetastatic effects. Thus, melatonin has been reported to play an antiangiogenetic role in blocking ROS-activated extracellular regulated protein kinases (ERKs) and Ak pathways in oral cancers [92]. Melatonin also suppresses hypoxia-induced cancer cell migration and invasion through the inhibition of HIF-1α due to its antioxidant impact on hypoxia [68,93].

All these findings point to the dual impact of melatonin, depending on its regulation of ROS homeostasis [94]. Although one possible explanation for this dual effect is the differences in experimental procedures and cancer models used in the different studies, further research is required to better understand the contradictory activities of melatonin in cancer treatment.

7. Conclusions

The oncostatic effects of melatonin through its pro-oxidant actions have been widely described [17–22,95]. Its mechanism of action, however, remains unclear. This review described the possible pathways involved in ROS production by melatonin in cancer cells. These mechanisms include, besides melatonin receptors, sirtuins and Akt pathways and melatonin's anti-Warburg activity. Furthermore, this review suggests that melatonin does not only induce ROS production by inducing oxidative stress but also by decreasing antioxidant defenses [19,20,23,31,40,71–73]. Finally, in view of new experimental data, we propose that melatonin induces ROS production in cancer cells, activating mitochondrial reverse electron transport [83]. Therefore, we describe and clarify a new mechanism of action of melatonin to induce ROS production in cancer cells, a finding that may be considered for its anti-neoplastic clinical applications.

Author Contributions: Writing—original draft preparation, J.F.; writing—review and editing, G.E.; formal analysis, G.E.; resources, J.F., C.R.-S., L.M.-R. and A.L.-R.; supervision, G.E., D.A.-C. and I.R.; project administration, G.E. and D.A.-C.; funding acquisition, G.E. All authors have read and agreed to the published version of the manuscript.

Funding: This review was funded by grants from the MCIN/AEI/10.13039/501100011033, Spain, the ERDF (SAF2017-85903-P, ID2020-115112RB-I00), the Consejería de Innovación, Ciencia y Empresa, Junta de Andalucía (P07-CTS-03135, P10-CTS-5784, and CTS-101), and the University of Granada (Grant "UNETE," UCE-PP2017-05), Spain. J.F. and L.M. are recipients of FPU fellowships from the Ministerio de Educación Cultura y Deporte, Spain.

Acknowledgments: We wish to thank Michael O'Shea for proofreading the paper.

Conflicts of Interest: The authors declare no conflict to interest.

Abbreviations

5-LOX	5-lipoxygenase
AA	Arachidonic acid
aMT	Melatonin
CaM	Calmodulin
COX	Cyclooxygenases
ERK	Extracellular-regulated protein kinases
GSH-Px	Glutathione peroxidase
GSK-3β	Glycogen synthase kinase-3β
H_2O_2	Hydrogen peroxide
HNSCC	Head and neck squamous cell carcinoma
HIF-1α	Hypoxia-inducible factor-1α
iPLA2	Ca^{2+}-independent phospholipase A2
LOX	Lipoxygenases
NAD	Nicotinamide adenine dinucleotide
O_2^-	Superoxide anion
PCNA	Proliferating cell nuclear antigen
PDC	Pyruvate dehydrogenase complex
PDH	Pyruvate dehydrogenase
PINK1	PTEN-induced putative kinase 1
PKB	Protein kinase B
PLA2	Phospholipase A2
PrPC	Cellular prion proteins
pO_2	Oxygen partial pressure
RET	Reverse electron transport
ROS	Reactive oxygen species
SHK	Shikonin
SIRT3	Sirtuin-3
SOD2	Superoxide dismutase 2

References

1. Sung, H.; Ferlay, J.; Siegel, R.L.; Laversanne, M.; Soerjomataram, I.; Jemal, A.; Bray, F. Global Cancer Statistics 2020: GLOBOCAN Estimates of Incidence and Mortality Worldwide for 36 Cancers in 185 Countries. *CA Cancer J. Clin.* **2021**, *71*, 209–249. [CrossRef] [PubMed]
2. Zorov, D.B.; Juhaszova, M.; Sollott, S.J. Mitochondrial Reactive Oxygen Species (ROS) and ROS-Induced ROS Release. *Physiol. Rev.* **2014**, *94*, 909–950. [CrossRef]
3. Moloney, J.N.; Cotter, T.G. ROS signalling in the biology of cancer. *Semin. Cell Dev. Biol.* **2018**, *80*, 50–64. [CrossRef]
4. Suofu, Y.; Li, W.; Jean-Alphonse, F.G.; Jia, J.; Khattar, N.K.; Li, J.; Baranov, S.V.; Leronni, D.; Mihalik, A.C.; He, Y.; et al. Dual role of mitochondria in producing melatonin and driving GPCR signaling to block cytochrome c release. *Proc. Natl. Acad. Sci. USA* **2017**, *114*, E7997–E8006. [CrossRef]
5. Venegas, C.; García, J.A.; Escames, G.; Ortiz, F.; López, A.; Doerrier, C.; García-Corzo, L.; López, L.C.; Reiter, R.J.; Acuña-Castroviejo, D. Extrapineal melatonin: Analysis of its subcellular distribution and daily fluctuations. *J. Pineal Res.* **2012**, *52*, 217–227. [CrossRef]
6. NaveenKumar, S.K.; Hemshekhar, M.; Jagadish, S.; Manikanta, K.; Vishalakshi, G.J.; Kemparaju, K.; Girish, K.S. Melatonin restores neutrophil functions and prevents apoptosis amid dysfunctional glutathione redox system. *J. Pineal Res.* **2020**, *69*, e12676. [CrossRef] [PubMed]
7. Sartorelli, L.S.; Neto, R.J.B.; Moscheta-Pinheiro, M.G.; de Castro, T.B. Blood melatonin level can serve as a potential biomarker for prostate and hepatocellular carcinomas. *Melatonin Res.* **2021**, *4*, 253–269. [CrossRef]

8. Tan, D.-X.; Manchester, L.C.; Reiter, R.J.; Qi, W.-B.; Karbownik, M.; Calvo, J.R. Significance of Melatonin in Antioxidative Defense System: Reactions and Products. *Neurosignals* **2000**, *9*, 137–159. [CrossRef]
9. Zhang, H.-M.; Zhang, Y.-Q.; Zhang, B.-X. The role of mitochondrial complex III in melatonin-induced ROS production in cultured mesangial cells. *J. Pineal Res.* **2010**, *50*, 78–82. [CrossRef]
10. Reiter, R.J. Melatonin: Lowering the High Price of Free Radicals. *N. Physiol. Sci.* **2000**, *15*, 246–250. [CrossRef]
11. Martín, M.; Macías, M.; Escames, G.; León, J.; Acuña-Castroviejo, D. Melatonin but not vitamins C and E maintains glutathione homeostasis in t-butyl hydroperoxide-induced mitochondrial oxidative stress. *FASEB J.* **2000**, *14*, 1677–1679. [CrossRef] [PubMed]
12. Acuña-Castroviejo, D.; Martín, M.; Macías, M.; Escames, G.; León, J.; Khaldy, H.; Reiter, R.J. Melatonin, mitochondria, and cellular bioenergetics. *J. Pineal Res.* **2001**, *30*, 65–74. [CrossRef] [PubMed]
13. Ortiz, F.; Acuña-Castroviejo, D.; Doerrier, C.; Dayoub, J.C.; Lopez, L.C.; Venegas, C.; García, J.A.; López, A.; Volt, H.; Sánchez, M.L.; et al. Melatonin blunts the mitochondrial/NLRP3 connection and protects against radiation-induced oral mucositis. *J. Pineal Res.* **2015**, *58*, 34–49. [CrossRef] [PubMed]
14. López, A.; Ortiz, F.; Doerrier, C.; Venegas, C.; Fernández-Ortiz, M.; Aranda, P.; Díaz-Casado, M.E.; Fernandez-Gil, B.; Barriocanal-Casado, E.; Escames, G.; et al. Mitochondrial impairment and melatonin protection in parkinsonian mice do not depend of inducible or neuronal nitric oxide synthases. *PLoS ONE* **2017**, *12*, e0183090. [CrossRef] [PubMed]
15. Sun, T.C.; Liu, X.C.; Yang, S.H.; Song, L.L.; Zhou, S.J.; Deng, S.L.; Tian, L.; Cheng, L.Y. Melatonin inhibits oxidative stress and apoptosis in cryopreserved ovarian tissues via Nrf2/HO-1 signaling pathway. *Front. Mol. Biosci.* **2020**, *7*, 176. [CrossRef]
16. Mortezaee, K.; Najafi, M.; Farhood, B.; Ahmadi, A.; Potes, Y.; Shabeeb, D.; Musa, A.E. Modulation of apoptosis by melatonin for improving cancer treatment efficiency: An updated review. *Life Sci.* **2019**, *228*, 228–241. [CrossRef]
17. Wölfler, A.; Caluba, H.-C.; Abuja, P.M.; Dohr, G.; Schauenstein, K.; Liebmann, P.M. Prooxidant activity of melatonin promotes fas-induced cell death in human leukemic Jurkat cells. *FEBS Lett.* **2001**, *502*, 127–131. [CrossRef]
18. Moreira, A.J.; Ordoñez, R.; Cerski, C.T.; Picada, J.N.; Garcia-Palomo, A.; Marroni, N.P.; Mauriz, J.L.; González-Gallego, J. Melatonin Activates Endoplasmic Reticulum Stress and Apoptosis in Rats with Diethylnitrosamine-Induced Hepatocarcinogenesis. *PLoS ONE* **2015**, *10*, e0144517. [CrossRef]
19. Fernandez-Gil, B.I.; Guerra-Librero, A.; Shen, Y.-Q.; Florido, J.; Ruiz, L.M.; García-López, S.; Adan, C.; Rodríguez-Santana, C.; Acuña-Castroviejo, D.; Quiñones-Hinojosa, A.; et al. Melatonin Enhances Cisplatin and Radiation Cytotoxicity in Head and Neck Squamous Cell Carcinoma by Stimulating Mitochondrial ROS Generation, Apoptosis, and Autophagy. *Oxidative Med. Cell. Longev.* **2019**, *2019*, 7187128. [CrossRef]
20. Guerra-Librero, A.; Fernandez-Gil, B.; Florido, J.; Martinez-Ruiz, L.; Rodríguez-Santana, C.; Shen, Y.-Q.; García-Verdugo, J.; López-Rodríguez, A.; Rusanova, I.; Quiñones-Hinojosa, A.; et al. Melatonin Targets Metabolism in Head and Neck Cancer Cells by Regulating Mitochondrial Structure and Function. *Antioxidants* **2021**, *10*, 603. [CrossRef]
21. Chen, K.; Zhu, P.; Chen, W.; Luo, K.; Shi, X.-J.; Zhai, W. Melatonin inhibits proliferation, migration, and invasion by inducing ROS-mediated apoptosis via suppression of the PI3K/Akt/mTOR signaling pathway in gallbladder cancer cells. *Aging* **2021**, *13*, 22502–22515. [CrossRef] [PubMed]
22. Yun, S.-M.; Woo, S.H.; Oh, S.T.; Hong, S.-E.; Choe, T.-B.; Ye, S.-K.; Kim, E.-K.; Seong, M.K.; Kim, H.-A.; Noh, W.C.; et al. Melatonin enhances arsenic trioxide-induced cell death via sustained upregulation of Redd1 expression in breast cancer cells. *Mol. Cell. Endocrinol.* **2016**, *422*, 64–73. [CrossRef] [PubMed]
23. Bułdak, R.J.; Pilc-Gumuła, K.; Bułdak, Ł.; Witkowska, D.; Kukla, M.; Polaniak, R.; Zwirska-Korczala, K. Effects of ghrelin, leptin and melatonin on the levels of reactive oxygen species, antioxidant enzyme activity and viability of the HCT 116 human colorectal carcinoma cell line. *Mol. Med. Rep.* **2015**, *12*, 2275–2282. [CrossRef] [PubMed]
24. Lamson, D.W.; Brignall, M.S. Antioxidants in cancer therapy; their actions and interactions with oncologic therapies. *Altern. Med. Rev. A J. Clin. Ther.* **1999**, *4*, 304–329.
25. Reppert, S.M.; Weaver, D.R.; Ebisawa, T. Cloning and characterization of a mammalian melatonin receptor that mediates reproductive and circadian responses. *Neuron* **1994**, *13*, 1177–1185. [CrossRef]
26. Wiesenberg, I.; Missbach, M.; Carlberg, C. The potential role of the transcription factor RZR/ROR as a mediator of nuclear melatonin signaling. *Restor. Neurol. Neurosci.* **1998**, *12*, 143–150.
27. León, J.; Macías, M.; Escames, G.; Camacho, E.; Khaldy, H.; Martín, M.; Espinosa, A.; Gallo, M.A.; Acuña-Castroviejo, D. Structure-Related Inhibition of Calmodulin-Dependent Neuronal Nitric-Oxide Synthase Activity by Melatonin and Synthetic Kynurenines. *Mol. Pharmacol.* **2000**, *58*, 967–975. [CrossRef]
28. Macías, M.; Escames, G.; Leon, J.; Coto, A.; Sbihi, Y.; Osuna, A.; Acuña-Castroviejo, D. Calreticulin-melatonin: An unexpected relationship. *Eur. J. Biochem.* **2003**, *270*, 832–840. [CrossRef]
29. Radogna, F.; Paternoster, L.; De Nicola, M.; Cerella, C.; Ammendola, S.; Bedini, A.; Tarzia, G.; Aquilano, K.; Ciriolo, M.; Ghibelli, L. Rapid and transient stimulation of intracellular reactive oxygen species by melatonin in normal and tumor leukocytes. *Toxicol. Appl. Pharmacol.* **2009**, *239*, 37–45. [CrossRef]
30. Liu, L.; Labani, N.; Cecon, E.; Jockers, R. Melatonin Target Proteins: Too Many or Not Enough? *Front. Endocrinol.* **2019**, *10*, 791. [CrossRef]

31. Shen, Y.; Guerra-Librero, A.; Fernandez-Gil, B.I.; Florido, J.; García-López, S.; Martinez-Ruiz, L.; Mendivil-Perez, M.; Soto-Mercado, V.; Acuña-Castroviejo, D.; Ortega-Arellano, H.; et al. Combination of melatonin and rapamycin for head and neck cancer therapy: Suppression of AKT/mTOR pathway activation, and activation of mitophagy and apoptosis via mitochondrial function regulation. *J. Pineal Res.* **2018**, *64*, e12461. [CrossRef] [PubMed]
32. Laothong, U.; Hiraku, Y.; Oikawa, S.; Intuyod, K.; Murata, M.; Pinlaor, S. Melatonin induces apoptosis in cholangiocarcinoma cell lines by activating the reactive oxygen species-mediated mitochondrial pathway. *Oncol. Rep.* **2015**, *33*, 1443–1449. [CrossRef]
33. Gonzalez-Periz, A.; Claria, J. New approaches to the modulation of the cyclooxygenase-2 and 5-lipoxygenase pathways. *Curr. Top. Med. Chem.* **2007**, *7*, 297–309. [CrossRef] [PubMed]
34. Radogna, F.; Sestili, P.; Martinelli, C.; Paolillo, M.; Paternoster, L.; Albertini, M.; Accorsi, A.; Gualandi, G.; Ghibelli, L. Lipoxygenase-mediated pro-radical effect of melatonin via stimulation of arachidonic acid metabolism. *Toxicol. Appl. Pharmacol.* **2009**, *238*, 170–177. [CrossRef] [PubMed]
35. Singh, C.K.; Chhabra, G.; Ndiaye, M.A.; Garcia-Peterson, L.M.; Mack, N.J.; Ahmad, N. The Role of Sirtuins in Antioxidant and Redox Signaling. *Antioxid. Redox Signal.* **2018**, *28*, 643–661. [CrossRef]
36. Mayo, J.C.; Sainz, R.M.; Menéndez, P.G.; Cepas, V.; Tan, D.-X.; Reiter, R.J. Melatonin and sirtuins: A 'not-so unexpected' relationship. *J. Pineal Res.* **2017**, *62*, e12391. [CrossRef]
37. Weir, H.J.; Lane, J.D.; Balthasar, N. SIRT3: A Central Regulator of Mitochondrial Adaptation in Health and Disease. *Genes Cancer* **2013**, *4*, 118–124. [CrossRef]
38. Rangarajan, P.; Karthikeyan, A.; Lu, J.; Ling, E.-A.; Dheen, S. Sirtuin 3 regulates Foxo3a-mediated antioxidant pathway in microglia. *Neuroscience* **2015**, *311*, 398–414. [CrossRef]
39. Kong, J.; Wang, L.; Ren, L.; Yan, Y.; Cheng, Y.; Huang, Z.; Shen, F. Triptolide induces mitochondria-mediated apoptosis of Burkitt's lymphoma cell via deacetylation of GSK-3β by increased SIRT3 expression. *Toxicol. Appl. Pharmacol.* **2018**, *342*, 1–13. [CrossRef]
40. Li, M.; Wu, C.; Muhammad, J.S.; Yan, D.; Tsuneyama, K.; Hatta, H.; Cui, Z.H.; Inadera, H. Melatonin sensitises shikonin-induced cancer cell death mediated by oxidative stress via inhibition of the SIRT3/SOD2-AKT pathway. *Redox Biol.* **2020**, *36*, 101632. [CrossRef]
41. Lu, J.; Fu, L.; Tang, Z.; Zhang, C.; Qin, L.; Wang, J.; Yu, Z.; Shi, D.; Xiao, X.; Xie, F.; et al. Melatonin inhibits AP-2β/hTERT, NF-κB/COX-2 and Akt/ERK and activates caspase/Cyto C signaling to enhance the antitumor activity of berberine in lung cancer cells. *Oncotarget* **2016**, *7*, 2985–3001. [CrossRef] [PubMed]
42. Chen, X.; Hao, B.; Li, D.; Reiter, R.J.; Bai, Y.; Abay, B.; Chen, G.; Lin, S.; Zheng, T.; Ren, Y.; et al. Melatonin inhibits lung cancer development by reversing the Warburg effect via stimulating the SIRT3/PDH axis. *J. Pineal Res.* **2021**, *71*, e12755. [CrossRef] [PubMed]
43. Naaz, S.; Mishra, S.; Pal, P.K.; Chattopadhyay, A.; Das, A.R.; Bandyopadhyay, D. Activation of SIRT1/PGC 1α/SIRT3 pathway by melatonin provides protection against mitochondrial dysfunction in isoproterenol induced myocardial injury. *Heliyon* **2020**, *6*, e05159. [CrossRef]
44. Ding, M.; Feng, N.; Tang, D.; Feng, J.; Li, Z.; Jia, M.; Liu, Z.; Gu, X.; Wang, Y.; Fu, F.; et al. Melatonin prevents Drp1-mediated mitochondrial fission in diabetic hearts through SIRT1-PGC1α pathway. *J. Pineal Res.* **2018**, *65*, e12491. [CrossRef] [PubMed]
45. Cheng, Y.; Cai, L.; Jiang, P.; Wang, J.; Gao, C.; Feng, H.; Wang, C.; Pan, H.; Yang, Y. SIRT1 inhibition by melatonin exerts antitumor activity in human osteosarcoma cells. *Eur. J. Pharmacol.* **2013**, *715*, 219–229. [CrossRef]
46. Jung-Hynes, B.; Schmit, T.L.; Reagan-Shaw, S.R.; Siddiqui, I.A.; Mukhtar, H.; Ahmad, N. Melatonin, a novel Sirt1 inhibitor, imparts antiproliferative effects against prostate cancer in vitro in culture and in vivo in TRAMP model. *J. Pineal Res.* **2010**, *50*, 140–149. [CrossRef]
47. Nitulescu, G.M.; van de Venter, M.; Nitulescu, G.; Ungurianu, A.; Juzenas, P.; Peng, Q.; Olaru, O.T.; Grădinaru, D.; Tsatsakis, A.; Tsoukalas, D.; et al. The Akt pathway in oncology therapy and beyond (Review). *Int. J. Oncol.* **2018**, *53*, 2319–2331. [CrossRef]
48. Fang, X.; Yu, S.X.; Lu, Y.; Bast, R.C.; Woodgett, J.R.; Mills, G.B. Phosphorylation and inactivation of glycogen synthase kinase 3 by protein kinase A. *Proc. Natl. Acad. Sci. USA* **2000**, *97*, 11960–11965. [CrossRef]
49. Phiboonchaiyanan, P.P.; Puthongking, P.; Chawjarean, V.; Harikarnpakdee, S.; Sukprasansap, M.; Chanvorachote, P.; Priprem, A.; Govitrapong, P. Melatonin and its derivative disrupt cancer stem-like phenotypes of lung cancer cells via AKT downregulation. *Clin. Exp. Pharmacol. Physiol.* **2021**, *48*, 1712–1723. [CrossRef]
50. Song, J.; Ma, S.; Luo, J.; Liu, H.; Li, L.; Zhang, Z.; Chen, L.; Zhou, R. Downregulation of AKT and MDM2, Melatonin Induces Apoptosis in AGS and MGC803 Cells. *Anat. Rec.* **2019**, *302*, 1544–1551. [CrossRef]
51. Liu, W.; Yu, M.; Xie, D.; Wang, L.; Ye, C.; Zhu, Q.; Liu, F.; Yang, L. Melatonin-stimulated MSC-derived exosomes improve diabetic wound healing through regulating macrophage M1 and M2 polarization by targeting the PTEN/AKT pathway. *Stem Cell Res. Ther.* **2020**, *11*, 259. [CrossRef] [PubMed]
52. Gou, Z.; Su, X.; Hu, X.; Zhou, Y.; Huang, L.; Fan, Y.; Li, J.; Lu, L. Melatonin improves hypoxic-ischemic brain damage through the Akt/Nrf2/Gpx4 signaling pathway. *Brain Res. Bull.* **2020**, *163*, 40–48. [CrossRef] [PubMed]
53. Perdomo, J.; Quintana, C.; González, I.; Hernández, I.; Rubio, S.; Loro, J.; Reiter, R.; Estévez, F.; Quintana, J. Melatonin Induces Melanogenesis in Human SK-MEL-1 Melanoma Cells Involving Glycogen Synthase Kinase-3 and Reactive Oxygen Species. *Int. J. Mol. Sci.* **2020**, *21*, 4970. [CrossRef]
54. Ciotti, S.; Iuliano, L.; Cefalù, S.; Comelli, M.; Mavelli, I.; Di Giorgio, E.; Brancolini, C. GSK3β is a key regulator of the ROS-dependent necrotic death induced by the quinone DMNQ. *Cell Death Dis.* **2020**, *11*, 2. [CrossRef] [PubMed]

55. Icard, P.; Shulman, S.; Farhat, D.; Steyaert, J.-M.; Alifano, M.; Lincet, H. How the Warburg effect supports aggressiveness and drug resistance of cancer cells? *Drug Resist. Updat.* **2018**, *38*, 1–11. [CrossRef]
56. Wang, Q.; Guo, X.; Li, L.; Gao, Z.; Su, X.; Ji, M.; Liu, J. N6-methyladenosine METTL3 promotes cervical cancer tumorigenesis and Warburg effect through YTHDF1/HK2 modification. *Cell Death Dis.* **2020**, *11*, 911. [CrossRef]
57. Hou, Y.; Zhang, Q.; Pang, W.; Hou, L.; Liang, Y.; Han, X.; Luo, X.; Wang, P.; Zhang, X.; Li, L.; et al. YTHDC1-mediated augmentation of miR-30d in repressing pancreatic tumorigenesis via attenuation of RUNX1-induced transcriptional activation of Warburg effect. *Cell Death Differ.* **2021**, *28*, 3105–3124. [CrossRef]
58. Zong, W.-X.; Rabinowitz, J.D.; White, E. Mitochondria and Cancer. *Mol. Cell* **2016**, *61*, 667–676. [CrossRef]
59. Movahed, Z.G.; Rastegari-Pouyani, M.; Mohammadi, M.H.; Mansouri, K. Cancer cells change their glucose metabolism to overcome increased ROS: One step from cancer cell to cancer stem cell? *Biomed. Pharmacother.* **2019**, *112*, 108690. [CrossRef]
60. McFate, T.; Mohyeldin, A.; Lu, H.; Thakar, J.; Henriques, J.; Halim, N.D.; Wu, H.; Schell, M.J.; Tsang, T.M.; Teahan, O.; et al. Pyruvate Dehydrogenase Complex Activity Controls Metabolic and Malignant Phenotype in Cancer Cells. *J. Biol. Chem.* **2008**, *283*, 22700–22708. [CrossRef]
61. Wang, Y.; Qi, H.; Liu, Y.; Duan, C.; Liu, X.; Xia, T.; Chen, D.; Piao, H.-L.; Liu, H.-X. The double-edged roles of ROS in cancer prevention and therapy. *Theranostics* **2021**, *11*, 4839–4857. [CrossRef] [PubMed]
62. Ghasemishahrestani, Z.; Mattos, L.M.M.; Tilli, T.M.; dos Santos, A.L.S.; Pereira, M.D. Pieces of the Complex Puzzle of Cancer Cell Energy Metabolism: An Overview of Energy Metabolism and Alternatives for Targeted Cancer Therapy. *Curr. Med. Chem.* **2021**, *28*, 3514–3534. [CrossRef] [PubMed]
63. Yeung, S.J.; Pan, J.; Lee, M.H. Roles of p53, MYC and HIF-1 in regulating glycolysis—The seventh hallmark of cancer. *Cell. Mol. Life Sci.* **2008**, *65*, 3981–3999. [CrossRef] [PubMed]
64. Stacpoole, P.W. Therapeutic Targeting of the Pyruvate Dehydrogenase Complex/Pyruvate Dehydrogenase Kinase (PDC/PDK) Axis in Cancer. *JNCI J. Natl. Cancer Inst.* **2017**, *109*, djx071. [CrossRef] [PubMed]
65. Reiter, R.; Sharma, R.; Rosales-Corral, S. Anti-Warburg Effect of Melatonin: A Proposed Mechanism to Explain its Inhibition of Multiple Diseases. *Int. J. Mol. Sci.* **2021**, *22*, 764. [CrossRef] [PubMed]
66. Carbajo-Pescador, S.; Ordoñez, R.; Benet, M.; Jover, R.; García-Palomo, A.; Mauriz, J.L.; González-Gallego, J. Inhibition of VEGF expression through blockade of Hif1α and STAT3 signalling mediates the anti-angiogenic effect of melatonin in HepG2 liver cancer cells. *Br. J. Cancer* **2013**, *109*, 83–91. [CrossRef]
67. Park, J.-W.; Hwang, M.-S.; Suh, S.-I.; Baek, W.-K. Melatonin down-regulates HIF-1α expression through inhibition of protein translation in prostate cancer cells. *J. Pineal Res.* **2009**, *46*, 415–421. [CrossRef]
68. Park, S.-Y.; Jang, W.-J.; Yi, E.-Y.; Jang, J.-Y.; Jung, Y.; Jeong, J.-W.; Kim, Y.-J. Melatonin suppresses tumor angiogenesis by inhibiting HIF-1α stabilization under hypoxia. *J. Pineal Res.* **2010**, *48*, 178–184. [CrossRef]
69. Manchester, L.C.; Tan, D.-X. Neurotoxins: Free Radical Mechanisms and Melatonin Protection. *Curr. Neuropharmacol.* **2010**, *8*, 194–210. [CrossRef]
70. Reiter, R.J.; Mayo, J.C.; Tan, D.-X.; Sainz, R.M.; Alatorre-Jimenez, M.; Qin, L. Melatonin as an antioxidant: Under promises but over delivers. *J. Pineal Res.* **2016**, *61*, 253–278. [CrossRef]
71. Mihanfar, A.; Yousefi, B.; Darband, S.G.; Sadighparvar, S.; Kaviani, M.; Majidinia, M. Melatonin increases 5-flurouracil-mediated apoptosis of colorectal cancer cells through enhancing oxidative stress and downregulating survivin and XIAP. *BioImpacts* **2020**, *11*, 253–261. [CrossRef] [PubMed]
72. Ammar, O.A.; El-Missiry, M.A.; Othman, A.I.; Amer, M.E. Melatonin is a potential oncostatic agent to inhibit HepG2 cell proliferation through multiple pathways. *Heliyon* **2022**, *8*, e08837. [CrossRef] [PubMed]
73. Abadi, S.H.M.H.; Shirazi, A.; Alizadeh, A.M.; Changizi, V.; Najafi, M.; Khalighfard, S.; Nosrati, H. The Effect of Melatonin on Superoxide Dismutase and Glutathione Peroxidase Activity, and Malondialdehyde Levels in the Targeted and the Non-targeted Lung and Heart Tissues after Irradiation in Xenograft Mice Colon Cancer. *Curr. Mol. Pharmacol.* **2018**, *11*, 326–335. [CrossRef] [PubMed]
74. Lee, J.H.; Yun, C.W.; Lee, S.H. Cellular Prion Protein Enhances Drug Resistance of Colorectal Cancer Cells via Regulation of a Survival Signal Pathway. *Biomol. Ther.* **2018**, *26*, 313–321. [CrossRef]
75. Westergard, L.; Christensen, H.M.; Harris, D.A. The cellular prion protein (PrPC): Its physiological function and role in disease. *Biochim. Biophys. Acta-Mol. Basis Dis.* **2007**, *1772*, 629–644. [CrossRef]
76. Brown, D.R.; Nicholas, R.S.; Canevari, L. Lack of prion protein expression results in a neuronal phenotype sensitive to stress. *J. Neurosci. Res.* **2001**, *67*, 211–224. [CrossRef]
77. Klamt, F.; Dal-Pizzol, F.; da Frota, M.L.C., Jr.; Walz, R.; Andrades, M.E.; da Silva, E.G.; Brentani, R.R.; Izquierdo, I.N.; Fonseca Moreira, J.C. Imbalance of antioxidant defense in mice lacking cellular prion protein. *Free Radic. Biol. Med.* **2001**, *30*, 1137–1144. [CrossRef]
78. Loh, D.; Reiter, R.J. Melatonin: Regulation of Prion Protein Phase Separation in Cancer Multidrug Resistance. *Molecules* **2022**, *27*, 705. [CrossRef]
79. Lee, J.H.; Yoon, Y.M.; Han, Y.-S.; Yun, C.W.; Lee, S.H. Melatonin Promotes Apoptosis of Oxaliplatin-resistant Colorectal Cancer Cells Through Inhibition of Cellular Prion Protein. *Anticancer Res.* **2018**, *38*, 1993–2000. [CrossRef]
80. Yun, C.W.; Kim, S.; Lee, J.H.; Lee, S.H. Melatonin Promotes Apoptosis of Colorectal Cancer Cells via Superoxide-mediated ER Stress by Inhibiting Cellular Prion Protein Expression. *Anticancer Res.* **2018**, *38*, 3951–3960. [CrossRef]

81. Onukwufor, J.O.; Berry, B.J.; Wojtovich, A.P. Physiologic Implications of Reactive Oxygen Species Production by Mitochondrial Complex I Reverse Electron Transport. *Antioxidants* **2019**, *8*, 285. [CrossRef] [PubMed]
82. Scialò, F.; Fernández-Ayala, D.J.; Sanz, A. Role of Mitochondrial Reverse Electron Transport in ROS Signaling: Potential Roles in Health and Disease. *Front. Physiol.* **2017**, *8*, 428. [CrossRef] [PubMed]
83. Florido, J.; Martinez-Ruiz, L.; Rodriguez-Santana, C.; López-Rodríguez, A.; Hidalgo-Gutiérrez, A.; Cottet-Rousselle, C.; Lamarche, F.; Schlattner, U.; Guerra-Librero, A.; Aranda-Martínez, P.; et al. Melatonin drives apoptosis in head and neck cancer by increasing mitochondrial ROS generated via reverse electron transport. *J. Pineal Res.* **2022**. [CrossRef]
84. Fuhrmann, D.C.; Brüne, B. Mitochondrial composition and function. under the control of hypoxia. *Redox Biol.* **2017**, *12*, 208–215. [CrossRef] [PubMed]
85. Yang, L.; Zheng, J.; Xu, R.; Zhang, Y.; Gu, L.; Dong, J.; Zhu, Y.; Zhou, R.; Zheng, L.; Zhang, X.; et al. Melatonin Suppresses Hypoxia-Induced Migration of HUVECs via Inhibition of ERK/Rac1 Activation. *Int. J. Mol. Sci.* **2014**, *15*, 14102–14121. [CrossRef]
86. He, M.; Zhou, C.; Lu, Y.; Mao, L.; Xi, Y.; Mei, X.; Wang, X.; Zhang, L.; Yu, Z.; Zhou, Z. Melatonin antagonizes nickel-induced aerobic glycolysis by blocking ROS-mediated HIF-1 α /miR210/ISCU axis activation. *Oxid. Med. Cell. Longev.* **2020**, *2020*, 5406284. [CrossRef]
87. Vriend, J.; Reiter, R.J. Melatonin and the von Hippel–Lindau/HIF-1 oxygen sensing mechanism: A review. *Biochim. Biophys. Acta* **2016**, *1865*, 176–183. [CrossRef]
88. Asghari, M.H.; Moloudizargari, M.; Ghobadi, E.; Fallah, M.; Abdollahi, M. Melatonin as a multifunctional anti-cancer molecule: Implications in gastric cancer. *Life Sci.* **2017**, *185*, 38–45. [CrossRef]
89. Zhang, H.-M.; Zhang, Y. Melatonin: A well-documented antioxidant with conditional pro-oxidant actions. *J. Pineal Res.* **2014**, *57*, 131–146. [CrossRef]
90. Reiter, R.J.; Rosales-Corral, S.A.; Tan, D.-X.; Acuna-Castroviejo, D.; Qin, L.; Yang, S.-F.; Xu, K. Melatonin, a Full Service Anti-Cancer Agent: Inhibition of Initiation, Progression and Metastasis. *Int. J. Mol. Sci.* **2017**, *18*, 843. [CrossRef]
91. Galano, A.; Tan, D.-X.; Reiter, R.J. Melatonin: A Versatile Protector against Oxidative DNA Damage. *Molecules* **2018**, *23*, 530. [CrossRef] [PubMed]
92. Liu, R.; Wang, H.; Deng, M.; Wen, X.; Mo, Y.; Chen, F.; Zou, C.; Duan, W.; Li, L.; Nie, X. Melatonin inhibits reactive oxygen species-driven proliferation, epithelial-mesenchymal transition, and vasculogenic mimicry in oral cancer. *Oxid. Med. Cell. Longev.* **2018**, *2018*, 3510970. [CrossRef] [PubMed]
93. Zhang, Y.; Liu, Q.; Wang, F.; Ling, E.-A.; Liu, S.; Wang, L.; Yang, Y.; Yao, L.; Chen, X.; Wang, F.; et al. Melatonin antagonizes hypoxia-mediated glioblastoma cell migration and invasion via inhibition of HIF-1α. *J. Pineal Res.* **2013**, *55*, 121–130. [CrossRef]
94. Saeidnia, S.; Abdollahi, M. Toxicological and pharmacological concerns on oxidative stress and related diseases. *Toxicol. Appl. Pharmacol.* **2013**, *273*, 442–455. [CrossRef] [PubMed]
95. Targhazeh, N.; Reiter, R.J.; Rahimi, M.; Qujeq, D.; Yousefi, T.; Shahavi, M.H.; Mir, S.M. Oncostatic activities of melatonin: Roles in cell cycle, apoptosis, and autophagy. *Biochimie* **2022**, *200*, 44–59. [CrossRef] [PubMed]

antioxidants

MDPI

Article

Melatonin Nuclear Receptors Mediate Green-and-Blue-Monochromatic-Light-Combinations-Inhibited B Lymphocyte Apoptosis in the Bursa of Chickens via Reducing Oxidative Stress and Nfκb Expression

Yijia Zhang, Zixu Wang, Yulan Dong, Jing Cao and Yaoxing Chen *

Laboratory of Anatomy of Domestic Animals, College of Veterinary Medicine, China Agricultural University, Beijing 100193, China; bs20193050473@cau.edu.cn (Y.Z.); zxwang@cau.edu.cn (Z.W.); ylbcdong@cau.edu.cn (Y.D.); caojing@cau.edu.cn (J.C.)

* Correspondence: yxchen@cau.edu.cn; Tel.: +86-10-6273-3778; Fax: +86-10-6273-3199

Abstract: Previous studies found that melatonin modulates a combination of green-and-blue-light-induced B-lymphocyte proliferation via its membrane receptors Mel1a and Mel1c. However, in addition to its membrane-bound receptors, melatonin also functions through binding to nuclear receptors RORα/RORβ/RORγ. In this study, we raised 120 chicks under 400–700 nm white (WW), 660 nm red (RR), 560 nm green (GG) and 480 nm blue light (BB) from P0 to P26. From P27 to P42, half of the chickens in green, blue and red were switched to blue (G→B), green (B→G) and red (R→B), respectively. We used immunohistochemistry, Western blotting, qRT-PCR, Elisa and MTT to investigate the influence of various monochromatic light combinations on the bursal B lymphocyte apoptosis and oxidative stress levels as well as estimate whether melatonin and its nuclear receptors were involved in this process. Consistent with the increase in the plasma melatonin concentration and antioxidant enzyme activity, we observed that G→B significantly decreased the *RORα*, *RORγ* mRNA level, inhibited Bax, Caspase-3 and p-iκb, p-p65 protein expression, increased the IL-10 level and Nrf2, HO-1 protein expression, down-regulated the MDA and pro-inflammatory IL-6, TNF-α and IFN-γ levels in the bursa compared with WW, RR, GG, BB and R→B, respectively. Our in vitro results showed exogenous melatonin supplementation inhibited B-lymphocyte apoptosis, decreased IL-6, TNF-α, IFN-γ and ROS production, down-regulated *RORα*, *RORγ* mRNA level and p-iκb and p-p65 protein expression, whereas it improved the IL-10 level and Nrf2 and the HO-1 protein expression in bursal B lymphocyte. Moreover, these responses were abrogated by RORα agonist SR1078 but were mimicked by RORα antagonist SR3335 or RORγ antagonist GSK2981278. In addition, p65 antagonist BAY reversed RORα/RORγ-mediated G→B-inhibited bursal B lymphocyte apoptosis. Overall, we concluded that melatonin nuclear RORα/RORγ mediates G→B-inhibited bursal B lymphocyte apoptosis via reducing oxidative stress and Nfκb expression.

Keywords: monochromatic light combination; melatonin; oxidative stress; nuclear receptor; B lymphocyte; apoptosis; chick

check for updates

Citation: Zhang, Y.; Wang, Z.; Dong, Y.; Cao, J.; Chen, Y. Melatonin Nuclear Receptors Mediate Green-and-Blue-Monochromatic-Light-Combinations-Inhibited B Lymphocyte Apoptosis in the Bursa of Chickens via Reducing Oxidative Stress and Nfκb Expression. *Antioxidants* **2022**, *11*, 748. https://doi.org/10.3390/antiox11040748

Academic Editors: Marina Garcia-Macia and Adrián Santos-Ledo

Received: 8 March 2022
Accepted: 7 April 2022
Published: 8 April 2022

1. Introduction

Birds are sensitive to light because of their highly developed visual systems. In addition to the light intensity and photoperiod, the wavelength of light affects the growth and development [1,2], immune [3], reproduction [4] and productive performance [5] in chicks and hens. Most visible light of different colors is a mixture of three monochromatic lights in proportion. These three monochromatic lights are: 660 nm red light, 560 nm green light and 480 nm blue light. Our lab has shown that, compared to white light, 660 nm red light could result in exacerbating the oxidative stress level [6] and inhibit the proliferation of spleen T lymphocyte [7]. By contrast, the 560 nm green light enhanced the mesor and amplitude of the melatonin and all kinds of the liver clock gene [8] and promoted muscle growth [9], meat quality properties [10] and T/B lymphocyte proliferation activity [6,11] during the early growth stage, while the 480 nm blue light is more effective during the later growth stage of chicks [10,12,13]. Therefore, we suggested that, compared with white light, different monochromatic lights have profound effects on the behavior, physiology, production performance and welfare of poultry. Further research found a combination of green and blue monochromatic light (G→B) effectively enhanced the proliferation activity of the bursal B lymphocyte [14]. These studies suggested that different light wavelengths have important impacts on the lymphocyte development in chickens. In addition to the B lymphocyte proliferation activity, the B lymphocyte apoptosis level, as another important indicator, can reflect bursal B lymphocyte development. On the contrary, whether different monochromatic light combinations will have impacts on bursal B lymphocyte apoptosis is unclear. Previous research reported that, when chicks in red light switched to other monochromatic lights (R→W, R→R, R→G, R→B), the body weight increased slowly, but the body weight in R→B was greater than R→R [9]. This result indicated that blue light can ameliorate the adverse effects of red light on the performance and oxidative stress of chicks. Therefore, we chose the R→B group to compare with G→B and B→G, and to detect the difference in the effects of different monochromatic light combinations on bursal B lymphocytes' apoptosis.

Melatonin, as one neuroendocrine hormone secreted by the pineal gland, could regulate the function of innate immunity [15] and adaptive immunity [16]. At present, melatonin mainly functions through membrane receptors, such as Mel1a, Mel1b and Mel1c. However, melatonin has the ability to penetrate the cell membrane, and the binding sites of melatonin have been detected in purified spleen and thymus lymphocyte nucleus [17]. Thus, in addition to membrane receptors, melatonin also functions through its nuclear receptor pathway. The melatonin nuclear receptor family consists of three members: RORα, RORβ and RORγ. However, the expression patterns of the three members are tissue-specific. RORα can be expressed in the thymus, skin, kidney, muscle and adipose tissue and can regulate a variety of physiological and pathological processes, including immune response, nervous system development, circadian rhythm and oxidative stress. The Mel–RORα pathway is essential for T lymphocyte proliferation, autoimmune function regulation [18] and anti-inflammatory function [19]. RORβ mainly exists in the central nervous system, and its function is mainly related to the circadian rhythm and the development of the nervous system sensory organs [20]. The expression pattern of RORγ is similar to that of RORα, and it is mainly found in immune cells, muscle, adipose tissue, liver and kidney [21]. RORγ is critical for the development and function of immune cells. Previous studies found that RORγ is a key regulatory factor of the differentiation and development of Th17 cells [22] and ILC3 cells [23]. Moreover, a prior study in our lab reported melatonin nuclear receptors mediate green light-activated T lymphocyte proliferation [7]. However, it is unclear whether the ROR nuclear receptors are involved in the melatonin-dependent inhibition of apoptosis in the bursal B lymphocyte.

In addition to affecting the activity of immune cells, melatonin also plays a role in regulating immune function by disrupting the balance between secretion levels of pro-inflammatory and anti-inflammatory cytokines. When the body is in the chronic inflammatory phase, melatonin can protect the body by inhibiting the secretion of TNF-α, IL-1β and IL-6, which are inducers of inflammation, as well as promoting the secretion of IL-2 and IL-10, which are protective factors that reduce inflammation [24]. The phenomenon that melatonin can enhance the immune response through influencing the concentration of cytokines at the inflammation stage is called "cytoprotective effects". Moreover, melatonin is also involved in the reduction in oxidative stress levels. Firstly, melatonin itself can be used as a strong antioxidant to directly remove oxygen free radicals generated during normal metabolism [25]. Secondly, melatonin can block the iNOS activity in various tissues and reduce the production of reactive oxygen species [26]. Thirdly, melatonin can improve the expression and enzyme activity of Gsh-Px, GSH-RD, SOD, CAT and other antioxidant enzymes, and improve the antioxidant level of tissues through the above aspects so as to protect lymphocytes from oxidative stress [27].

In the present work, we investigated the influence of various monochromatic light combinations on the bursal B lymphocyte apoptosis, as well as further exploring the mechanisms by which melatonin could exert its effects on B lymphocytes to inhibit their apoptosis as well as which melatonin nuclear receptors are involved in this process.

2. Materials and Methods

2.1. Animals and Treatments

A total of 120 male Arbor Acre chicks (post-hatching Day 0, Beijing Hua du Breeding Co., Ltd., Beijing, China) were randomly divided into four light treatment groups with 30 replicates (at a temperature of 32 °C in the first week and then kept at a temperature of 30 °C, the relative humidity of 60%). Unlike the 400 to 700 nm white light group (WW), the chicks in red light group were raised under 660 nm monochromatic light by an LED system (RR); the chicks in green light group were raised under 560 nm monochromatic light by an LED system; the chicks in blue light group were raised under 480 nm monochromatic light by an LED system. On the P26 at 23:00, we transferred 15 chicks in 560 nm green light group to 480 nm blue light group (G→B), 15 chicks in 480 nm blue light group to 560 nm green light group (B→G) and 15 chicks in 660 nm red light group to 480 nm blue light group (R→B) for breeding until P42. Therefore, the WW, RR, GG, BB groups (before P27) were changed into WW, RR, GG, BB, G→B, B→G, R→B until P42. For more details, refer to those described previously [14]. The light parameters are shown in Table 1.

Table 1. Light parameters.

Items	Light Treatments						
	WW	RR	GG	BB	G→B	B→G	R→B
Light wavelength [nm] (1–26 days)	400–700	660	560	480	560	480	660
Light wavelength [nm] (27–42 days)	400–700	660	560	480	480	560	480
Light intensity [W/m^2]	0.19	0.19	0.19	0.19	0.19	0.19	0.19
Photoperiod [Light: Dark]	23:1	23:1	23:1	23:1	23:1	23:1	23:1

2.2. Sampling

At P42, all chicks were euthanized, then their blood samples were collected and centrifuged for 30 min at 3000× *g*. Then, the plasma was decanted for melatonin measurement. We randomly selected 5 chicks from each light treatment group for subsequent immunohistochemistry, Western blot analysis and qRT-PCR. Another five chicks from each of the remaining chicks in the seven light treatment groups were selected and used for inflammatory factors, antioxidant capacity assay.

2.3. Lymphocyte Proliferation Assay

To determine the mechanism of action of melatonin, B lymphocyte was obtained aseptically from bursa of the G→B group, cultured in RPMI 1640 medium and stimulated with 25 µg/mL LPS and 250 pg/mL melatonin, which were both supplied from Sigma (St. Louis, MO, USA). Then, the B lymphocyte was incubated in the automatic-controlled incubator for 44 h. The temperature in the incubator was kept at 37 °C and the CO_2 concentration was 5%. We used methyl thiazolyl tetrazolium (MTT) assay and stimulation index to evaluate B lymphocyte proliferative activity. The stimulation index was calculated as the optical density (OD) values in stimulated cells compared to optical density values in unstimulated cells. The optical density was measured by using a microplate reader (Synergy HT; BioTek, Winooski, VT, USA) at 570 nm.

In addition, bursal B lymphocytes of the G→B group were prepared with either RORα antagonist (SR3335, 5 µM, MCE, Weehawken, NJ, USA), RORα agonist (SR1078, 10 µM, MCE, Weehawken, NJ, USA), RORγ antagonist (GSK2981278, 1 µM, MCE, Weehawken, NJ, USA), Nrf2 antagonist (ML385, 5 µM, MCE, Weehawken, NJ, USA), Nfκb antagonist (BAY, 1 µM,

MCE, Weehawken, NJ, USA) for 30 min before the addition of LPS and melatonin. After 48 h, each plate of treated cells was collected for determining the mechanism of melatonin action and the inflammatory cytokine. We used 6 wells as replicates in each assay.

2.4. Elisa Assay

Plasma (n = 5) melatonin were measured using the competitive inhibition Enzyme-linked Immunosorbent Assay Kit (USCN Life Science Inc., Wuhan, China) for melatonin. The detection range of the assay was 4.49–1000 pg/mL, and the intra- and inter assay coefficients of variation were <10% and <12%, respectively. Briefly, according to the manufacturer's protocol, 50 μL serial dilutions of melatonin standard (1000 pg/mL, 333.33 pg/mL, 111.11 pg/mL, 37.04 pg/mL, 12.35 pg/mL and 0 pg/mL) and samples were incubated with 50 μL Detection Reagent A for 1 h and then 100 μL Detection Reagent B for 30 min at 37 °C. Then, 90 μL Substrate Solution (TMB, 3,3′,5,5′-Tetramethylbenzidine) was micropipetted into each well of the microplate and reacted for 20 min at 37 °C, and the reaction was terminated by the addition of 50 μL stop solution. OD values were immediately measured using an ELISA analyzer (Bio-Rad, Model 680, Hercules, CA, USA) at 450 nm. Each sample was measured in triplicate. We created a standard curve with the log of melatonin concentration of the standard on the y-axis and the OD values of the standard on the x-axis. The concentration of the sample was calculated according to the standard curve.

Plasma (n = 5) GSH-Px, CAT, SOD, T-AOC and MDA levels were measured using commercial kits (Beyotime, Beijing, China). GSH-Px, CAT and SOD are well-known scavenger enzymes that protect cells from oxidative stress. SOD was detected by the xanthine oxidase method, and GSH-Px was determined by the rate at which it was converted to the enzymatic reaction of oxidized glutathione disulfide (GSSG). Those values were expressed as units/mL of plasma. T-AOC was detected by converting Fe^{3+} to Fe^{2+}. Those values were expressed as mmol/L of plasma. MDA is responsible for inducing oxidative stress, and it reacts with thiobarbituric acid to form a red complex and is expressed as μmol/L of plasma. These results were detected at specific wavelengths (GSH-Px: 340 nm, CAT: 520 nm, SOD: 450 nm, T-AOC: 593 nm and MDA: 532 nm). Five samples were included in each group, and each sample was tested in triplicate.

The levels of reactive oxygen species (ROS) in the bursal B lymphocyte were detected using a commercial assay kit (Nanjing Jiancheng Co., Ltd., Nanjing, China). The cell concentration was then adjusted to 1×10^5 cells/mL and the cells were loaded with 2′-7′-dichlorofluorescein diacetate (DCFH-DA) (10 μM) for 30 min at 37 °C in the dark. Fluorescence intensity was detected at 502/530 nm (excitation/emission) using a fluorescence microscope reader (Synergy HT; BioTek, Winooski, VT, USA) and expressed as the fluorescent intensity normalized to controls for cultured bursal B lymphocyte. Each sample was assayed three times.

Bursa tissues (n = 5) were homogenized in ice-cold PBS (pH = 7.4). The supernatants were then extracted by centrifugation (2000× g for 10 min) at 4 °C and stored at −80 °C to allow assay of TNF-α, IFN-γ, IL-6 and IL-10 levels using competitive ELISA (Uscn Life Science, Inc., Wuhan, China). The intra-assay CV was <10%, and the inter-assay CV was <12%. The protein concentration was determined using the BCA protein assay kit (Beyotime, Beijing, China). All tests were performed according to the manufacturer's instructions. Briefly, according to the manufacturer's protocol, 100 μL serial dilutions of standard (500 pg/mL, 250 pg/mL, 125 pg/mL, 62.5 pg/mL, 31.2 pg/mL, 15.6 pg/mL, 7.8 pg/mL and 0 pg/mL) and samples were incubated for 60 min at 37 °C. After that, added 100 μL Detection Reagent A for 1 h and then 100 μL Detection Reagent B for 60 min at 37 °C. Then, 90 μL Substrate Solution (TMB, 3,3′,5,5′-Tetramethylbenzidine) was micropipetted into each well of the microplate and reacted for 20 min at 37 °C, and the reaction was terminated by the addition of 50 μL stop solution. OD values were immediately measured using an ELISA analyzer (Bio-Rad, Model 680, Hercules, CA, USA) at 450 nm. Each sample was measured in triplicate. We created a standard curve with the log of inflammatory factors concentration of the standard on the *y*-axis and the OD values of the standard on the *x*-axis. The concentration of the sample was calculated according to the standard curve. The data were expressed as pg/mg.

2.5. Immunohistochemical Staining

For immunohistochemical staining, the primary antibodies (rabbit anti–Bcl-2,1:1000, Biorbyt, Cambridge, UK; rabbit anti–Caspase-3, 1:1000, CST, Boston, MA, USA) incubated with the sections overnight at 4 °C and visualized by incubating 0.05% 3,3–diaminobenzidine tetrahydrochloride (DAB, Sigma, St. Louis, MO, USA) and 0.003% hydrogen peroxide. A total of 25 fields were randomly selected in each sample. The integrated optical densities (IODs) of positive cells were measured by using Image–Pro Plus software.

2.6. Real-Time Reverse Transcription-Polymerase Chain Reaction (qRT-PCR)

Total RNA was extracted from bursa (n = 5) with TRIzol testing agent (CoWin Biotech Co., Inc., Beijing, China). The method was modified according to the standard procedure [28]. In brief, cDNA was reverse-transcribed and amplified using the Revertaid™ first strand cDNA synthesis kit (Fermentas Life Sciences, Burlington, Ontario, Canada). Each 12 μL transcription system included 2 μg total RNA, oligo (dT) 18 primer and nuclease-free water. Each reaction was incubated for 5 min at 65 °C and then mixed with 4 μL 5 × reaction buffer, 1 μL RNase Inhibitor 2 μL 10 mM dNTP mix and 1 μL reverse transcriptase. The mixtures were then incubated for 1 h at 42 °C and then 15 min at 70 °C. Real-time polymerase chain reaction (PCR) was performed with AceQ® qPCRSYBR® Green Master Mix (Q141-02, Vazyme, Nanjing, China). Briefly, 2 μL cDNA was mixed with 10 μL 2 × SYBR Green Master Mix, 1 μL forward primer (20 μM) and 1 μL reverse primer (20 μM) in a final volume of 20 μL per reaction. The PCR amplification protocol was 95 °C for 10 min, 40 cycles of 95 °C for10 s, 57 °C for 30 s and 72 °C for 30 s. Melt curve analyses were performed with the default program of the Light Cycler® 480 (Roche, LightCycler® 480 System, Roche Diagnostics GmbH, Bavaria, Germany). The relative mRNA levels were normalized to GAPDH and calculated using the formula 2^{-Ct}. The Ct value was calculated by the formula Ct = Ct target gene –Ct reference gene. Table 2 showed primers sequences of RORα, RORβ, RORγ and GAPDH used in the present research, and each sample was repeated in triplicate.

Table 2. Sequences of primers used for RT-PCR.

Gene	Product Size	Primer Sequences (5′–3′)	Accession No.
RORα	140	F: TGG GCATACCCCTGAAGGTA<break/>R: CCG ATGCTGGTGTGTAGTCA	XM_413763.2
RORβ	270	F: AAA TCG TTG CCA ACA CTG CC<break/>R: AGG TCA ATG ACG TGC CCA TT	NM_205093.1
RORγ	90	F: GTG GGGTAATATCGGGAGCG<break/>R: CTT ATCGGGACAACCTGCGT	XM_015280013.1
GAPDH	124	F: ATCACAGCCACACAGAAGACG<break/>R: TGACTTTCCCCACAGCCTTA	NM_204305

2.7. Western Blot Analysis

The proteins (n = 5) of bursa were extracted with RIPA lysis buffer and determined concentration with the bicinchoninic acid (BCA) kit (Beyotime, Shanghai, China). Then, the equal amount of protein in each group was added to the SDS–polyacrylamide gel and transferred onto PVDF membranes and blocked for 1 h using 5% skimmed milk. Subsequently, the primary antibodies, including goat anti–Bax (1:1000, Biorbyt, Cambridge, UK), rabbit anti–Bcl–2 (1:1000, Biorbyt, Cambridge, UK), rabbit anti–Caspase–3 (1:1000, CST, Boston, MA, USA), rabbit anti–Nrf2 (1:1000, Proteintech Group, Inc., Wuhan, China), rabbit anti–HO-1 (1:1000, BIOSS, Beijing, China), rabbit anti–p-iκb (1:1000, Abcam, Cambridge, UK), rabbit anti–p-p65 (1:1000, Abcam, Cambridge, UK) or rabbit anti–β–actin (1:4000; Co Win Biotech Co., Inc., Beijing, China), incubated with the membranes overnight at 4 °C. Then, the membranes were washed with TBST and incubated with horseradish peroxidase–conjugated goat anti–mouse/rabbit IgG (1:8000; Co Win Biotech Co., Inc., Beijing, China) for 2 h. The target bands obtained in the blots were scanned and measured using ImageJ 4.0.2 software (Scion Corp., Frederick, MD, USA) The data were expressed as the IOD of the target bands and compared to the corresponding β–actin values. Repeat the test three times for each sample.

2.8. Statistical Analysis

The results of each group were expressed as the mean ± standard error and analyzed using SPSS 25.0 software (SPSS, Chicago, IL, USA). The one-way ANOVA was used to evaluate the effects of different light treatment. The significant differences between seven light treatment groups were considered at $p < 0.05$.

3. Results

3.1. Caspase-3, Bcl-2, Bax, Protein in the Bursa of Chickens at P42

As shown in Figure 1A–H, the IOD of Caspase-3 positive cells in bursa was 55.19–123.98% lower in G→B than in WW, RR, GG, BB and R→B ($p < 0.001$). There was no significant difference between G→B and B→G ($p > 0.05$), but G→B was lower than B→G by 4.30%. Similarly, the Western blot analysis showed that the expression of Caspase-3 in the bursa was 7.05–13.13% lower in G→B than WW, RR, GG and R→B (Figure 1I, p = 0.000–0.025). In addition, the Caspase-3 protein level of G→B was lower by 1.13–4.62% than that of BB and B→G, but there was no significant difference among G→B, BB and B→G ($p > 0.05$). Next, we tested the expression of Bcl-2 protein, which plays a crucial role in inhibiting cell apoptosis. As shown in Figure 1J, the level of Bcl-2 protein in G→B was 17.66–73.61% higher than WW, RR, GG and R→B (p = 0.000–0.017). However, there was no significant difference between G→B, B→G and BB ($p > 0.05$), but G→B was higher than B→G and BB by 1.90–9.12%. G→B also significantly decreased the protein expression of Bax (9.38–51.27%, p = 0.000–0.012) and the ratio of Bax/Bcl-2 (18.70–169.47%, $p < 0.001$) compared with WW, RR, GG and R→B, respectively. In contrast to G→B, RR decreased the expression of Bcl-2 protein and increased the expression of Bax and Caspase-3 protein. These results indicated that G→B could inhibit bursal B lymphocyte apoptosis, and RR has the opposite results.

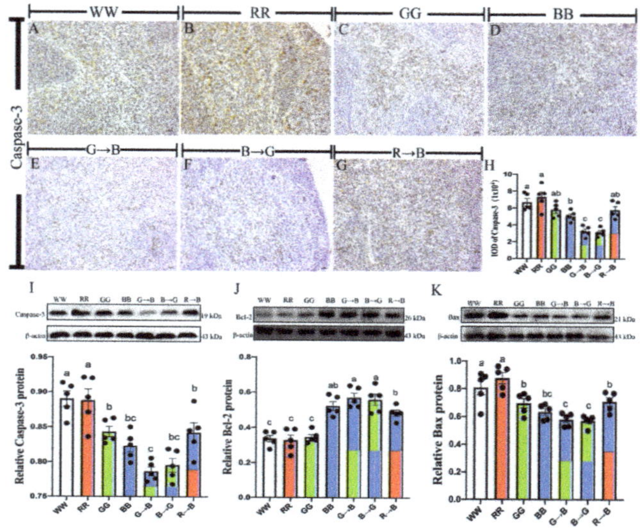

Figure 1. Immunohistochemical staining of Caspase-3 (scale bar = 50μm) in WW (**A**), RR (**B**), GG (**C**), BB (**D**), G→B (**E**), B→G (**F**), R→B (**G**), IOD of Caspase-3 positive cells (**H**), Caspase-3 protein (**I**), Bcl-2 protein (**J**), Bax protein (**K**) in the bursa at P42. The results of each group were expressed as the mean ± standard error. Differences between seven light treatment groups were evaluated by one-way ANOVA, and values with no common letters differ significantly ($p < 0.05$).

3.2. Plasma GSH-Px, CAT, SOD, T-AOC, MDA and Bursal IL-6, TNF-α, IFN-γ, IL-10 Concentration in Chickens at P42

Then, to evaluate the impact of different light wavelengths upon the oxidative stress level, we determined the change in five antioxidant indices in plasma. As shown in Figure 2A–E, G→B significantly increased the GSH-Px (10.83–48.51%, p = 0.000–0.004), CAT (35.07–100.47%, p = 0.000), SOD (47.83–115.30%, p = 0.000) and T-AOC (22.79–96.45%, p = 0.000) levels and decreased the plasma lipid metabolite and MDA levels (78.71–453.59%, p = 0.000–0.023) compared with WW, RR, GG, BB and R→B. There was no significant difference between G→B and B→G ($p > 0.05$), but the antioxidant enzymes and T-AOC were higher in G→B than in B→G by 1.93–10.01%, and the MDA level was lower in G→B than in B→G by 29.39%.

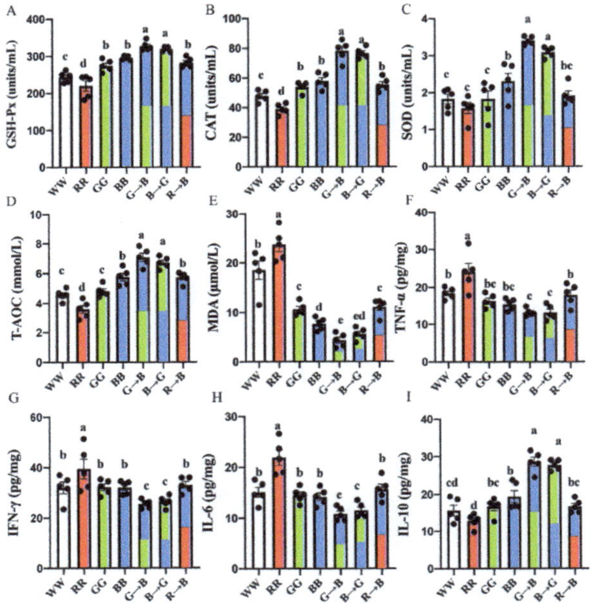

Figure 2. Plasma GSH-Px (**A**), CAT (**B**), SOD (**C**), T-AOC (**D**), MDA level (**E**), bursal TNF-α (**F**), IFN-γ (**G**), IL-6 (**H**), IL-10 (**I**) level in the chick at P42. The results of each group were expressed as the mean ± standard error. Differences between seven light treatment groups were evaluated by one-way ANOVA, and values with no common letters differ significantly ($p < 0.05$).

In addition, as shown in Figure 2F, G→B significantly reduced the levels of pro-inflammatory cytokine TNF-α (16.89–82.10%, p = 0.001–0.007) in the bursa compared with WW, RR and R→B. There was no significant difference between G→B, B→G, BB and GG ($p > 0.05$), but the TNF-α level was lower in G→B than in BB and GG by 16.89–23.73%. Consistent with this result, the IFN-γ and IL-6 levels in the bursa of G→B chicks were significantly lower than WW, RR, GG, BB and R→B by (25.89–54.24%, p = 0.000–0.022) and (32.05–103.00%, p = 0.000–0.012). Additionally, there was no significant difference between G→B and B→G. However, RR increased the levels of IL-6, TNF-α, and IFN-γ in the bursa compared with WW, GG, BB, G→B, B→G and R→B (p = 0.000–0.014). On the contrary, the level of anti-inflammatory factor IL-10 was the highest in G→B but had the lowest level in RR. The IL-10 level was higher in G→B than in WW, RR, GG, BB and R→B (47.72–122.58%, $p < 0.001$) but had no difference between G→B and B→G. In the meanwhile, the GSH-Px, CAT, SOD and T-AOC were the lowest in RR. Therefore, these results suggested that G→B can significantly reduce the oxidative stress level, while RR can significantly aggravate the level of oxidative stress.

3.3. *Plasma Melatonin Concentration and Melatonin Nuclear Expression in the Bursa of Chickens at P42*

First, we explored the influences of various monochromatic light combinations on the plasma melatonin concentration of chicks at P42. As shown in Figure 2A, the plasma melatonin concentration was higher in G→B than in WW, RR, GG, BB and R→B by 6.34–82.26% (p = 0.000–0.300). There was no significant difference between G→B and B→G ($p > 0.05$), but G→B was higher than that of B→G by 6.34%. Additionally, there was a strong negative correlation between the melatonin concentration in the plasma and the pro-apoptosis protein expression of Bax ($r = -0.97$, $p < 0.001$) and Caspase-3 ($r = -0.99$, $p < 0.001$).

In addition, as shown in Figure 3B, G→B significantly decreased the bursal *RORα* mRNA level (71.58–271.35%, p = 0.000–0.008) compared with WW, RR, GG, BB and R→B, and there was no significant difference between G→B and B→G. Similarly, as shown in Figure 3D, G→B significantly decreased the bursal *RORγ* mRNA level (15.85–135.69%, $p = 0.000$) compared with WW, RR, GG and R→B. However, the *RORβ* mRNA level in the bursa had no significant difference between WW, RR, GG, BB, G→B, B→G and R→B ($p > 0.05$). In addition, the Pearson's correlation analysis showed that the *RORα* mRNA ($r = -0.98$, $p < 0.001$) and *RORγ* mRNA levels ($r = -0.92$, $p = 0.003$) were negatively correlated with the plasma melatonin concentration. However, there was no significant correlation between the *RORβ* mRNA level and plasma melatonin concentration ($p > 0.05$). Therefore, these results suggested that melatonin nuclear receptors RORα and RORγ may play important roles in mediating G→B-inhibited bursal B lymphocyte proliferation.

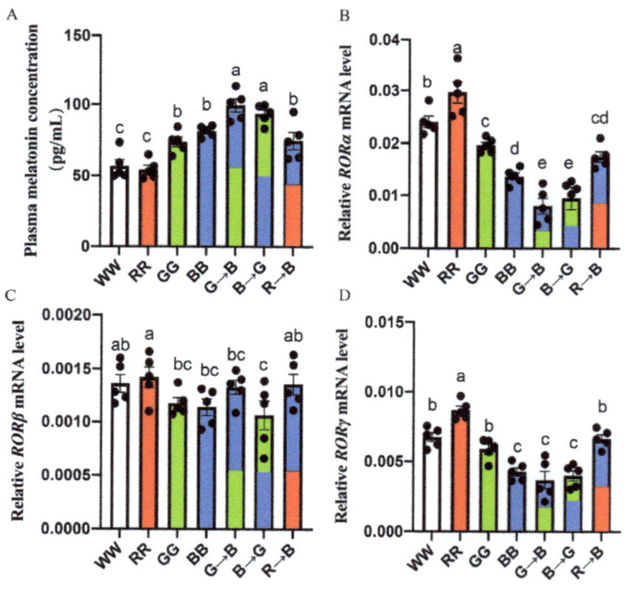

Figure 3. Plasma melatonin concentration (**A**), *RORα* mRNA level (**B**), *RORβ* mRNA level (**C**), *RORγ* mRNA level (**D**) in the bursa at P42. The results of each group were expressed as the mean ± standard error. Differences between seven light treatment groups were evaluated by one-way ANOVA, and values with no common letters differ significantly ($p < 0.05$).

3.4. *Nrf2, HO-1, p-iκb, p-p65 Protein Level in the Bursa of Chickens at P42*

In order to explore the mechanism by which G→B-inhibited bursal B lymphocyte apoptosis, we investigated the expression of the Nrf2/HO-1 signaling pathway and Nfκb-signaling-pathway-related proteins in the bursa of chickens. As shown in Figure 4A,B, the Nrf2 and HO-1 protein levels of the bursa in G→B were significant higher by 46.90–163.49% (Nrf2, $p < 0.001$) and 36.44–89.41% (HO-1, $p < 0.001$) than WW, RR, GG, BB and R→B, while the p-iκb and p-p65 protein levels of the bursa in G→B were significant lower by 26.54–100.81% (p-iκb, $p < 0.001$) and 48.98–179.41% (p-p65, $p = 0.000$–0.007) than WW, RR, GG, BB and R→B. However, there were no significant differences between G→B and B→G ($p > 0.05$), but the Nrf2 and HO-1 protein levels of the bursa were higher in G→B than in B→G by 3.54–7.10%, whereas the p-iκb and p-p65 protein levels in G→B were significant lower by 6.52–14.59% than B→G. These results indicated that Nrf2, HO-1, p-iκb and p-p65 proteins may play important roles in G→B-inhibited bursal B lymphocyte apoptosis.

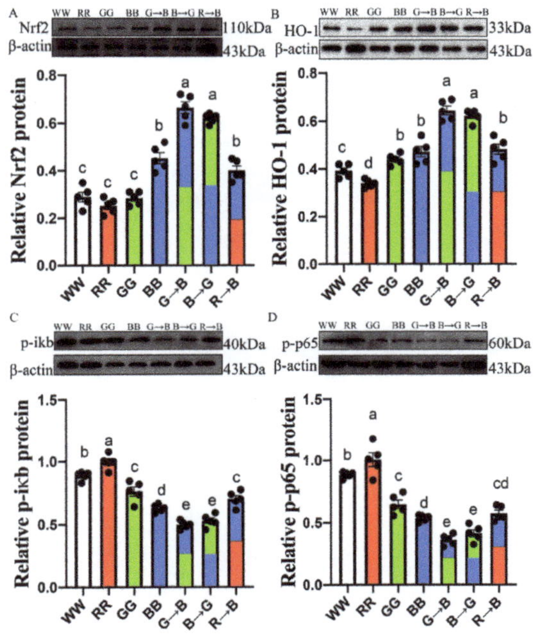

Figure 4. Nrf2 (**A**), HO-1 (**B**), p-iκb (**C**), p-p65 (**D**) protein level in bursa at P42. The results of each group were expressed as the mean ± standard error. Differences between seven light treatment groups were evaluated by one-way ANOVA, and values with no common letters differ significantly ($p < 0.05$).

3.5. Melatonin Nuclear Receptors RORα/RORγ-Mediated G→B-Inhibited B Lymphocyte Apoptosis in the Bursa of Chickens via Nrf2/HO-1 and Nfκb Pathway

Next, we further verified whether exogenous melatonin could mediate G→B-inhibited bursal B-lymphocytes apoptosis in vitro. We isolated aseptically the bursa B lymphocytes from G→B and detected the expression of apoptosis-related proteins to determine the effect of melatonin on B-lymphocyte apoptosis after 44 h of culture. As shown in Figure 5A–D, LPS + melatonin could significantly increase the anti-apoptosis protein Bcl-2 expression ($p < 0.001$) but significantly decreased the pro-apoptosis protein Bax ($p < 0.001$) and Caspase-3 ($p = 0.009$) levels compared with the control group, respectively. In addition, as shown in Figure 5E, exogenous melatonin supplement could down-regulate ROS production in B lymphocyte compared with the LPS group ($p < 0.001$). To determine the involvement of melatonin nuclear receptors RORα and RORγ on B lymphocyte apoptosis, primary cultures were pretreated with SR3335 (an antagonist of RORα), SR1078 (an agonist of RORα) and GSK2981278 (an antagonist of RORγ). After the pretreatment of B lymphocyte with RORα agonist SR1078, the Western blotting results showed that SR1078 markedly inhibited melatonin-induced Bcl-2 protein expression up-regulation ($p = 0.011$), promoted Bax ($p < 0.001$) and Caspase-3 ($p < 0.001$) protein down-regulation and induced ROS production ($p = 0.000$). In contrast, in SR3335 + LPS + melatonin or GSK2981278 + LPS + melatonin-treated B lymphocyte, the Bcl-2 protein was 28.78–20.17% ($p = 0.000$–0.009) higher than in the LPS + melatonin group, while the Bax, Caspase-3 and ROS level was 18.83–31.32% (Bax, $p = 0.000$–0.021), 15.22–28.75% (Caspase-3, $p < 0.001$) and 5.55–9.26% (ROS, $p = 0.000$–0.005) lower than in the LPS + melatonin group, respectively. Similar to SR3335 and GSK2981278 pretreatment, Nrf2 antagonist ML385 could also markedly reduce the Bcl-2 protein level ($p = 0.033$) and enhance the Bax ($p < 0.001$), Caspase-3 protein ($p < 0.001$) and ROS level ($p = 0.001$). However, when treating bursal B lymphocyte with p65 antagonist BAY, the protein expression of Bax, Caspase-3 and Bcl-2 were opposite of ML385. These results suggested that melatonin mediates G→B-inhibited bursal B lymphocyte apoptosis through nuclear receptor RORα/RORγ and the Nrf2/HO-1 and Nfκb signaling pathway.

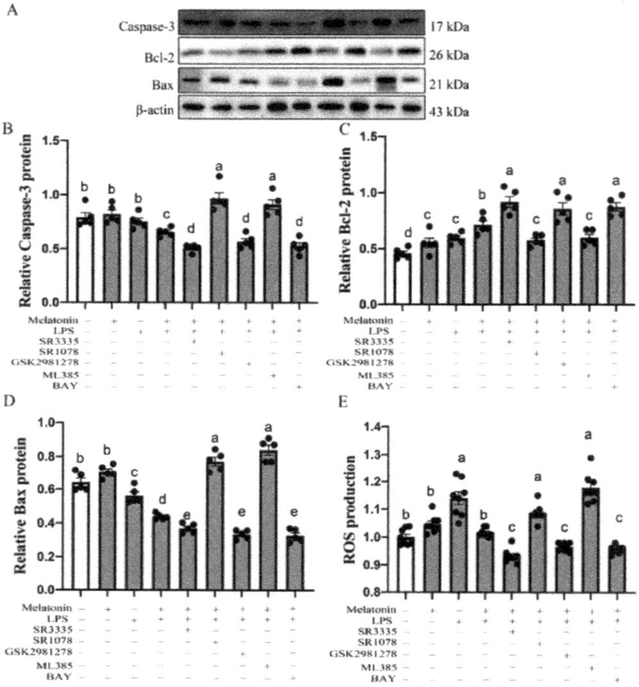

Figure 5. Effects of RORα antagonists, RORα agonist, RORγ antagonist, Nrf2 antagonists and p65 antagonists on Caspase-3 protein (**A**,**B**), Bcl-2 protein (**C**), Bax protein (**D**), ROS level (**E**) of bursal B-lymphocyte in response to LPS in the G→B group. SR3335 is a RORα antagonist; SR1078 is a RORα agonist; GSK1981278 is a RORγ antagonist; ML385 is an NRF2 antagonist; BAY is a p65 antagonist. Differences between seven light treatment groups were evaluated by one-way ANOVA, and values with no common letters differ significantly ($p < 0.05$).

To further clarify the respective roles of melatonin nuclear receptors in the Nrf2/HO-1 and Nfκb pathway, we pretreated B lymphocyte with melatonin nuclear receptor antagonist. As shown in Figure 6A,B, the observed melatonin-induced up-regulation in the Nrf2 and HO-1 protein was abrogated by RORα antagonist SR3335 or RORγ antagonist GSK1981278 ($p = 0.000$–0.017). Additionally, melatonin-induced down-regulation of p-iκb and p-p65 protein was also abrogated by the RORα antagonist SR3335 or RORγ antagonist GSK2981278 ($p = 0.000$–0.025). However, the protein levels of Nrf2 and HO-1 in the LPS + melatonin + SR1078 group were significantly decreased compared with the LPS + melatonin group ($p < 0.001$), and the p-iκb and p-p65 protein expression levels in the "LPS+ melatonin +SR1078" group were significantly higher than those in the "LPS + melatonin" group ($p < 0.001$).

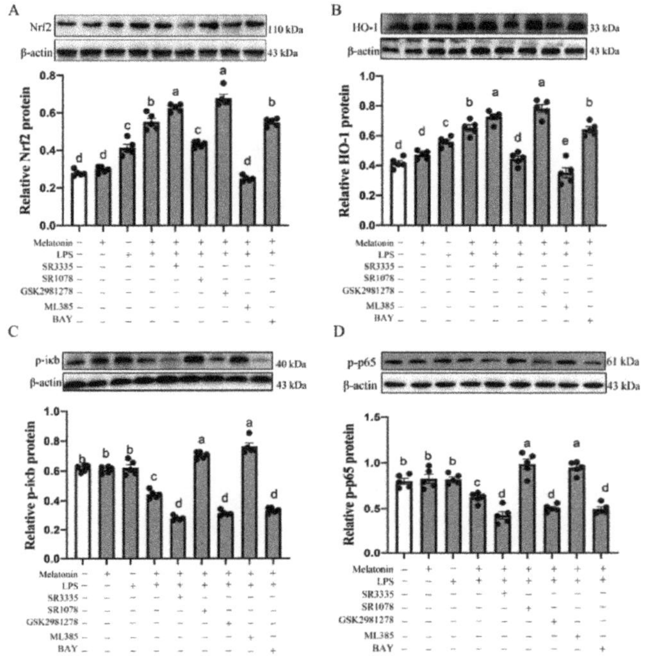

Figure 6. Effects of RORα antagonists, RORα agonist, RORγ antagonist, Nrf2 antagonists and p65 antagonists on Nrf2 (**A**), HO-1 (**B**), p-iκb (**C**), p-p65 (**D**) protein level of bursal B-lymphocyte in response to LPS in the G→B group. SR3335 is a RORα antagonist; SR1078 is a RORα agonist; GSK1981278 is a RORγ antagonist; ML385 is an NRF2 antagonist; BAY is a p65 antagonist. Differences between seven light treatment groups were evaluated by one-way ANOVA, and values with no common letters differ significantly ($p < 0.05$).

Next, we explored further the relationship between the Nrf2/HO-1 and Nfκb pathway in melatonin-inhibited B lymphocyte apoptosis. As shown in Figure 6C,D, the LPS + melatonin + ML385 could significantly increase the p-iκb protein expression in the cytoplasm and p-p65 protein expression in the nuclear receptor compared with the LPS + melatonin group ($p < 0.001$). However, the co-addition of p65 antagonist with LPS and melatonin had no significant effect on the expression of Nrf2 and HO-1 protein level ($p > 0.05$). Therefore, these results suggested that melatonin activates the downstream Nrf2/HO-1 signaling pathway by binding RORα and RORγ receptors, and then the activated Nrf2/HO-1 signaling pathway leads to the activity of p-iκb and p-p65 down-regulation, thereby inhibiting B lymphocyte apoptosis.

As shown in Figure 7A,B, the pretreatment of B lymphocyte with melatonin nuclear receptor RORα antagonist SR3335 or RORγ antagonist GSK2981278 significantly reduced the pro-inflammatory factor TNF-α (p = 0.000–0.001), IFN-γ (p = 0.000–0.002) and IL-6 (p = 0.024–0.044) level while improving the anti-inflammatory factor IL-10 level (p = 0.009–0.024) compared with the LPS + melatonin group. These responses were reversed by RORα agonists, SR1078. In addition, the addition of the Nrf2 pathway blocker also increased the levels of pro-inflammatory cytokines IL-6 ($p < 0.001$), TNF-α ($p < 0.001$) and IFN-γ ($p < 0.001$) in cell supernatant. However, the results of the pretreatment of B lymphocyte with SR1078 showed the opposite outcomes. These results indicated that melatonin can activate the Nrf2/HO-1 signaling pathway and inhibit the Nfκb pathway by RORα and RORγ nuclear receptors, resulting in anti-inflammatory cytokines' level enhancing and pro-inflammatory cytokines decreasing.

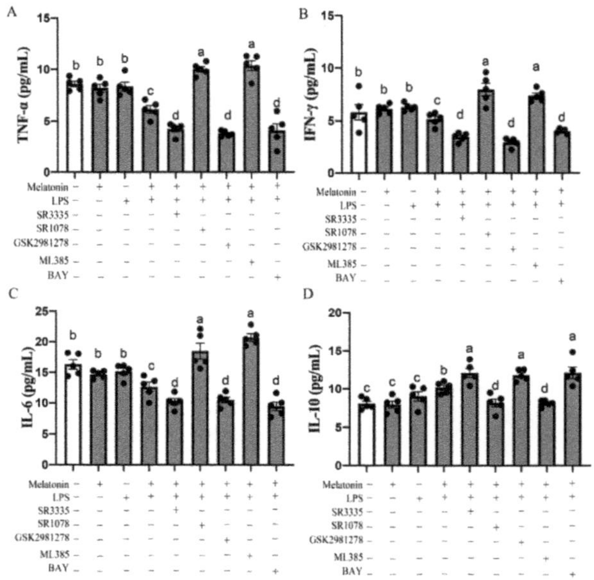

Figure 7. Effects of RORα antagonists, RORα agonist, RORγ antagonist, Nrf2 antagonists and p65 antagonists on pro-inflammatory cytokine TNF-α (**A**), IFN-γ (**B**), IL-6 (**C**), anti-inflammatory cytokine IL-10 (**D**) level of bursal B-lymphocyte in response to LPS in the G→B group. SR3335 is a RORα antagonist; SR1078 is a RORα agonist; GSK1981278 is a RORγ antagonist; ML385 is an NRF2 antagonist; BAY is a p65 antagonist. Differences between seven light treatment groups were evaluated by one-way ANOVA, and values with no common letters differ significantly ($p < 0.05$).

4. Discussion

With the increase in the broiler age, the bursa of Fabricius, which is the primary immunity organ in the chick, will gradually shrink. The bursa volume reached the maximum at sexual maturity and began to degenerate after sexual maturity. Atrophy of the bursa of Fabricius is related to B lymphocyte apoptosis. Previous research found that G→B promoted bursa morphological development and induced B lymphocyte proliferation. However, whether the apoptosis of B lymphocytes in the bursa of Fabricius is affected by different monochromatic light remains to be further explored. In this research, we found that the bursal pro-apoptosis protein Bax and Caspase-3 level were significantly lower in G→B than WW, as well as that the anti-apoptosis protein Bcl-2 level was higher in G→B than WW, indicating G→B could inhibit bursal B lymphocyte apoptosis. Previous research also found G→B could better induce bursal B lymphocyte proliferation [14] and elevate the level of anti-Newcastle disease virus (NDV) and anti-bovine serum albumin (BSA) IgG in plasma compared to WW [3]. These results showed that a combination of green and blue light can promote bursa morphological development as well as improve the humoral immunity of chicks better than white light. However, the bursal Bax and Caspase-3 protein level were the highest in the RR group. Xiong et al. showed that chickens reared under red light have lower T lymphocyte proliferation activity and higher apoptosis protein expression in the thymus, suggesting different light wavelengths have a various effect on the cell vitality in lymphocytes [7,29]. In addition, we observed G→B improved the bursal anti-inflammatory cytokine IL-10 level and decreased the bursal pro-inflammatory IL-6, TNF-α, IFN-γ level as well as enhanced the antioxidant capacity in plasma compared with WW. By contrast, RR improved the plasma pro-inflammatory level and increased lipid peroxidation production of MDA. These results mentioned above indicate that G→B effectively relieves the bursal inflammatory response and protects the body from oxidative stress, thereby inhibiting bursal B lymphocyte apoptosis. However, 660 nm red light improved the IL-6, TNF-α and IFN-γ secretion level to induce the inflammation response, thus leading to the imbalance between the proliferation of the B lymphocyte and apoptosis. Consistent with our results, green light showed the highest IL-2 activity in the spleen during the early growth stage, while the splenic IL-2 activity was the highest under blue light during the later growth stage [30,31]; blue light can effectively reduce the heat stress response in commercial broilers [32], significantly decrease lipid peroxidation and improve the antioxidant activities in the breast and thigh muscles of chicks compared to white light [10].

In addition, G→B improved the plasma melatonin concentration in chickens. Melatonin is a wide-spread hormone, mainly secreted by the pineal gland at night, and its secretion could be affected by environmental light information. For example, the melatonin concentration and release duration in sheep depend on the light cycle, with the highest melatonin concentration in short days (winter) and lowest in long days [33,34]. Interestingly, the Pearson's correlation analysis indicated that there was a significant negative correlation between the plasma melatonin concentration and apoptosis of B lymphocytes in the bursa of chickens. Previous research reported that melatonin-mediate 560 nm green light inhibited thymus T lymphocyte apoptosis [29], and an exogenous melatonin supplement can inhibit the development and maturation of mouse bone marrow B lymphocytes [35]. These studies indicated that melatonin, as an immune mediator, is crucial for B lymphocyte development. In addition, melatonin has been reported to inhibit pro-inflammatory cytokines IL-8 and TNF-α production in neutrophils [36], suggesting that melatonin helps alleviate acute and chronic inflammatory responses, as well as prevents cell apoptosis through inhibiting the production of pro-inflammatory cytokines [37]. The antioxidant ability of melatonin also accounts for its anti-apoptotic actions on immune cells [37]. Espino et al. found melatonin alleviated aging-induced apoptosis in neutrophils and lymphocytes by neutralizing free radicals and counteracting oxidative stress at the cellular level [38].

Moreover, the functions of melatonin, which exhibits anti-inflammatory, antioxidant and immunity regulation, are often mediated through its binding to the melatonin membrane-bound receptor (Mel1a/Mel1b). In addition to its membrane-bound receptors, studies have found the melatonin binding sites in purified cell nuclear receptors and activated the nuclear receptors in the liver, spleen and thymus [39]. Another study further confirmed the interaction of melatonin and ROR through a co-immunoprecipitation and co-localization assay [40]. In our study, we observed the *ROR*α mRNA level was the lowest in the G→B group, whereas it was the highest in the RR group. Additionally, the *ROR*γ mRNA level was similar to the RORα results. However, the *ROR*β mRNA level was markedly lower than *ROR*α and *ROR*γ. The Pearson's correction results showed there existed a strongly negative correction between the *ROR*α or *ROR*γ mRNA level and plasma melatonin concentration, indicating melatonin inhibited B lymphocyte apoptosis by negatively regulating the expression of nuclear receptors RORα or RORγ. It has been confirmed in many studies that melatonin can negatively regulate the expression of nuclear receptors. Zhao et al. demonstrated that the regulation of RORα expression by melatonin is dose-dependent, so, as the exogenous melatonin concentration increases, the *ROR*α mRNA expression decreases [41]. In addition, Wang et al. proved that melatonin can significantly reduce the mRNA and protein expression levels of RORγ [42]. In in vitro experiments, our results showed that the inhibitory effect of melatonin on B-lymphocyte apoptosis was reversed by SR1078 but enhanced by SR3335. This result corroborates the previous report that melatonin can inhibit apoptosis in mouse Leydig cells via RORα and suppressed Th17 cell differentiation via the inhibition of ROR-γ expression [24].

In our study, we found G→B significantly promoted Nrf2 and its downstream HO-1 protein expression compared with WW, RR, GG, BB and R→B. Melatonin inhibited B lymphocyte apoptosis, and induced pro-apoptosis protein down-regulation was blocked by Nrf2 antagonist ML385, implying the Nrf2 and HO-1 protein may participate in G→B-inhibited B lymphocyte apoptosis. Additionally, melatonin-induced Nrf2 and HO-

1activation were blocked by RORα agonist SR1078 and enhanced by RORα antagonist SR3335 or RORγ antagonist GSK2981278, implying that Nrf2 and HO-1 may participate in G→B-inhibited B-lymphocyte apoptosis by RORα/RORγ. Nrf2 (NF-ER-related factor2) is a transcription factor that mediates a broad-based set of adaptive responses to environmental and endogenous stresses, and the absence or inhibition of HO-1 is related to the level of inflammatory response [43]. The Nrf2/HO-1 pathway, which can be activated by the direct binding of the transcription factor to antioxidant response elements, could protect cells from oxidative stress. However, disruption of the Nrf2/HO-1 pathway will exacerbate oxidative stress [44]. Our in vitro results showed that the pretreatment of B lymphocyte with Nrf2 antagonist ML385 not only improved the pro-apoptosis protein expression but also relieved the inhibition of ROS produced by melatonin. Previous research found ROS production resulted in enhancing Nfκb translocation, leading to elevated angiogenic and pro-inflammatory mediators in endometriosis patients [45]. In our study, the protein expression of p-iκb and p-p65 was the lowest in G→B while the highest in RR. In an in vitro experiment, the pretreatment of B lymphocyte with p65 antagonist BAY resulted in reducing Bax and Caspase-3 protein but did not affect the Nrf2 and HO-1 activation. Interestingly, supplementing with ML385 significantly increased the p-iκb protein in the cytoplasm as well as improved the p-p65 protein expression in the nuclear receptor. These results showed that activation of Nrf2/HO-1 can inhibit the activity of Nfκb, thereby inhibiting the inflammatory response, reducing the production of ROS and, ultimately, leading to decreased apoptosis of B lymphocytes. A series of studies have demonstrated that Nrf2 could interact with Nfκb and directly affect the cellular activity of skeletal muscle satellite cells [46], endothelial cells [47] and macrophages [48], supporting our speculation.

Although the modulation of light information on animal immune response has been reported in recent years, there is still a lack of research data on the effect of different monochromatic lights' combinations on immune function in chicks, especially the neuromodulation mechanism of how the body converts external environmental light stimuli into internal molecular signals to influence a series of physiological changes, including the oxidative stress level and bursal development. In this study, we found that a combination of green and blue light could effectively decrease bursal pro-apoptosis protein Bax, Caspase-3, pro-inflammatory factor TNF-α, IFN-γ, IL-6 and lipid metabolites' MDA level, increase bursal anti-apoptosis protein Bcl-2, anti-inflammatory factor IL-10 and antioxidant enzyme activity, which promoted bursal development and reduced the stress level. Furthermore, we also explored the signaling pathway of bursal B lymphocyte apoptosis inhibited by melatonin. We considered that our results provide another lighting solution for the rational use of artificial light to reduce the stress levels, enhance health conditions and improve the survival rate of chicks. Therefore, we strongly recommend using 560 nm green light during the early stage (P0–P26) and then transferring to blue light during the later stage (P27–P42) in poultry houses.

5. Conclusions

In summary, a combination of green and blue light increased melatonin secretion but inhibited bursal B-lymphocyte apoptosis and the *RORα/RORγ* mRNA level. Melatonin nuclear receptors RORα/RORγ mediate G→B-inhibited bursal B lymphocyte apoptosis via reducing oxidative stress, activating Nrf2/HO-1 and inhibiting the Nfκb pathway in the bursa of chickens.

Author Contributions: Conceptualization, Y.C. and Y.Z.; methodology, Y.Z.; software, Z.W.; validation, Y.D.; formal analysis, J.C.; investigation, Y.Z.; resources, Y.C.; data curation, Y.Z.; writing—original draft preparation, Y.Z.; writing—review and editing, Y.C.; visualization, Y.D.; supervision, Z.W.; project administration, J.C.; funding acquisition, Y.C. All authors have read and agreed to the published version of the manuscript.

Funding: This research was funded by the "Chinese National Natural Science Foundation, grant number 32172801 and 31873000", and the "Beijing Natural Science Foundation, grant number 6222019".

Institutional Review Board Statement: The study was conducted in accordance with the Declaration of Helsinki and approved by the Animal Welfare and Ethics Committee of China Agricultural University (No. CAU 20171114–2.).

Informed Consent Statement: Not applicable.

Data Availability Statement: Data is contained within the article.

Acknowledgments: The authors would like to thank all members of the neurobiology laboratory.

Conflicts of Interest: The authors declare no conflict of interest.

Abbreviations

Bax	Bcl-2 associated x protein
Bcl-2	B-cell lymphoma-2
BB	Blue light
B→G	A combination of blue and green monochromatic light
Caspases	Cysteinyl aspartate specific proteinase
CAT	Catalase
DAB	3,3′-diaminobenzidine-4HCl
GSH-Px	Glutathione peroxidase
GG	Green light
G→B	A combination of green and blue monochromatic light
HRP	Horseradish-peroxidase
IOD	Integral optical density
INF-γ	Interferon-γ
IL-6	Interleukin-6
IL-10	Interleukin-10
LED	Light Emitting Diode
LPS	Lipopolysaccharide
Mel	Melatonin
Mel1a	Melatonin receptor 1a
Mel1b	Melatonin receptor 1b
Mel1c	Melatonin receptor 1c
MTT	Methyl thiazolyl tetrazolium
MDA	Malondialdehyde
NF-κB	Nuclear factor kappa B
OD	Optical density
PBS	Phosphate-buffered saline
PCNA	Proliferating cell nuclear antigen
RR	Red light
R→B	A combination of red and blue monochromatic light
ROR	Retinoic acid receptor related Orphan Receptor
RORα	Retinoic acid receptor related Orphan Receptor α
RORβ	Retinoic acid receptor related Orphan Receptor β
RORγ	Retinoic acid receptor related Orphan Receptor γ
ROS	Reactive oxygen species
SOD	Superoxide dismutase
SDS	Sodium dodecyl sulfate
T-AOC	Total antioxidant capacity
TNFα	Tumor necrosis factor α
WW	White light

References

1. Cao, J.; Liu, W.; Wang, Z.; Xie, D.; Jia, L.; Chen, Y. Green and blue monochromatic lights promote growth and development of broilers via stimulating testosterone secretion and myofiber growth. *J. Appl. Poult. Res.* **2008**, *17*, 211–218. [CrossRef]
2. Rozenboim, I.; Biran, I.; Uni, Z.; Robinzon, B.; Halevy, O. The effect of monochromatic light on broiler growth and development. *Poult. Sci.* **1999**, *78*, 135–138. [CrossRef] [PubMed]

3. Zhang, Z.; Cao, J.; Wang, Z.; Dong, Y.; Chen, Y. Effect of a combination of green and blue monochromatic light on broiler immune response. *J. Photochem. Photobiol. B Biol.* **2014**, *138*, 118–123. [CrossRef] [PubMed]
4. Yang, Y.F.; Jiang, J.S.; Pan, J.M.; Ying, Y.B.; Wang, X.S.; Zhang, M.L.; Lu, M.S.; Chen, X.H. The relationship of spectral sensitivity with growth and reproductive response in avian breeders (Gallus gallus). *Sci. Rep.* **2016**, *6*, 19291. [CrossRef]
5. Er, D.; Wang, Z.; Cao, J.; Chen, Y. Effect of monochromatic light on the egg quality of laying hens. *J. Appl. Poult. Res.* **2007**, *16*, 605–612. [CrossRef]
6. Li, J.; Cao, J.; Wang, Z.; Dong, Y.; Chen, Y. Melatonin plays a critical role in inducing B lymphocyte proliferation of the bursa of fabricius in broilers via monochromatic lights. *J. Photochem. Photobiol. B Biol.* **2015**, *142*, 29–34. [CrossRef]
7. Xiong, J.; Wang, Z.; Cao, J.; Dong, Y.; Chen, Y. Effect of the melatonin nuclear receptor RORα on monochromatic light-induced t-lymphocyte proliferation in chicken thymus. *Immunol. Lett.* **2019**, *213*, 21–29. [CrossRef]
8. Liu, L.; Wang, Z.; Cao, J.; Dong, Y.; Chen, Y. Effect of melatonin on monochromatic light-induced changes in clock gene circadian expression in the chick liver. *J. Photochem. Photobiol. B Biol.* **2019**, *197*, 111537. [CrossRef]
9. Cao, J.; Wang, Z.; Dong, Y.; Zhang, Z.; Li, J.; Li, F.; Chen, Y. Effect of combinations of monochromatic lights on growth and productive performance of broilers. *Poult. Sci.* **2012**, *91*, 3013–3018. [CrossRef]
10. Ke, Y.; Liu, W.; Wang, Z.; Chen, Y. Effects of monochromatic light on quality properties and antioxidation of meat in broilers. *Poult. Sci.* **2011**, *90*, 2632–2637. [CrossRef]
11. Guo, Q.; Wang, Z.; Dong, Y.; Cao, J.; Chen, Y. Physiological crosstalk between the ac/pka and plc/pkc pathways modulates melatonin-mediated, monochromatic-light-induced proliferation of t-lymphocytes in chickens. *Cell Tissue Res.* **2017**, *369*, 555–565. [CrossRef]
12. Xie, D.; Li, J.; Wang, Z.; Cao, J.; Li, T.; Chen, J.; Chen, Y. Effects of monochromatic light on mucosal mechanical and immunological barriers in the small intestine of broilers. *Poult. Sci.* **2011**, *90*, 2697–2704. [CrossRef] [PubMed]
13. Zhang, Y.; Wang, Z.; Dong, Y.; Cao, J.; Chen, Y. Blue light alters the composition of the jejunal microbiota and promotes the development of the small intestine by reducing oxidative stress. *Antioxidants* **2022**, *11*, 274. [CrossRef] [PubMed]
14. Zhang, Y.; Wang, Z.; Cao, J.; Dong, Y.; Chen, Y. A green and blue monochromatic light combination therapy reduces oxidative stress and enhances b-lymphocyte proliferation through promoting melatonin secretion. *Oxid. Med. Cell Longev.* **2021**, *2021*, 5595376. [CrossRef] [PubMed]
15. Currier, N.L.; Sun, Z.Y.; Miller, S.C. Exogenous melatonin: Quantitative enhancement in vivo of cells mediating non-specific immunity. *J. Neuroimmunol.* **2000**, *104*, 101–108. [CrossRef]
16. Ren, W.; Liu, G.; Chen, S.; Yin, J.; Wang, J.; Tan, B.; Wu, G.; Bazer, F.W.; Peng, Y.; Li, T.; et al. Melatonin signaling in t cells: Functions and applications. *J. Pineal Res.* **2017**, *62*, e12394. [CrossRef]
17. Ma, H.; Kang, J.; Fan, W.; He, H.; Huang, F. ROR: Nuclear receptor for melatonin or not? *Molecules* **2021**, *26*, 2693. [CrossRef]
18. Ding, S.; Lin, N.; Sheng, X.; Zhao, Y.; Su, Y.; Xu, L.; Tong, R.; Yan, Y.; Fu, Y.; He, J.; et al. Melatonin stabilizes rupture-prone vulnerable plaques via regulating macrophage polarization in a nuclear circadian receptor rorα-dependent manner. *J. Pineal Res.* **2019**, *67*, e12581. [CrossRef]
19. Farez, M.F.; Calandri, I.L.; Correale, J.; Quintana, F.J. Anti-inflammatory effects of melatonin in multiple sclerosis. *BioEssays* **2016**, *38*, 1016–1026. [CrossRef]
20. Liu, H.; Aramaki, M.; Fu, Y.; Forrest, D. Retinoid-related orphan receptor β and transcriptional control of neuronal differentiation. *Curr. Top. Dev. Biol.* **2017**, *125*, 227–255. [CrossRef]
21. Jetten, A.M.; Cook, D.N. (inverse) agonists of retinoic acid-related orphan receptor γ: Regulation of immune responses, inflammation, and autoimmune disease. *Annu. Rev. Pharmacol. Toxicol.* **2020**, *60*, 371–390. [CrossRef] [PubMed]
22. Zhang, S.; Takaku, M.; Zou, L.; Gu, A.D.; Chou, W.C.; Zhang, G.; Wu, B.; Kong, Q.; Thomas, S.Y.; Serody, J.S.; et al. Reversing ski-smad4-mediated suppression is essential for t(h)17 cell differentiation. *Nature* **2017**, *551*, 105–109. [CrossRef] [PubMed]
23. Kasal, D.N.; Bendelac, A. Multi-transcription factor reporter mice delineate early precursors to the ilc and lti lineages. *J. Exp. Med.* **2021**, *218*, e20200487. [CrossRef] [PubMed]
24. Farez, M.F.; Mascanfroni, I.D.; Méndez-Huergo, S.P.; Yeste, A.; Murugaiyan, G.; Garo, L.P.; Balbuena Aguirre, M.E.; Patel, B.; Ysrraelit, M.C.; Zhu, C.; et al. Melatonin contributes to the seasonality of multiple sclerosis relapses. *Cell* **2015**, *162*, 1338–1352. [CrossRef]
25. Reiter, R.J.; Mayo, J.C.; Tan, D.X.; Sainz, R.M.; Alatorre-Jimenez, M.; Qin, L. Melatonin as an antioxidant: Under promises but over delivers. *J. Pineal Res.* **2016**, *61*, 253–278. [CrossRef]
26. Leon, J.; Acua-Castroviejo, D.; Sainz, R.M.; Mayo, J.C.; Tan, D.X.; Reiter, R.J. Melatonin and mitochondrial function. *Life Sci.* **2004**, *75*, 765–790. [CrossRef]
27. Shin, Y.H.; Lee, S.H.; Mun, K.C. Effect of melatonin on the antioxidant enzymes in the kidneys of cyclosporine-treated rats. *Transplant. Proc.* **2002**, *34*, 2650–2651. [CrossRef]
28. Bian, J.; Wang, Z.; Dong, Y.; Cao, J.; Chen, Y. Effect of pinealectomy on the circadian clock of the chick retina under different monochromatic lights. *Chronobiol. Int.* **2019**, *36*, 548–563. [CrossRef]
29. Xiong, J.; Wang, Z.; Cao, J.; Dong, Y.; Chen, Y. Melatonin mediates monochromatic light—Induced proliferation of T/B lymphocytes in the spleen via the membrane receptor or nuclear receptor. *Poult. Sci.* **2020**, *99*, 4294–4302. [CrossRef]
30. Xie, D.; Wang, Z.; Dong, Y.; Cao, J.; Wang, J.; Chen, J.; Chen, Y. Effects of monochromatic light on immune response of broilers. *Poult. Sci.* **2008**, *87*, 1535–1539. [CrossRef]

31. Xie, D.; Wang, Z.; Cao, J.; Dong, Y.; Chen, Y. Effects of monochromatic light on proliferation response of splencyte in broilers. *Anat. Histol. Embryol.* **2008**, *37*, 332–337. [CrossRef] [PubMed]
32. Abdo, S.E.; El-Kassas, S.; El-Nahas, A.F.; Mahmoud, S. Modulatory effect of monochromatic blue light on heat stress response in commercial broilers. *Oxid. Med. Cell Longev.* **2017**, *2017*, 1351945. [CrossRef] [PubMed]
33. Hill, S.M.; Belancio, V.P.; Dauchy, R.T.; Xiang, S.; Brimer, S.; Mao, L.; Hauch, A.; Lundberg, P.W.; Summers, W.; Yuan, L.; et al. Melatonin: An inhibitor of breast cancer. *Endocr. Relat. Cancer* **2015**, *22*, R183–R204. [CrossRef] [PubMed]
34. Wojtulewicz, K.; Tomaszewska-Zaremba, D.; Herman, A.P. Endotoxin-induced inflammation suppresses the effect of melatonin on the release of lh from the ovine pars tuberalis explants—Ex vivo study. *Molecules* **2017**, *22*, 1933. [CrossRef]
35. Yu, Q.; Miller, S.C.; Osmond, D.G. Melatonin inhibits apoptosis during early b-cell development in mouse bone marrow. *J. Pineal Res.* **2000**, *29*, 86–93. [CrossRef]
36. Silva, S.O.; Rodrigues, M.R.; Ximenes, V.F.; Bueno-da-Silva, A.E.; Amarante-Mendes, G.P.; Campa, A. Neutrophils as a specific target for melatonin and kynuramines: Effects on cytokine release. *J. Neuroimmunol.* **2004**, *156*, 146–152. [CrossRef]
37. Sainz, R.M.; Mayo, J.C.; Uría, H.; Kotler, M.; Antolfn, I.; Rodriguez, C.; Menendez-Pelaez, A. The pineal neurohormone melatonin prevents in vivo and in vitro apoptosis in thymocytes. *J. Pineal Res.* **1995**, *19*, 178–188. [CrossRef]
38. Espino, J.; Bejarano, I.; Paredes, S.D.; Barriga, C.; Reiter, R.J.; Pariente, J.A.; Rodríguez, A.B. Melatonin is able to delay endoplasmic reticulum stress-induced apoptosis in leukocytes from elderly humans. *Age* **2011**, *33*, 497–507. [CrossRef]
39. Acunacastroviejo, D.; Reiter, R.J.; Menendezpelaez, A.; Pablos, M.I.; Burgos, A. Characterization of high-affinity melatonin binding-sites in purified cell nuclei of rat liver. *J. Pineal Res.* **1994**, *16*, 100–112. [CrossRef]
40. Lardone, P.J.; Guerrero, J.M.; Fernandez-Santos, J.M.; Rubio, A.; Martin-Lacave, I.; Carrillo-Vico, A. Melatonin synthesized by t lymphocytes as a ligand of the retinoic acid-related orphan receptor. *J. Pineal Res.* **2011**, *51*, 454–462. [CrossRef]
41. Zhao, Y.H.; Liu, Z.H.; Wang, L.; Xiao, H.M.; Du, C.G.; Zhang, Y.J.; Su, R.; Li, J.Q. Expression of the rorα gene in inner mongolian cashmere goat hair follicles. *Genet. Mol. Res.* **2015**, *14*, 380–388. [CrossRef] [PubMed]
42. Wang, R.X.; Liu, H.; Xu, L.; Zhang, H.; Zhou, R.X. Melatonin downregulates nuclear receptor rzr/rorγ expression causing growth-inhibitory and anti-angiogenesis activity in human gastric cancer cells in vitro and in vivo. *Oncol. Lett.* **2016**, *12*, 897–903. [CrossRef] [PubMed]
43. Xiao, Y.; Xia, J.; Wu, S.; Lv, Z.; Huang, S.; Huang, H.; Su, X.; Cheng, J.; Ke, Y. Curcumin inhibits acute vascular inflammation through the activation of heme oxygenase-1. *Oxid. Med. Cell Longev.* **2018**, *2018*, 3295807. [CrossRef]
44. Wakabayashi, N.; Slocum, S.L.; Skoko, J.J.; Shin, S.; Kensler, T.W. When nrf2 talks, who's listening? *Antioxid. Redox Signal.* **2010**, *13*, 1649–1663. [CrossRef] [PubMed]
45. Lousse, J.C.; Van Langendonckt, A.; Defrere, S.; Ramos, R.G.; Colette, S.; Donnez, J. Peritoneal endometriosis is an inflammatory disease. *Front. Biosci.* **2012**, *4*, 23–40. [CrossRef]
46. Al-Sawaf, O.; Fragoulis, A.; Rosen, C.; Kan, Y.W.; Sönmez, T.T.; Pufe, T.; Wruck, C.J. Nrf2 protects against tweak-mediated skeletal muscle wasting. *Sci. Rep.* **2014**, *4*, 3625. [CrossRef] [PubMed]
47. Liu, Y.C.; Hsieh, C.W.; Wu, C.C.; Wung, B.S. Chalcone inhibits the activation of nf-kappab and stat3 in endothelial cells via endogenous electrophile. *Life Sci.* **2007**, *80*, 1420–1430. [CrossRef] [PubMed]
48. Pan, M.H.; Lin-Shiau, S.Y.; Lin, J.K. Comparative studies on the suppression of nitric oxide synthase by curcumin and its hydrogenated metabolites through down-regulation of ikappab kinase and nfkappab activation in macrophages. *Biochem. Pharmacol.* **2000**, *60*, 1665–1676. [CrossRef]

antioxidants

Article

Functional Characterization of Serotonin *N*-Acetyltransferase in Archaeon *Thermoplasma volcanium*

Kyungjin Lee [1], Geun-Hee Choi [2] and Kyoungwhan Back [1,*]

1 Department of Biotechnology, College of Agriculture and Life Sciences, Chonnam National University, Gwangju 61186, Korea; nicekj7@hanmail.net
2 Nakdonggang National Institute of Biological Resources, 137 Donam 2-gil, Sangju-si 37242, Korea; ghchoi@nnibr.re.kr
* Correspondence: kback@chonnam.ac.kr; Tel.: +82-62-530-2165

Abstract: Serotonin *N*-acetyltransferase is the penultimate enzyme in the melatonin biosynthetic pathway that catalyzes serotonin into *N*-acetylserotonin. Many *SNAT* genes have been cloned and characterized from organisms ranging from bacteria to plants and mammals. However, to date, no *SNAT* gene has been identified from Archaea. In this study, three archaeal *SNAT* candidate genes were synthesized and expressed in *Escherichia coli*, and SNAT enzyme activity was measured using their purified recombinant proteins. Two *SNAT* candidate genes, from Methanoregulaceae (Archaea) and *Pyrococcus furiosus*, showed no SNAT enzyme activity, whereas a *SNAT* candidate gene from *Thermoplasma volcanium* previously named *TvArd1* exhibited SNAT enzyme activity. The substrate affinity and the maximum reaction rate of TvSNAT toward serotonin were 621 μM and 416 pmol/min/mg protein, respectively. The highest amine substrate was tyramine, followed by tryptamine, serotonin, and 5-methoxytryptamine, which were similar to those of plant SNAT enzymes. Homologs of *TvSNAT* were found in many Archaea families. Ectopic overexpression of *TvSNAT* in rice resulted in increased melatonin content, antioxidant activity, and seed size in conjunction with the enhanced expression of seed size-related gene. This study is the first to report the discovery of *SNAT* gene in Archaea. Future research avenues include the cloning of *TvSNAT* orthologs in different phyla, and identification of their regulation and functions related to melatonin biosynthesis in living organisms.

Keywords: archaea; Ard1; *N*-acetylserotonin; *N*-acetyltyramine; synthetic genes; melatonin; rice seed size

Citation: Lee, K.; Choi, G.-H.; Back, K. Functional Characterization of Serotonin *N*-Acetyltransferase in Archaeon *Thermoplasma volcanium*. *Antioxidants* **2022**, *11*, 596. https://doi.org/10.3390/antiox11030596

Academic Editors: Marina Garcia-Macia and Adrián Santos-Ledo

Received: 24 February 2022
Accepted: 18 March 2022
Published: 21 March 2022

1. Introduction

Melatonin is a universal molecule; it has been found in nearly all living organisms that have been tested for its presence [1,2]. Melatonin functions as a neurohormone, influencing circadian rhythms and seasonal behavioral changes, and acting as a potent antioxidant in animals. In plants, melatonin functions as an antioxidant as well as a signaling molecule, orchestrating various growth and defense processes, possibly via protein quality control and reactive oxygen species (ROS) scavenging mechanisms, although its role in circadian rhythm remains elusive [3–7]. In contrast to animals, which show a nocturnal peak in melatonin content, plants show nocturnal peaks in the content of melatonin metabolites, such as 2-hydroxymelatonin and cyclic 3-hydroxymelaotnin in rice plants [8,9]. Recently, 2-hydroxymelatonin was found to trigger ROS production in *Arabidopsis thaliana* leaves in a manner dependent on respiratory bursts of NADPH (nicotinamide adenine dinucleotide) oxidase [10], indicating the physiological significance of melatonin metabolites in plants. As in animals, these biological functions of melatonin are thought to be mediated by phytomelatonin receptors [11,12], whereas its antioxidant activity appears to be independent of phytomelatonin receptors [1,13].

The final two enzymes in the melatonin biosynthesis pathway are serotonin *N*-acetyltransferase (SNAT; also called arylalkyl *N*-acetyltransferase) and *N*-acetylserotonin *O*-methyltransferase [13,14]. SNAT is thought to be a rate-limiting enzyme in melatonin synthesis in both animals and plants [1]. In animals, *SNAT* evolved from Gram-positive bacteria [15], whereas in plants, *SNAT* evolved from cyanobacteria [2]. These independent evolutionary pathways have led a lack of significant homology among *SNAT* genes between animals and plants [16]. SNAT belongs to the GCN5-related *N*-acetyltransferase (GNAT) superfamily, members of which share a common acetyl-CoA binding domain and acetyl transfer mechanism; however, each family exhibits dramatically different substrate preferences [15,17]. In plants, it has been reported that the ectopic overexpression of *SNAT* either from animals or plants led to enhanced melatonin production followed by a series of physiological effects such as increased tolerance against biotic and abiotic stresses as well as yield increase [6,14].

Despite the functional characterization of a number of *SNAT* genes from bacteria [18], cyanobacteria [19], yeast [20], animals [21,22], and plants [23,24], *SNAT* from archaea have not been characterized to date. In common with other organisms, archaea are estimated to possess at least 200 GNAT proteins [25], of which two archaeal *GNAT* genes have been cloned and characterized according to crystallographic analysis of protein acetyltransferase activity [25,26]. However, SNAT enzyme activity has not been characterized in archaea to date. In this study, we employed two archaeal *GNAT* genes from *Pyrococcus furiosus* and *Thermoplasma volcanium* and one methanogenic archaea *GNAT* gene exhibiting similarity to rice *SNAT* [27] to investigate whether these archaeal GNAT proteins exhibit SNAT enzyme activity. This study is the first to report the identification of archaeal *SNAT* gene, providing a new avenue for investigating melatonin biosynthetic pathways in all kingdoms of living organisms.

2. Materials and Methods

2.1. Synthesis of Archaeal GNAT Genes

On the basis of protein information for three archaeal *GNAT* genes from *Pyrococcus furiosus* (GenBank accession no. NC_003413), *Thermoplasma volcanium* (NC_002689 or WP_010916271), and *Methanoregulaceae archaeon* PtaB.Bin152 (MVQF01000146), we manually designed corresponding nucleotide sequences with reference to the rice *SNAT2* codon [27]. Three synthetic *GNAT* genes were custom synthesized at Bioneer (Daejeon, South Korea).

2.2. Vector Construction and Purification of Recombinant Proteins

The full-length *GNAT* genes of *M. archaeon* (*MaGNAT*), *P. furiosus* (*PfGNAT*), and *T. volcanium* (*TvGNAT*) were initially amplified using polymerase chain reaction (PCR), and the primer sets are listed in Table S1. These were used with template plasmids containing each synthetic *GNAT* DNA (pBHA-MaGNAT, pBHA-PfGNAT, and pBHA-TvGNAT), which were provided by Bioneer. The initial *GNAT* PCR products were further amplified using a primer set containing the *attB* recombination sequence described in Table S1. These *GNAT* PCR products were cloned using Gateway recombination reactions into the pDONR221 vector (Invitrogen, Carlsbad, CA, USA), and then into the destination vector pET300/NT-DEST (Invitrogen) according to the manufacturer's procedure. The three plasmids, pET300-MaGNAT, pET300-PfGNAT, and pET300-TvGNAT, were then transformed into *Escherichia coli* strain BL21(DE3) (Invitrogen). The seed culture (10 mL) was cultured overnight in the presence of the antibiotic ampicillin (50 mg/L) and inoculated into 100 mL of Terrific Broth medium, consisting of 20 g/L bacto-tryptone, 24 g/L bacto-yeast extract, 4 mL/L glycerol, and phosphate buffer (0.017 M KH_2PO_4 and 0.072 M K_2HPO_4). The culture was incubated at 37 °C until the optical density of the *E. coli* culture at 600 nm reached 1.0. After the addition of 1 mM isopropyl-β-d-1-thiogalactopyranoside (Sigma, St. Louis, MO, USA), the culture was grown at 28 °C and shaken at 180 rpm for 5 h. Purification procedures were performed using affinity (Ni^{2+}) chromatography (Qiagen, Tokyo, Japan), according to the manufacturer's recommendations.

2.3. Homology and Phylogenetic Analysis

Amino acid (aa) sequence homology analysis was performed using the BLASTp tool (National Library of Medicine, Bethesda, MD, USA), and non-redundant protein sequences databases of the National Center for Biotechnology Information (http://www.ncbi.nlm.nih.gov/, accessed on 24 April 2019). Phylogenetic analysis was performed using the BLAST-Explorer program (version 2, Information Genomique & Structurale, Marseille, France) [28].

2.4. SNAT Enzyme Kinetics Measurements

Each purified recombinant GNAT protein was incubated in a total volume of 100 μL containing 0.5 mM serotonin (or other substrates) and 0.5 mM acetyl-CoA in 100 mM potassium phosphate at pH 8.8 and 45 °C for 1 h; pH and temperature were varied in some samples. Reaction products such as *N*-acetylserotonin and melatonin were subjected to high-performance liquid chromatography (HPLC) as described previously [23]. Substrate affinity (K_m) and the maximum reaction rate (V_{max}) were calculated using Lineweaver–Burk plots. Protein concentrations were determined using the Bradford method and a protein assay dye (Bio-Rad, Hercules, CA, USA). These analyses were performed in triplicate.

2.5. Transgenic Rice Plants Overexpressing *TvSNAT*

To obtain transgenic rice plants overexpressing the synthetic *TvSNAT* gene, their full-length sequences were amplified by PCR using specific primers (Table S1) using the synthetic cDNA gene as a template. The initial PCR products were further amplified using a second primer set containing 14 nt of the *attB* sequence using the initial PCR product as a template. Secondary PCR products were gel purified and cloned into the pDONR221 gateway vector (Invitrogen) via BP recombination. The pDONR221-TvSNAT gene entry vector was then recombined with the pIPKb002 destination vector [29] via LR recombination to yield the pIPKb002-TvSNAT binary plasmid. Constitutive expression in rice transgenic lines was ensured by the maize ubiquitin promoter.

The binary vector plasmid was transformed into *Agrobacterium tumefaciens* LBA4404, followed by transformation into rice as described previously [30].

2.6. Quantification of Melatonin and Radical Scavenging Activity Using the DPPH Method

Melatonin levels in rice seedlings were quantified by HPLC using a fluorescence detection system (Waters, Milford, MA, USA) as described previously [9]. As for seed melatonin quantification, rough rice seeds were first imbibed for 24 h followed by chloroform extraction as performed in seedling samples. As for antioxidant activity measurement, rough rice seeds were employed. The radical scavenging activity was measured using the 1,1-diphenyl-2-picrylhydrazyl (DPPH) method as described previously [31]. In brief, rough rice seeds (0.2 g), which were imbibed for 24 h, were extracted with 1 mL methanol and spun down for 5 min at 12,000× *g*. The resulting supernatants of methanolic extracts (0.1 mL) were mixed with 0.9 mL of 0.15 mM DPPH solution dissolved in methanol at 27 °C for 20 min. The control was prepared as above without any extract. Radical scavenging activity was expressed as a percentage of inhibition and was calculated using the following numerical formula: % radical scavenging activity = (control optical density (OD) − sample OD/control OD) × 100.

2.7. Statistical Analyses

All data analyses were performed using the IBM SPSS Statistics 23 software (IBM Corp., Armonk, NY, USA). Data are reported as means ± standard deviation; significant differences were evaluated at $p < 0.05$ according to Tukey's post hoc honest significant difference test.

3. Results

3.1. Codon-Optimized Synthesis of Three Archaeal GNAT Genes

Three archaeal *GNAT* genes were synthesized on the basis of aa sequence data from the GenBank database (http://www.ncbi.nlm.nih.gov/, accessed on 27 May 2019; Table S2).

Codon selection was performed on the basis of rice *SNAT2* codon preference for further overexpression in the rice genome. Because the rice *SNAT2* gene (GenBank accession no. AK068156) contains high G+C content (70%), the G+C contents of the synthetic genes of the three archaeal *GNAT* genes increased by 17% on average (Figure 1A). The *GNAT* gene from *M. archaeon* PtaB.Bin152 (*MaGNAT*) was selected due to its high aa sequence identity to rice SNAT2 (>45%) in the acetyl-CoA binding region, whereas there is little homology in the arylalkylamine binding region of GNAT family members [15]. Two other *GNAT* genes from *Pyrococcus furiosus* (*PvGNAT*) and *Thermoplasma volcanium* (*TvGNAT* or *TvArd1* (*arrest-defective-1*)) were chosen on the basis of their *N*-acetyltransferase (NAT) enzyme activity and successful expression in *Escherichia coli* [25,26]. The three codon-optimized archaeal *GNAT* genes were expressed as fusion proteins for the N-terminal hexa-histidine tag, followed by Ni^{2+} affinity purification (Figure 1B). All three GNAT proteins were successfully purified and subjected to SNAT enzyme assay analysis. MaGNAT and PfGNAT showed no SNAT enzyme activity, whereas TvGNAT exhibited high SNAT enzyme activity (6 pkat/mg protein) at 45 °C and pH 8.8 (Figure 1C). TvGNAT was previously annotated as an NAT Ard1 protein (TvArd1), which acetylates N-terminal residues of proteins such as *T. volcanium* Alba [26]. These data indicate that TvArd1 (now named TvSNAT) exhibits acetylation activity in transferring an acetyl group from acetyl-coenzyme A into both protein and serotonin. The optimum growth temperature of *T. volcanium* is approximately 60 °C, which is the lowest reported among Archaea [32].

(A)

	Wild-type *GNAT* gene			Codon-modified *GNAT* gene		
	M. archaeon	*P. furious*	*Tv*Ard1	M. archaeon	*P. furious*	*Tv*Ard1
No. of total aa	147	176	151	147	176	151
G+C content (%)	62	37	43	72	57	64
No. of modified codon	0	0	0	71 (48%)	118 (67%)	105 (70%)

(B)

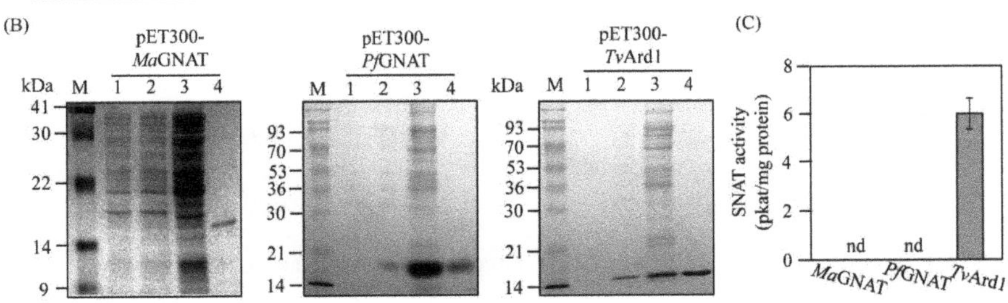

Figure 1. Summary of synthetic archaeal *GNAT* genes and recombinant GNAT protein purification. (**A**) Modification of archaeal *GNAT* genes. (**B**) Purification of N-terminal His × 6-tagged GNAT proteins. (**C**) SNAT activity measurements. Bacterial host strain BL21 (DE3) cells harboring the pET30-GNAT plasmids were incubated with isopropyl β-d-1-thiogalactopyranoside (IPTG) for 5 h at 28 °C. M, molecular mass standards; lane 1, total proteins in 15-μL aliquots of bacterial culture without IPTG; lane 2, total proteins in 15-μL aliquots of bacterial culture with IPTG; lane 3, 20 μg soluble protein; lane 4, 5 μg protein purified by affinity chromatography. Protein samples were separated using 12% sodium dodecyl sulfate–polyacrylamide gel electrophoresis and stained with Coomassie blue. SNAT enzyme activity was measured according to *N*-acetylserotonin production in the presence of 0.5 mM serotonin at 45 °C and pH 8.8.

3.2. Enzyme Kinetics of TvSNAT

To investigate the features of the TvSNAT enzyme (TvArd1), we performed experiments to determine their optimum pH and temperature. The highest SNAT activity was observed at pH 8.8 (Figure 2), similar to those of many plant SNAT proteins [23,27,33,34]. TvSNAT also showed optimum activity at temperatures ranging from 45 to 55 °C, similar to plant SNAT proteins [24]. By contrast, animal SNAT proteins have an optimum pH of 6.7 [21,22] and optimal temperature of approximately 37 °C [21,22]. The K_m and V_{max} values of TvSNAT were 621 μM and 416 pmol/min/mg protein, respectively. The K_m value of TvSNAT was similar to those of many other plant SNAT proteins [24], but different from those of SNAT proteins in animals such as human (1350 μM) and sheep (125 μM) [22]. On the basis of these data enzymatic features, we conclude that the TvSNAT protein is more closely related to plant SNAT proteins than to animal SNAT proteins. The high optimum temperature of TvSNAT enzyme activity was closer to the high optimum growth temperature of *T. volcanium* than to the low optimum temperature (25 °C) of protein NAT activity of TvArd1 [26], suggesting that TvArd1 shows SNAT activity in preference to NAT activity in vitro, although in vivo evidence is required.

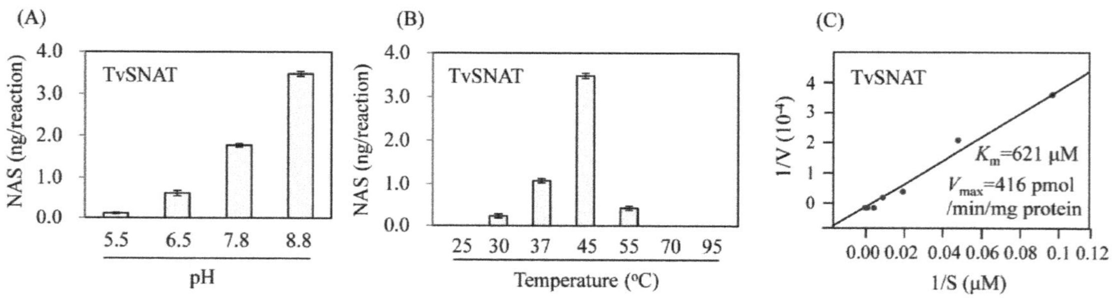

Figure 2. SNAT enzyme activity catalyzing the conversion of serotonin to *N*-acetylserotonin as a function of (**A**) pH and (**B**) temperature. (**C**) Determination of substrate affinity (K_m) and maximum reaction rate (V_{max}) values of TvSNAT (also called TvArd1). TvSNAT (1 μg) was incubated at a range of pH values and temperatures for 30 min. K_m and V_{max} values were determined using Lineweaver–Burk plots. In vitro enzymatic *N*-acetylserotonin products were measured by high-performance liquid chromatography. Data are the means ± standard deviation (n = 3).

3.3. Substrate Specificity

SNAT enzymes can accept many other substrates, including phenylethylamines such as tyramine and indolethylamines such as tryptamine, serotonin, and 5-methoxytryptamine [35]. To determine whether TvSNAT can also acetylate other substrates, we measured SNAT enzyme activity in several other candidate substrates (Figure 3). The best substrates for the TvSNAT enzyme were tyramine (16.9 pkat/mg protein) and tryptamine (7.5 pkat/mg protein), followed by serotonin (6.7 pkat/mg protein) and 5-methoxytryptamine (0.7 pkat/mg protein) (Figure 3B). Substrate preference toward tyramine was also reported in sheep SNAT (20,413 pkat/mg protein) and rice SNAT2 (140 pkat/mg protein) [27]. TvSNAT exhibited the lowest enzyme activity toward 5-methoxytryptamine, whereas sheep and yeast SNAT showed the highest enzyme activity toward 5-methoxytryptamine [20], suggesting that melatonin biosynthesis via serotonin to 5-methoxytryptamine to melatonin is a less likely pathway in Archaea. In addition to these preferred arylalkylamines, other arylakylamines (dopamine, octopamine, 2-phenylethylamine, and histamine) and polyamines (spermidine and putrescine) were tested for possible acceptance as substrates for TvSNAT. Due to the lack of commercially available standard compounds of those acetylated substrates, we performed a SNAT inhibition assay (0.5 mM serotonin) in the presence of each substrate (0.5 mM) to determine whether SNAT activity is inhibited by co-incubation with one of these compounds. Spermidine strongly inhibited TvSNAT activity, followed by octopamine (Figure 4A). However, other compounds had no inhibitory effect on TvSNAT enzyme activity. These findings indirectly indicate that TvSNAT can acetylate spermidine and octopamine into *N*-acetylspermidine and *N*-acetyloctopamine, respectively. A dose-dependent inhibition assay showed that SNAT enzyme activity decreased to 50% in the presence of 0.1 mM spermidine; the same inhibition effect was observed at 0.5 mM octopamine (Figure 4B). Thus, TvSNAT appears to prefer spermidine to serotonin as a substrate. TvSNAT shares 9% sequence identity with human spermidine/spermine N^1-acetyltransferase-1 [36], suggesting the bifunctional activity of TvSNAT toward serotonin and spermidine. This enhanced SNAT activity in the presence of dopamine was unexpected because recombinant fish SNAT enzyme activity was inhibited by co-incubation with dopamine [37]. The detailed enzymatic inhibition kinetics of TvSNAT require further study.

(A)

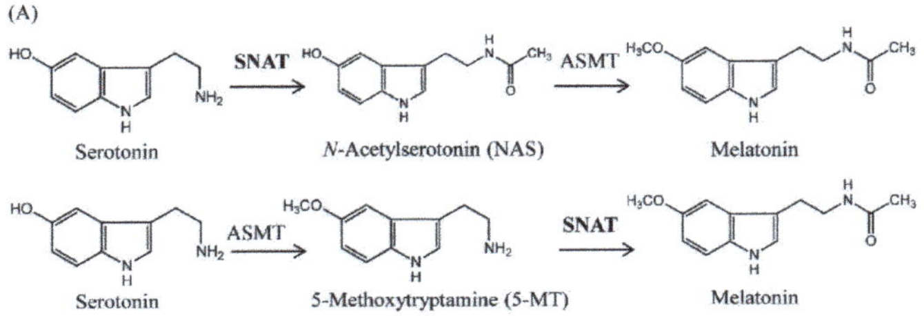

(B)

Substrate	Product	NAT activity (pkat/mg protein)
Serotonin	*N*-Acetylserotonin	6.7 ± 0.3
Tryptamine	*N*-Acetyltryptamine	7.5 ± 0.1
Tyramine	*N*-Acetyltyramine	16.9 ± 2.6
5-Methoxytryptamine	Melatonin	0.7 ± 0.1

Figure 3. Schematic diagram of the SNAT reaction and substrate preference. (**A**) Enzymatic reaction of SNAT in two different melatonin biosynthetic pathways. (**B**) TvSNAT enzyme activity measurements for various substrates. SNAT enzyme activity was measured in the presence of 0.5 mM of each substrate at 45 °C and pH 8.8. Data are the means ± standard deviation (n = 3). ASMT, *N*-acetylserotonin *O*-methyltransferase; SNAT, serotonin *N*-acetyltransferase.

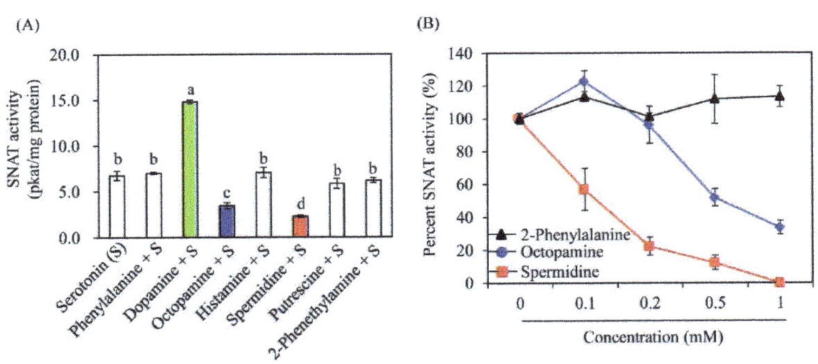

Figure 4. Effects of various substrates on SNAT activity. (**A**) SNAT enzyme activity of recombinant purified TvSNAT in the presence of serotonin (0.5 mM) and various amines (0.5 mM). Different letters indicate significant differences [$p < 0.05$; analysis of variance (ANOVA), followed by Tukey's honest significant difference (HSD) post hoc tests]. The green, blue and red colors emphasize compounds which alter SNAT activity. (**B**) Dose-dependent inhibition of SNAT enzyme activity by spermidine or octopamine. SNAT activity was assayed in the presence of polyamines at various concentrations and expressed as a percentage relative to that in the absence of polyamines (Figure 4B). Data are the means ± standard deviation (n = 3).

3.4. Phylogenetic Analysis

A phylogenetic tree constructed on the basis of SNAT aa sequences from different organisms revealed that TvSNAT orthologs are widely distributed in many other Archaea families, including Picrophilaceae (*Picrophilus torridus*), Ferroplasmaceae (*Ferroplasma acidiphilum*), DHVE2 (*Aciduliprofundum boonei*), Cuniculiplasmataceae (*Cuniculiplasma divulgatum*), and Thermoprotei (*Desulfurococcales archaeon*) (Figure 5A). A non-redundant search of the National Center for Biotechnology Information and National Institutes of Health protein sequence databases (http://www.ncbi.nlm.nih.gov/, accessed on 24 April 2019) using the BLASTp program revealed that TvSNAT had the highest homology (33%) to *Drosophila novamexicana* (GenBank accession no. XP_030561899) among the kingdom Animalia, whereas among the kingdom Plantae, *Dendrobium catenatum* exhibited 31% identity to TvSNAT (Figure 5B). Further study is required to determine whether these animal and plant homologs of TvSNAT exhibit SNAT enzyme activity.

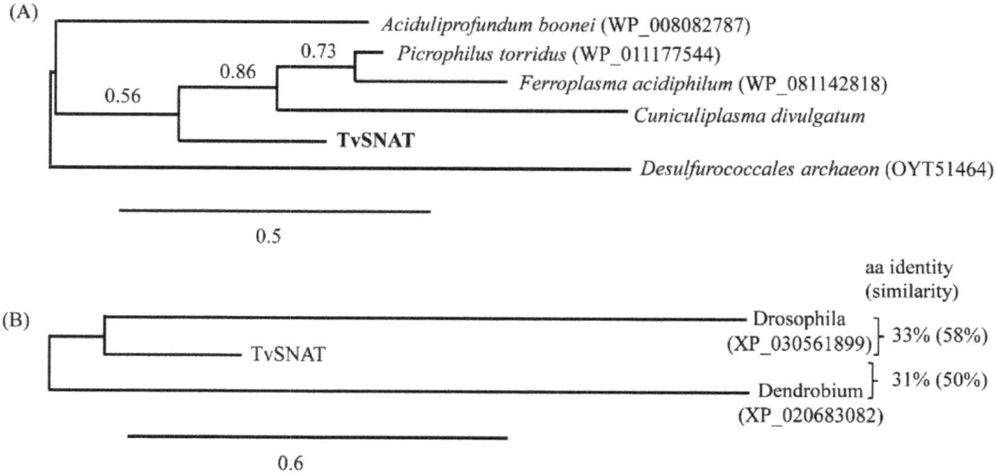

Figure 5. Phylogenetic analysis of TvSNAT constructed using the neighbor-joining method for (**A**) Archaea and (**B**) Animalia and Plantae. Scale bars in (**A**,**B**) represent 0.3 and 0.6 substitutions per site, respectively. Numbers in parentheses are GenBank accession numbers of corresponding genes. Alignments and the phylogenetic analyses were performed using the BLAST-Explorer tool (www.phylogeny.fr, accessed on 6 November 2019)).

3.5. Characterization of Transgenic Rice Plants

To gain further insight into the role of *TvSNAT* in melatonin biosynthesis in vivo, we generated transgenic rice plants overexpressing the *TvSNAT* gene under the control of the maize ubiquitin promoter (Figure 6A). Eleven independent T_1 transgenic rice lines were initially screened on half-strength Murashige and Skoog medium containing 50 µg/mL hygromycin. Among these lines, one-copy transgene insertion lines with a 3:1 hygromycin segregation ratio were further selected and selfed to produce T_2 seeds. Three homozygous transgenic rice plants overexpressing *TvSNAT* (TvSNAT-OE) were used for further analysis. Transgenic plants showed ectopic overexpression of transgenes in rice plants according to reverse-transcription (RT)-PCR analyses (Figure 6B). Three 7-day-old TvSNAT-OE transgenic rice seedlings produced more melatonin than wild-type (WT) seedlings (Figure 6C,D), indicating that *TvSNAT* gene overexpression was functionally coupled to enhanced melatonin production in rice plants. To see whether melatonin increase is closely coupled to increased antioxidant activity, the radical scavenging activities were investigated from the transgenic and wild-type seeds. As expected, the TvSNAT-OE lines exhibited high antioxidant activities to quench DPPH radicals compared to wild type (Figure 6E). On average, the transgenic lines showed 30% radical scavenging activity, whereas wild type had 25% radical scavenging activity, indicative of 12% higher antioxidant activity in the TvSNAT-OE lines.

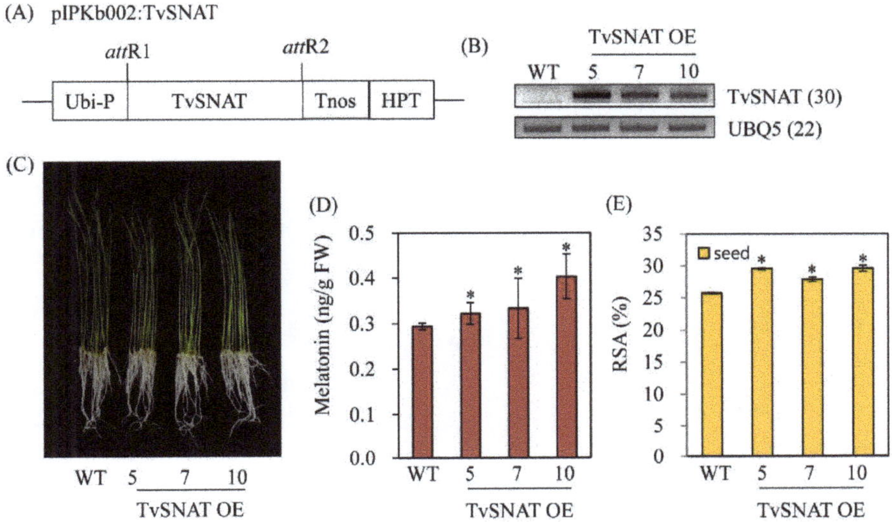

Figure 6. Schematic diagram of overexpression binary vectors, reverse-transcription polymerase chain reaction (RT-PCR) analyses, and melatonin content in transgenic rice. (**A**) Binary vector (GenBank accession no. EU161568) used for *TvSNAT* overexpression. (**B**) RT-PCR analyses of 7-day-old transgenic and wild-type (WT) rice seedlings. (**C**) Phenotypes of 7-day-old seedlings. (**D**) Melatonin content in 7-day-old seedlings. (**E**) Radical scavenging activity assessed by 1,1-diphenyl-2-picrylhydrazyl (DPPH) in seeds. *TvSNAT*, *Thermoplasma volcanium serotonin* N*-acetyltransferase*; *Ubi-P*, maize ubiquitin promoter; *HPT*, hygromycin phosphotransferase; *UBQ5*, rice ubiquitin 5 gene (GenBank accession no. Os03g13170). Numbers in parentheses indicate the numbers of PCR cycles performed. Asterisks indicate significant differences from the WT (Tukey's honest significant difference test; $p < 0.05$). RSA, radical scavenging activity.

Enhanced melatonin synthesis in the TvSNAT-OE transgenic rice plants led to increased grain size due to increased seed length and width compared with the WT control (Figure 7A,B). Similarly, 1000-seed weight was higher in transgenic seeds than in WT seeds (Figure 7C). We investigated two representative genes responsible for controlling rice seed size: *GRAIN INCOMPLETE FILLING 1* (*GIF1*) is a positive factor and *GRAIN WIDTH 2* (*GW2*) is a negative factor in controlling rice seed size. *GIF1* expression was higher in the TvSNAT-OE lines than in the WT, whereas *GW2* expression did not differ between the TvSNAT-OE and WT lines (Figure 7D). Melatonin levels were also increased in seeds of the TvSNAT-OE lines compared with the WT (Figure 7E). This is the first report of enhanced grain size in transgenic rice plants caused by an increase in endogenous melatonin, whereas small grain size was previously observed in various transgenic rice plants downregulating melatonin synthesis [38]. These data indicate that melatonin is positively associated with the control of grain size in plants. Further in-depth studies are required to investigate the roles of TvSNAT in rice growth, development, yield, and defense responses against environmental stresses.

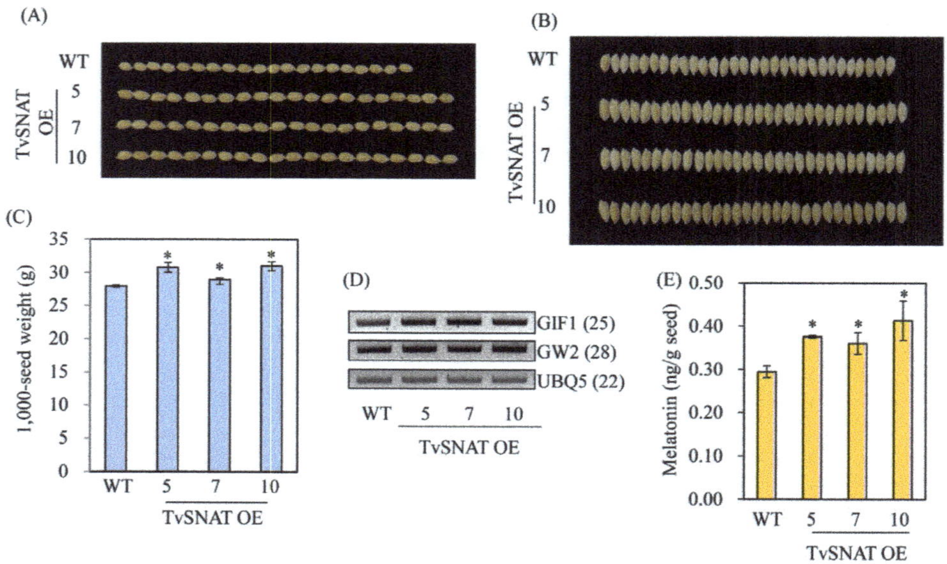

Figure 7. Rice grain morphology and expression levels of seed size-related genes. (**A**) Grain lengths, (**B**) grain widths, and (**C**) 1000-grain weights of the wild-type (WT) and TvSNAT-OE lines. (**D**) Expression levels of rice seed size-related genes determined by reverse-transcription polymerase chain reaction (RT-PCR) in 7-day-old seedlings. (**E**) Seed melatonin content levels. GenBank accession numbers for *GIF1*, *GW2*, and *UBQ5* are Os04g33740, Os02g14720, and Os03g13170, respectively. Numbers in parentheses indicate the numbers of PCR cycles performed. Asterisks indicate significant differences from the WT (Tukey's honest significant difference test; $p < 0.05$).

4. Discussion

It has long been postulated that archaeans may be capable of synthesizing melatonin; however, no evidence to support this hypothesis has been discovered to date. Similarly, no enzymatic evidence has been discovered to confirm the existence of archaean melatonin biosynthetic genes such as *SNAT*, although archaeans contain approximately 200 genes belonging to the GNAT family [25]. Among these, two *GNAT* genes were suggested to be potentially capable of melatonin synthesis on the basis of sequence homology [2,25,26]; however, no direct evidence has been obtained through either enzymatic or mutant analysis.

In this study, we initially searched for *SNAT* genes encoding SNAT enzyme activity among archaean GNAT family members. SNAT is the penultimate enzyme of the melatonin biosynthesis pathway; it catalyzes serotonin into *N*-acetylserotonin, followed by melatonin synthesis through *O*-methyltransferase enzymes [14]. As SNAT plays a rate-limiting role in melatonin biosynthesis in both animals and plants [1], SNAT identification from archaeans could provide strong evidence for the presence of melatonin in archaea. Commensurate with the pivotal role of *SNAT* genes, it is highly likely that Gram-positive bacterial *SNAT* genes are ancestral to present animal *SNAT* genes, while cyanobacterial *SNAT* genes are ancestral to plant *SNAT* genes, suggesting an essential role of *SNAT* genes in melatonin biosynthesis among all living organisms [15,35].

In this study, we employed three archaean *SNAT* candidate genes from the GNAT family, including two recommended genes possessing NAT activity [25,26]. Using recombinant GNAT enzymes, we discovered a *SNAT* gene in *T. volcanium* that was previously annotated as a TvArd1 [26]. Ard1 encodes NAT, which transfers an acetyl group from acetyl-CoA to the N-terminal of various proteins [39,40]. Although TvArd1 (or TvSNAT) exhibits NAT activity at 25 °C [26], aa sequence identity between TvArd1 and human Ard1 was <14%, and human Ard1 (235 aa) was larger than TvArd1 (154 aa), suggesting that *TvArd1* may not be an ortholog gene of human *Ard1*.

In common with TvArd1, *Arabidopsis thaliana* SNAT1 (AtSNAT1) exhibits acetyltransferase activity toward a series of substrates such as histone [41], chloroplast protein [42], and arylalkylamines [14,43]. Of note, AtSNAT1 exhibited SNAT activity within a broad range of temperatures (25–55 °C), as well as high serotonin affinity (K_m = 309 μM). The melatonin biosynthetic roles of AtSNAT1 were confirmed by gain- and loss-of-function analyses as well as exogenous melatonin treatment in view of pathogen defense [44], endoplasmic reticulum stress [5], high light stress [45], and flowering [23], suggesting that AtSNAT is more important to melatonin biosynthesis than to NAT or histone acetyltransferase.

TvSNAT-OE transgenic rice plants exhibited a large-seed phenotype, suggesting the involvement of melatonin in seed size regulation. Although larger seeds have not been observed in previous studies of transgenic rice plants producing higher melatonin levels than WT plants [46,47], other plants, including maize and cucumber, have shown increased seed or fruit size following exogenous melatonin treatment [48,49]. Intriguingly, transgenic rice plants overexpressing the rice *SNAT2* gene produce longer seeds than WT plants, although seed width remained unaffected [37]. These data suggest that melatonin is involved to some extent in rice seed size control. Several rice genes have been found to be responsible for controlling grain size [50,51] including *GRAIN SIZE 3* (*GS3*) [52], *GW2* [53], *GIF1* [54], and *RICE BIG GRAIN 1* (*RBG1*) [55]. GS3 and GW2 function as negative regulators of grain size, whereas *GIF1* and *RBG1* function as positive regulators. In this study, we found that melatonin-mediated larger grain size was mainly ascribed to the induction of *GIF1*, which encodes a cell wall invertase [54]. A previous study reported that melatonin treatment elevated cell wall invertase activity in *A. thaliana*, followed by pathogen resistance due to the increased cell wall reinforcement and callose deposition [56]. Further detailed mechanisms by which melatonin regulates seed size remain to be investigated in the near future. In addition to enhanced seed size, we anticipate that many other biological functions will be discovered through TvSNAT overexpression in rice plants, on the basis of the previous known roles of melatonin counteracting damages caused by biotic and abiotic stresses such as drought and salt [8,57–59].

5. Conclusions

This is the first study to discover the *SNAT* gene in Archaea. We demonstrated that the *TvSNAT* candidate gene from *Thermoplasma volcanium*, which was previously named *TvArd1*, exhibited SNAT enzyme activity. The highest amine substrate for TvSNAT was tyramine, followed by tryptamine, serotonin, and 5-methoxytryptamine, which were similar to those of plant SNAT enzymes. Homologs of *TvSNAT* have been found in many Archaea families. Ectopic overexpression of *TvSNAT* in rice was functionally coupled with the enhanced melatonin synthesis, resulting in the increased rice seed size. Consequently, one of key genes responsible for controlling rice seed size, *GIF1* was significantly overexpressed in the TvSNAT-OE lines compared to that in wild type. Our findings will open new avenues for research involving the cloning of *TvSNAT* orthologs in many different phyla, allowing the exploration of their functional roles and regulation of melatonin biosynthesis in living organisms.

Supplementary Materials: The following are available online at https://www.mdpi.com/article/10.3390/antiox11030596/s1, Table S1: Sequences of primers used in this study. Table S2: Full length nucleotide sequences of three *GNAT* genes of both native and synthetic forms.

Author Contributions: Conceptualization, K.B.; data curation, K.L. and K.B.; formal analysis, K.L. and G.-H.C.; writing—original draft, K.B.; and writing—review and editing, K.B. All authors have read and agreed to the published version of the manuscript.

Funding: This research was supported from grants by the Basic Science Research Program of the National Research Foundation of Korea (NRF-2021R1I1A2042237) funded by the Ministry of Education and by the Nakdonggang National Institute of Biological Resources (NNIBR) funded by the Ministry of Environment (MOE) of the Republic of Korea (NNIBR202202105).

Institutional Review Board Statement: Not applicable.

Informed Consent Statement: Not applicable.

Data Availability Statement: The data presented in this study are available within the article.

Conflicts of Interest: The authors declare no conflict of interest.

References

1. Zhao, D.; Yu, Y.; Shen, Y.; Liu, Q.; Zhao, Z.; Sharma, R.; Reiter, R.J. Melatonin synthesis and function: Evolutionary history in animals and plants. *Front. Endocrinol.* **2019**, *10*, 249. [CrossRef]
2. Hardeland, R. Melatonin in the evolution of plants and other phototrophs. *Melatonin Res.* **2019**, *2*, 10–36. [CrossRef]
3. Reiter, R.J.; Tan, D.X.; Sharma, R. Historical perspective and evaluation of the mechanisms by which melatonin mediates seasonal reproduction in mammals. *Melatonin Res.* **2018**, *1*, 59–77. [CrossRef]
4. Yu, Y.; Lv, Y.; Shi, Y.; Li, T.; Chen, Y.; Zhao, D.; Zhao, Z. The role of phyto-melatonin and related metabolites in response to stress. *Molecules* **2018**, *23*, 1887. [CrossRef]
5. Lee, H.Y.; Back, K. Melatonin plays a pivotal role in conferring tolerance against endoplasmic reticulum stress via mitogen-activated protein kinases and bZIP60 in *Arabidopsis thaliana*. *Melatonin Res.* **2018**, *1*, 93–107. [CrossRef]
6. Arnao, M.B.; Hernández-Ruiz, J. Melatonin: A new plant hormone and/or a plant master regulator? *Trends Plant Sci.* **2019**, *24*, 38–48. [CrossRef] [PubMed]
7. Lee, H.Y.; Hwang, O.J.; Back, K. Phytomelatonin as a signaling molecule for protein quality control via chaperone, autophagy, and ubiquitin–proteasome systems in plants. *J. Exp. Bot.* **2022**, in press. [CrossRef] [PubMed]
8. Choi, G.H.; Back, K. Choi GH, Back K. Cyclic 3-hydroxymelatonin exhibits diurnal rhythm and cyclic 3-hydroxymelatonin overproduction increases secondary tillers in rice by upregulating *MOC1* expression. *Melatonin Res.* **2019**, *2*, 120–138. [CrossRef]
9. Choi, G.H.; Back, K. Suppression of melatonin 2-hydroxylase increases melatonin production leading to the enhanced abiotic stress tolerance against cadmium, senescence, salt, and tunicamycin in rice plants. *Biomolecules* **2019**, *9*, 589. [CrossRef] [PubMed]
10. Lee, H.Y.; Back, K. 2-Hydroxymelatonin, rather than melatonin, is responsible for RBOH-dependent reactive oxygen species production leading to premature senescence in plants. *Antioxidant* **2021**, *10*, 1782. [CrossRef] [PubMed]
11. Wei, J.; Li, D.X.; Zhang, J.R.; Shan, C.; Rengel, Z.; Song, Z.B.; Chen, Q. Phytomelatonin receptor PMTR1-mediated signaling regulates stomatal closure in *Arabidopsis thaliana*. *J. Pineal Res.* **2018**, *65*, e12500. [CrossRef] [PubMed]
12. Lee, H.Y.; Back, K. The phytomelatonin receptor (PMRT1) *Arabidopsis* Cand2 is not a *bona fide* G protein-coupled melatonin receptor. *Melatonin Res.* **2020**, *3*, 177–186. [CrossRef]
13. Tan, D.-X.; Reiter, R.J. Mitochondria: The birth place, battle ground and site of melatonin metabolism in cells. *Melatonin Res.* **2019**, *2*, 44–66. [CrossRef]
14. Back, K. Melatonin metabolism, signaling and possible roles in plants. *Plant J.* **2021**, *105*, 376–391. [CrossRef]
15. Coon, S.L.; Klein, D.C. Evolution of arylalkylamine *N*-acetyltransferase: Emergence and divergence. *Mol. Cell. Endocrinol.* **2006**, *252*, 2–10. [CrossRef]
16. Tan, D.X.; Zheng, X.; Kong, J.; Manchester, L.C.; Hardeland, R.; Kim, S.J.; Xu, X.; Reiter, R.J. Fundamental issues related to the origin of melatonin and melatonin isomers during evolution: Relation to their biological functions. *Int. J. Mol. Sci.* **2014**, *15*, 15858–15890. [CrossRef]
17. Dyda, F.; Klein, D.C.; Hickman, A.B. GCN5-related *N*-acetyltransferases: A structural overview. *Annu. Rev. Biophys. Biomol. Struct.* **2000**, *29*, 81–103. [CrossRef]
18. Pan, Q.; Zhao, F.-L.; Ye, B.-C. Eis, a novel family of arylalkylamine *N*-acetyltransferase (EC 2.3.1.87). *Sci. Rep.* **2018**, *8*, 2435. [CrossRef]
19. Byeon, Y.; Lee, K.; Park, Y.I.; Park, S.; Back, K. Molecular cloning and functional analysis of serotonin *N*-acetyltransferase from the cyanobacterium *Synechocystis* sp. PCC 6803. *J. Pineal Res.* **2013**, *55*, 371–376. [CrossRef]
20. Ganguly, S.; Mummaneni, P.; Steinbach, P.J.; Klein, D.C.; Coon, S.L. Characterization of the *Saccharomyces cerevisiae* homolog of the melatonin rhythm enzyme arylalkylamine *N*-acetyltransferase (EC 2.3.1.87). *J. Biol. Chem.* **2001**, *276*, 47239–47247. [CrossRef]
21. De Angelis, J.; Gastel, J.; Klein, D.C.; Cole, P.A. Kinetic analysis of the catalytic mechanism of serotonin *N*-acetyltransferase (EC 2.3.1.87). *J. Biol. Chem.* **1998**, *273*, 3045–3050. [CrossRef] [PubMed]
22. Ferry, G.; Loynel, A.; Kucharczyk, N.; Bertin, S.; Rodriguez, M.; Delagrange, P.; Galizzi, J.-P.; Jacoby, E.; Volland, J.-P.; Lesieur, D.; et al. Substrate specificity and inhibition studies of human serotonin *N*-acetyltransferase. *J. Biol. Chem.* **2000**, *275*, 8794–8805. [CrossRef] [PubMed]
23. Lee, H.Y.; Lee, K.; Back, K. Knockout of *Arabidopsis serotonin N-acetyltransferase-2* reduces melatonin levels and delays flowering. *Biomolecules* **2019**, *9*, 712. [CrossRef] [PubMed]

24. Lee, H.Y.; Hwang, O.J.; Back, K. Functional characterization of tobacco (*Nicotiana benthamiana*) serotonin *N*-acetyltransferases (*NbSNAT1* and *NbSNAT2*). *Melatonin Res.* **2021**, *4*, 507–521.
25. Biarrotte-Sorin, S.; Mayer, C. Cloning, purification, crystallization and preliminary crystallographic analysis of a hypothetical acetyltransferase from *Pyrococcus furious*. *Acta Cryst.* **2005**, *F61*, 269–270.
26. Ma, C.; Pathak, C.; Jang, S.; Lee, S.J.; Nam, M.; Kim, S.J.; Im, H.; Lee, B.J. Structure of *Thermoplasma volcanium* Ard1 belongs to *N*-acetyltransferase family member suggesting multiple ligand binding modes with acetyl coenzyme A and coenzyme A. *Biochim. Biophys. Acta* **2014**, *1844*, 1790–1797. [CrossRef]
27. Byeon, Y.; Lee, H.Y.; Back, K. Cloning and characterization of the serotonin *N*-acetyltransferase-2 gene (*SNAT2*) in rice (*Oryza sativa*). *J. Pineal Res.* **2016**, *61*, 198–207. [CrossRef]
28. Dereeper, A.; Audic, S.; Claverie, J.M.; Blanc, G. BLAST-EXPLORER helps you building datasets for phylogenetic analysis. *BMC Evol. Biol.* **2010**, *10*, 8. [CrossRef]
29. Himmelbach, A.; Zierold, U.; Hensel, G.; Riechen, J.; Douchkov, D.; Schweizer, P.; Kumlehn, J. A set of modular binary vectors for transformation of cereals. *Plant Physiol.* **2007**, *145*, 1192–1200. [CrossRef]
30. Lee, H.J.; Lee, S.B.; Chung, J.S.; Han, S.U.; Han, O.; Guh, J.O.; Jeon, J.S.; An, G.; Back, K. Transgenic rice plants expressing a *Bacillus subtilis* protoporphyrinogen oxidase gene are resistant to diphenyl ether herbicide oxyfluorfen. *Plant Cell Physiol.* **2000**, *41*, 743–749. [CrossRef]
31. Kang, K.; Kim, Y.S.; Park, S.; Back, K. Senescence-induced serotonin biosynthesis and its role in delaying senescence in rice leaves. *Plant Physiol.* **2009**, *150*, 1380–1393. [CrossRef] [PubMed]
32. Kawashima, T.; Amano, N.; Koike, H.; Makino, S.I.; Higuchi, S.; Kawashima-Ohya, T.; Watanabe, K.; Yamazaki, M.; Kanehori, K.; Kawamoto, T.; et al. Archaeal adaptation to higher temperatures reveals by genomic sequence of *Thermoplasma volcanium*. *Proc. Natl. Acad. Sci. USA* **2000**, *97*, 14257–14262. [CrossRef]
33. Kang, K.; Lee, K.; Park, S.; Byeon, Y.; Back, K. Molecular cloning of rice serotonin *N*-acetyltransferase, the penultimate gene in plant melatonin biosynthesis. *J. Pineal Res.* **2013**, *55*, 7–13. [CrossRef]
34. Yu, Y.; Bian, L.; Jiao, Z.; Keke, Y.; Wan, Y.; Zhang, G.; Guo, D. Molecular cloning and characterization of a grapevine (*Vitis vinifera* L.) serotonin *N*-acetyltransferase (*VvSNAT2*) gene involved in plant defense. *BMC Genom.* **2019**, *20*, 880. [CrossRef] [PubMed]
35. Falcón, J.; Coon, S.L.; Besseau, L.; Cazamea-Catalan, D.; Fuentes, M.; Magnanou, E.; Paulin, C.H.; Boeuf, G.; Sauzet, S.; Jorgensen, E.H.; et al. Drastic neofunctionalization associated with evolution of the timezyme AANAT 500 Mya. *Proc. Natl. Acad. Sci. USA* **2014**, *111*, 314–319. [CrossRef] [PubMed]
36. Xiao, L.; Celano, P.; Mank, A.R.; Pegg, A.E.; Casero, R.A., Jr. Characterization of a full-length cDNA which codes for the human spermidine/spermine N^1-acetyltransferase. *Biochem. Biophys. Res. Commun.* **1991**, *179*, 407–415. [CrossRef]
37. Zilberman-Peled, B.; Bransburg-Zabary, S.; Klein, D.C.; Gothilf, Y. Molecular evolution of multiple arylalkylamine *N*-acetyltransferase (AANAT) in fish. *Mar. Drugs* **2011**, *9*, 906–921. [CrossRef] [PubMed]
38. Hwang, O.J.; Back, K. Melatonin deficiency confers tolerance to multiple abiotic stresses in rice via decreased brassinosteroid levels. *Int. J. Mol. Sci.* **2019**, *20*, 5173. [CrossRef]
39. Arnesen, T.; Anderson, D.; Torsvik, J.; Halseth, H.B.; Varhaug, J.E.; Lillehaug, J.R. Cloning and characterization of hNAT5/hSAN: An evolutionarily conserved component of the NatA protein *N*-α-acetyltransferase complex. *Gene* **2006**, *371*, 291–295. [CrossRef]
40. Linster, E.; Wirtz, M. N-terminal acetylation: An essential protein modification emerges as an important regulator of stress responses. *J. Exp. Bot.* **2018**, *69*, 4555–4568. [CrossRef]
41. McGarry, R.C.; Barron, Y.D.; Carvalho, M.F.; Hill, J.E.; Gold, D.; Cheung, E.; Kraus, W.L.; Lazarowitz, S.G. A novel Arabidopsis acetyltransferase interacts with the geminivirus movement protein NSP. *Plant Cell* **2003**, *15*, 1605–1618. [CrossRef] [PubMed]
42. Koskela, M.M.; Brünje, A.; Ivanauskaite, A.; Grabsztunowicz, M.; Lassowskat, I.; Neumann, U.; Dinh, T.V.; Sindlinger, J.; Schwarzer, D.; Wirtz, M.; et al. Chloroplast acetyltransferase NSI is required for state transition in *Arabidopsis thaliana*. *Plant Cell* **2018**, *30*, 1695–1709. [CrossRef] [PubMed]
43. Lee, H.Y.; Back, K. Melatonin regulates chloroplast protein quality control via a mitogen-activated protein kinase signaling pathway. *Antioxidants* **2021**, *10*, 511. [CrossRef] [PubMed]
44. Lee, H.Y.; Byeon, Y.; Tan, D.X.; Reiter, R.J.; Back, K. *Arabidopsis* serotonin *N*-acetyltransferase knockout plants exhibit decreased melatonin and salicylic acid levels resulting in susceptibility to an avirulent pathogen. *J. Pineal Res.* **2015**, *58*, 291–299. [CrossRef] [PubMed]
45. Lee, H.Y.; Back, K. Melatonin induction and its role in high light stress tolerance in *Arabidopsis thaliana*. *J. Pineal Res.* **2018**, *65*, e12504. [CrossRef] [PubMed]
46. Byeon, Y.; Back, K. An increase in melatonin in transgenic rice causes pleiotropic phenotypes, including enhanced seedling growth, delayed flowering, and low grain yield. *J. Pineal Res.* **2014**, *56*, 408–414. [CrossRef]
47. Lee, K.; Back, K. Overexpression of rice serotonin *N*-acetyltransferase 1 in transgenic rice plants confers resistance to cadmium and senescence and increases grain yield. *J. Pineal Res.* **2017**, *62*, e12392. [CrossRef]
48. Janas, K.M.; Posmyk, M.M. Melatonin, an underestimated natural substance with great potential for agricultural application. *Acta Physiol. Plant.* **2013**, *35*, 3285–3292. [CrossRef]
49. Ahmad, S.; Kamran, M.; Zhou, X.; Ahmad, I.; Meng, X.; Javed, T.; Iqbal, A.; Wang, G.; Su, W.; Wu, X.; et al. 2021. Melatonin improves the seed filling rate and endogenous hormonal mechanism in grains of summer maize. *Physiol. Plant.* **2021**, *172*, 1059–1072. [CrossRef]

50. Yan, S.; Zou, G.; Li, S.; Wang, H.; Liu, H.; Zhai, G.; Guo, P.; Song, H.; Yan, C.; Tao, Y. Seed size is determined by the combinations of the genes controlling different seed characteristics in rice. *Theor. Appl. Genet.* **2011**, *123*, 1173–1181. [CrossRef]
51. Huang, R.; Jiang, L.; Zheng, J.; Wang, T.; Wang, H.; Huang, Y.; Hong, Z. Genetic bases of rice grain shape: So many genes, so little known. *Trends Plant Sci.* **2013**, *18*, 218–226. [CrossRef] [PubMed]
52. Fan, C.; Xing, Y.; Mao, H.; Lu, T.; Han, B.; Xu, C.; Li, X.; Zhang, Q. GS3, a major QTL for grain length and weight and minor QTL for grain width and thickness in rice, encodes a putative transmembrane protein. *Theor. Appl. Genet.* **2006**, *112*, 1164–1171. [CrossRef] [PubMed]
53. Song, X.J.; Huang, W.; Shi, M.; Zhu, M.Z.; Lin, H.X. A QTL for rice grain width and weight encodes a previously unknown RING-type E3 ubiquitin ligase. *Nat. Genet.* **2007**, *39*, 623630. [CrossRef]
54. Wang, E.; Wang, J.; Zhu, X.; Hao, W.; Wang, L.; Li, Q.; Zhang, L.; He, W.; Lu, B.; Lin, H.; et al. Control of rice grain-filling and yield by a gene with a potential signature of domestication. *Nat. Genet.* **2008**, *40*, 1370–1374. [CrossRef] [PubMed]
55. Lo, S.F.; Cheng, M.L.; Hsing, Y.C.; Chen, Y.S.; Lee, K.W.; Hong, Y.F.; Hsiao, Y.; Hsiao, A.S.; Chen, P.J.; Wong, L.I.; et al. *Rice Big Grain 1* promotes cell division to enhance organ development, stress tolerance and grain yield. *Plant Biotechnol. J.* **2020**, *18*, 1969–1983. [CrossRef] [PubMed]
56. Zhao, H.; Xu, L.; Su, T.; Jiang, Y.; Hu, L.; Ma, F. Melatonin regulates carbohydrate metabolism and defenses against *Pseudomonas syringae pv. tomato DC3000* infection in *Arabidopsis thaliana*. *J. Pineal Res.* **2015**, *59*, 109–119. [CrossRef] [PubMed]
57. Sun, C.; Liu, L.; Wang, L.; Li, B.; Jin, C.; Lin, X. Melatonin: A master regulator of plant development and stress responses. *J. Integr. Plant Biol.* **2021**, *63*, 126–145. [CrossRef]
58. Bai, Y.; Xiao, S.; Zhang, Z.; Zhang, Y.; Sun, H.; Zhang, K.; Wang, X.; Bai, Z.; Li, C.; Liu, L. Melatonin improves the germination rate of cotton seeds under drought stress by opening pores in the seed coat. *Peer J.* **2020**, *8*, e9450. [CrossRef]
59. Li, D.; Batchelor, W.D.; Zhang, D.; Miao, H.; Li, H.; Song, S.; Li, R. Analysis of melatonin regulation of germination and antioxidant metabolism in different wheat cultivars under polyethylene glycol stress. *PLoS ONE* **2020**, *15*, e0237536. [CrossRef]

MDPI AG
Grosspeteranlage 5
4052 Basel
Switzerland
Tel.: +41 61 683 77 34

Antioxidants Editorial Office
E-mail: antioxidants@mdpi.com
www.mdpi.com/journal/antioxidants

www.ingramcontent.com/pod-product-compliance
Lightning Source LLC
Chambersburg PA
CBHW061557210926
51850CB00009B/2132
* 9 7 8 3 7 2 5 8 2 1 4 8 8 *